Cellular Potts Models
Multiscale Extensions and Biological Applications

CHAPMAN & HALL/CRC
Mathematical and Computational Biology Series

Aims and scope:
This series aims to capture new developments and summarize what is known over the entire spectrum of mathematical and computational biology and medicine. It seeks to encourage the integration of mathematical, statistical, and computational methods into biology by publishing a broad range of textbooks, reference works, and handbooks. The titles included in the series are meant to appeal to students, researchers, and professionals in the mathematical, statistical and computational sciences, fundamental biology and bioengineering, as well as interdisciplinary researchers involved in the field. The inclusion of concrete examples and applications, and programming techniques and examples, is highly encouraged.

Proposals for the series should be submitted to one of the series editors above or directly to:
CRC Press, Taylor & Francis Group
3 Park Square, Milton Park
Abingdon, Oxfordshire OX14 4RN
UK

Published Titles

Algorithms in Bioinformatics: A Practical Introduction
Wing-Kin Sung

Bioinformatics: A Practical Approach
Shui Qing Ye

Biological Computation
Ehud Lamm and Ron Unger

Biological Sequence Analysis Using the SeqAn C++ Library
Andreas Gogol-Döring and Knut Reinert

Cancer Modelling and Simulation
Luigi Preziosi

Cancer Systems Biology
Edwin Wang

Cell Mechanics: From Single Scale-Based Models to Multiscale Modeling
Arnaud Chauvière, Luigi Preziosi, and Claude Verdier

Cellular Potts Models: Multiscale Extensions and Biological Applications
Marco Scianna and Luigi Preziosi

Clustering in Bioinformatics and Drug Discovery
John D. MacCuish and Norah E. MacCuish

Combinatorial Pattern Matching Algorithms in Computational Biology Using Perl and R
Gabriel Valiente

Computational Biology: A Statistical Mechanics Perspective
Ralf Blossey

Computational Hydrodynamics of Capsules and Biological Cells
C. Pozrikidis

Computational Neuroscience: A Comprehensive Approach
Jianfeng Feng

Computational Systems Biology of Cancer
Emmanuel Barillot, Laurence Calzone, Philippe Hupé, Jean-Philippe Vert, and Andrei Zinovyev

Data Analysis Tools for DNA Microarrays
Sorin Draghici

Differential Equations and Mathematical Biology, Second Edition
D.S. Jones, M.J. Plank, and B.D. Sleeman

Dynamics of Biological Systems
Michael Small

Engineering Genetic Circuits
Chris J. Myers

Exactly Solvable Models of Biological Invasion
Sergei V. Petrovskii and Bai-Lian Li

Game-Theoretical Models in Biology
Mark Broom and Jan Rychtář

Gene Expression Studies Using Affymetrix Microarrays
Hinrich Göhlmann and Willem Talloen

Genome Annotation
Jung Soh, Paul M.K. Gordon, and Christoph W. Sensen

Glycome Informatics: Methods and Applications
Kiyoko F. Aoki-Kinoshita

Handbook of Hidden Markov Models in Bioinformatics
Martin Gollery

Introduction to Bioinformatics
Anna Tramontano

Introduction to Bio-Ontologies
Peter N. Robinson and Sebastian Bauer

Introduction to Computational Proteomics
Golan Yona

Introduction to Proteins: Structure, Function, and Motion
Amit Kessel and Nir Ben-Tal

An Introduction to Systems Biology: Design Principles of Biological Circuits
Uri Alon

Kinetic Modelling in Systems Biology
Oleg Demin and Igor Goryanin

Published Titles (continued)

Knowledge Discovery in Proteomics
Igor Jurisica and Dennis Wigle

Meta-analysis and Combining Information in Genetics and Genomics
Rudy Guerra and Darlene R. Goldstein

Methods in Medical Informatics: Fundamentals of Healthcare Programming in Perl, Python, and Ruby
Jules J. Berman

Modeling and Simulation of Capsules and Biological Cells
C. Pozrikidis

Niche Modeling: Predictions from Statistical Distributions
David Stockwell

Normal Mode Analysis: Theory and Applications to Biological and Chemical Systems
Qiang Cui and Ivet Bahar

Optimal Control Applied to Biological Models
Suzanne Lenhart and John T. Workman

Pattern Discovery in Bioinformatics: Theory & Algorithms
Laxmi Parida

Python for Bioinformatics
Sebastian Bassi

Quantitative Biology: From Molecular to Cellular Systems
Sebastian Bassi

Spatial Ecology
Stephen Cantrell, Chris Cosner, and Shigui Ruan

Spatiotemporal Patterns in Ecology and Epidemiology: Theory, Models, and Simulation
Horst Malchow, Sergei V. Petrovskii, and Ezio Venturino

Statistics and Data Analysis for Microarrays Using R and Bioconductor, Second Edition
Sorin Drăghici

Stochastic Modelling for Systems Biology, Second Edition
Darren J. Wilkinson

Structural Bioinformatics: An Algorithmic Approach
Forbes J. Burkowski

The Ten Most Wanted Solutions in Protein Bioinformatics
Anna Tramontano

Chapman & Hall/CRC Mathematical and Computational Biology Series

Cellular Potts Models

Multiscale Extensions and Biological Applications

Marco Scianna and Luigi Preziosi

CRC Press
Taylor & Francis Group
Boca Raton London New York

CRC Press is an imprint of the
Taylor & Francis Group, an **informa** business
A CHAPMAN & HALL BOOK

CRC Press
Taylor & Francis Group
6000 Broken Sound Parkway NW, Suite 300
Boca Raton, FL 33487-2742

ISBN 13: 978-1-4665-1478-2 (hbk)

Visit the Taylor & Francis Web site at
http://www.taylorandfrancis.com

and the CRC Press Web site at
http://www.crcpress.com

Contents

Preface xi

I Basic Cellular Potts Model and Applications 1

1 Basic CPM 3
1.1 The CPM Domain 3
1.2 The CPM Algorithm 6
1.3 The Hamiltonian 7
1.4 Evaluation of Some Kinematic Parameters 11
1.5 Some Illustrative Simulations 11

2 HGF-Induced Cell Scatter 17
2.1 Biological Introduction 17
2.2 Mathematical Model for ARO Aggregates 19
2.3 Scattering of ARO Aggregates 21
2.4 Mathematical Model for MLP-29 Aggregates 25
2.5 Scattering of MLP-29 Aggregates 27

3 Mesothelial Invasion of Ovarian Cancer 33
3.1 Biological Introduction 33
3.1.1 Single Cell Transmigration 34
3.1.2 Multicellular Spheroid Invasion 36
3.2 Mathematical Model 38
3.3 Single Cell Transmigration 40
3.4 Multicellular Spheroid Invasion 43

II Extended Cellular Potts Model and Applications 47

4 Extended Cellular Potts Model 49
4.1 Advantages and Limitations of the Basic CPM 49
4.2 Compartmentalization Approach 50
4.3 Nested Approach 56
4.4 Motility of Individuals 61

5 Wound Healing Assay **67**
5.1 Biological Introduction . . . 67
5.2 Mathematical Model . . . 70
5.2.1 Cell-Level Model . . . 70
5.2.2 Molecular-Level Model . . . 72
5.3 Simulations . . . 75

6 Effect of Calcium–Related Pathways on Single Cell Motility **81**
6.1 Biological Introduction . . . 81
6.2 Mathematical Model . . . 83
6.2.1 Cell-Level Model . . . 84
6.2.2 Molecular-Level Model . . . 88
6.3 Simulation Details and Parameter Estimates . . . 93
6.4 Simulations in Standard Conditions . . . 94
6.5 Interfering with Calcium Machinery . . . 99
6.6 Altering Cell Morphology . . . 109
6.7 Varying the Chemical Source . . . 111

7 Tumor-Derived Vasculogenesis **115**
7.1 Biological Introduction . . . 115
7.2 Mathematical Model . . . 119
7.2.1 Cell-Level Model . . . 119
7.2.2 Molecular-Level Model . . . 121
7.3 Simulations in Standard Conditions . . . 121
7.4 Varying Cell Density . . . 126
7.5 Testing Anti-Angiogenic Therapies . . . 128

8 Different Morphologies of Tumor Invasion Fronts **137**
8.1 Biological Introduction . . . 137
8.2 Mathematical Model . . . 139
8.2.1 Cell-Level Model . . . 139
8.2.2 Molecular-Level Model . . . 142
8.3 Simulations in Standard Conditions . . . 143
8.4 Varying Cell Adhesive Properties . . . 146
8.5 Varying Cell Elasticity . . . 149
8.6 Altering Cell-Substrate Interactions . . . 149
8.7 Effect of Cell Proliferation . . . 152
8.8 Early Stages of Tumor Spheroid Growth . . . 156
8.9 Mathematical Model . . . 157
8.10 Simulations . . . 158

9 Cell Migration in Extracellular Matrices **165**
9.1 Biological Introduction . . . 165
9.2 Mathematical Model . . . 167
9.2.1 Simulation Details . . . 169
9.3 Isotropic Matrices . . . 171

9.4 Anisotropic 2D and 3D Matrices 172
9.5 Varying Fiber Density 175
9.6 Varying Cell-Fiber Adhesiveness 180
9.7 Varying Fiber Elasticity of 3D Matrix Scaffold 182
9.8 Effect of Varying Nucleus Compressibility in 3D 184
9.9 Effect of Matrix Degradation in 3D 186

10 Cancer Cell Migration in Matrix Microchannels 189
10.1 Biological Introduction 189
10.2 Mathematical Model 190
10.3 Simulations 192
10.4 Migration Velocities 198
10.5 Migration Modes 200

III Appendix 203

A Computational Implementation 205

B Glossary 209

C Parameter Values 221

D Color Insert 231

Bibliography 241

Index 277

Preface

All biological phenomena emerge from an intricate interconnection of multiple processes occurring at different levels of organization: namely, at the molecular, the cellular and the tissue level, see Figure 1. These natural levels can approximately be connected to a microscopic, mesoscopic, and macroscopic scale, respectively. The microscopic scale refers to those processes that occur at the subcellular level, such as DNA synthesis and duplication, gene dynamics, activation of receptors, transduction of chemical signals, diffusion of ions and transport of proteins. The mesoscopic scale, on the other hand, can refer to cell-level phenomena, such as adhesive interactions between cells or between cells and ECM components, cell duplication and death and cell motion. The macroscopic scale finally corresponds to those processes that are typical of multicellular behavior, such as population dynamics, tissue mechanics and organ growth and development.

It is evident that research in biology and medicine needs to work in a multiscale fashion. This brings many challenging questions and a complexity that can not be addressed in the classical way, but can take advantage of the increasing collaboration between natural and exact sciences (for more detailed comments the reader is referred to [90, 262]). On the other hand, the recent literature provides evidence of the increasing attention of the mathematical, statistical, computational and physical communities toward biological and biomedical modeling, consequence of the successful results obtained by a multidisciplinary approach to the life sciences problems.

Though it is obvious that no single theoretical model is so perfect to incorporate each and every aspect of all the processes involved in a given natural phenomenon, a computational approach is able to simplify the biological problem, offering both a concise description of its essential features and the possibility of highlighting the most relevant mechanisms and parameters. It can also be used, in a predictive manner, to determine the consequences of experimental manipulations, providing a useful guide for possible future experiments.

Procedurally, a good approach is to build the simplest possible model, focused on a single scale and based on the minimal set of assumptions that are consistent with biological observations. After checking its validity with experimental results, it is then possible to gradually add more components and more levels of abstraction. This continuous feedback and feedforward between *in silico* and *in vitro* techniques, typical of a *system biology* approach,

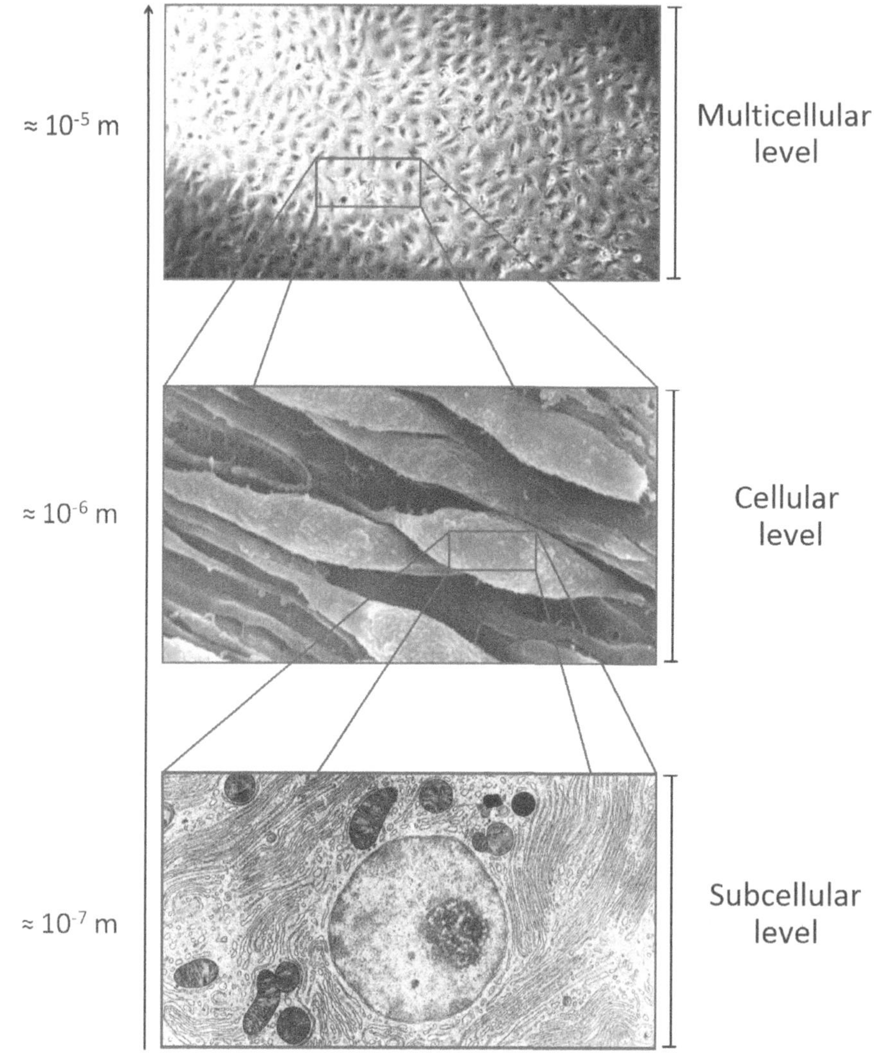

FIGURE 1: Multiscale view of biological systems. A cell population consists of a huge number of cells. A zoom allows the identification of single polarized cells adhering to each other. A further zoom permits the identification and localization of intracellular elements, such as the nucleus, the Golgi apparatus and some mitochondria. Representative images courtesy of the Department of Life Sciences and Systems Biology, Università degli Studi di Torino.

on one side may efficiently support the activity of experimental researchers, by adapting existing mathematical tools to their needs. On the other side it may contribute to the development of new and more realistic computational methods.

Mathematical approaches applied to biological problems embody a wide range of techniques, which depend on the particular spatio-temporal scale of interest, see Figure 2. However, most of them fall in two categories: *continuous* and *discrete* models.

Continuous models approach biological phenomena in terms of variation of fields. Characteristic of a macroscopic point of view, these methods represent populations of biological individuals as densities, which evolve satisfying sets of balance laws or diffusion equations. This type of approach includes the multiphase models, developed under the simple observation that biological systems are made of several constituents and are then treated with classical concepts of continuum mechanics.

On the other hand, discrete models, widely known as *Individual Cell-Based Models* (IBMs) or *Cellular Automata* (CA), approach the biological problem with a phenomenological point of view, focusing on the cell-level of abstraction and preserving the identity and the behavior of individual elements (for comprehensive reviews the reader is referred to [6, 11, 103, 113, 239]). Indeed, these techniques represent biological individuals, with the typical length scale of a cell, as one or a set of discrete units, with rules that describe their movements and interactions. The morphology of the elements is restricted according to some underlying discretizations of the simulation domain, which can be either regular (such as square or cubic grids) or irregular (Voronoi tassellations). These approaches can be further classified in two categories: those in which each individual is correlated to a single spatial unit of the domain, and those in which each element is instead constituted by a collection of spatial units. With respect to continuous methods, IBMs can more naturally capture detailed biophysical properties, such as cell shape, geometry and adhesion, and they are also able to handle local interactions between the simulated objects. In cell-based methods, the individuals behave according to a relatively small set of prescribed rules, which they execute depending on their type and on the signals received from the neighbors and the environment. In particular, these techniques are able to analyze the mechanisms by which the relatively simple behavior and interactions of individuals collectively manage macroscopic pattern formation and development, and, *vice versa*, to infer how phenomena that occur at the macroscopic level feed back in the phenomenology of single elements, as commented in [262]. Moreover, cell-based models help to unravel how abnormal cell behavior can produce abnormal patterns.

However, as it can be easily observed, the study even of simple biological phenomena does not directly and uniquely define the most suitable mathematical framework. In most cases, it is in fact hard to identify the best approach at all, because every method has its advantages and disadvantages. In particular, continuous techniques overlook the behavior of single individuals and also fail to describe their mutual interactions. They may therefore be unsatisfactory since, in order to deal with biological processes, it is fundamental what

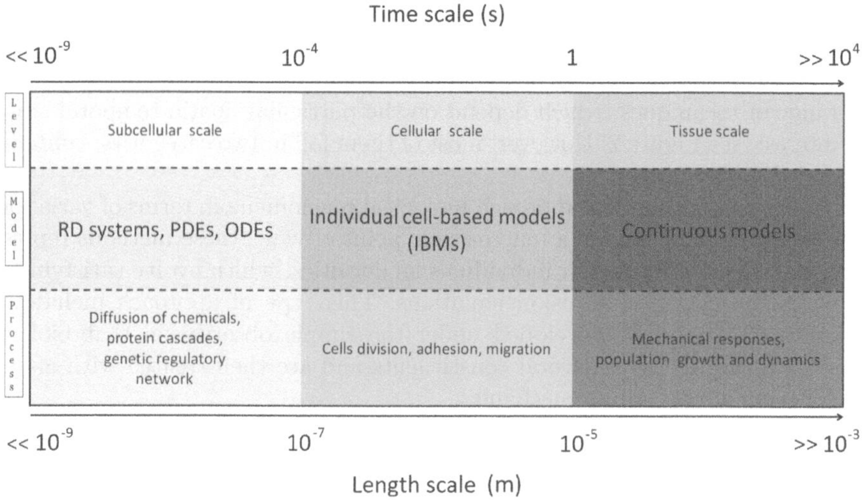

FIGURE 2: Hierarchies of biological length and temporal scales, and the corresponding modeling techniques.

occurs at the scale of the single element. On the other hand, by approaching biological phenomena from a mesoscopic point of view, IBMs usually do not describe molecular-level processes, and thus neglect the microscopic mechanisms that underly individual phenomenology. Furthermore, since they are computationally expensive, they can only simulate several individuals at once, being inefficient in providing a general outlook of the system as a whole.

In the last decade, we have therefore witnessed an increasing introduction of new modeling environments, that aim to overcome this *dichotomy* by integrating both approaches. The trend is to create computational frameworks able to span a wide range of spatio-temporal scales with a sufficient level of accuracy, offering the advantages brought by the different methods. These multilevel discrete-continuous approaches can be constructed building *nested* environments where the information needed by the cell-based model are obtained using another model model operating at the subcellular scale to describe for instance processed at the proteomic and/or genomic level. In many cases, one also need to interface discrete objects with concentration fields used to represent the evolution of nutrients and chemical factors diffusing in the environment, creating in this way *hybrid* models (the reader is referred to [317] and references therein for a useful overview).

Entering in more details, the nested models typically include sub-cellular mechanisms that, because of the smallness of the size of the molecules involved with respect to the size of the cell, can be often described with continuous models, into cell-level IBMs, i.e., they nest in a mesoscopic representation one or more modules accounting for processes characteristic of the microscopic scale, as reviewed in [11]. For example, in [324] accurate E-cadherin/β-catenin cascades are integrated in a cellular automata to obtain a deeper description of cell-cell adhesion, while in [299] a discrete model for vascularized tumors includes some aspects related to the VEGF and the p53 pathways, in addition to protein networks governing mitosis. Moreover, in some structured population models of cancer, the evolution of malignant individuals along the cell cycle is related to the activation of specific genetic dynamics [87, 147] or to the presence of specific control molecules, such as cyclin/CDC complexes [109]. It is useful to underline that nested approaches can be also developed by using only continuous techniques, i.e., by constitutively relating the parameters appearing in the evolutive PDEs to sub-cellular mechanisms. However, in these cases, the link between the different scales is more complex and often more qualitative than quantitative. However, formal and consistent results may be obtained by using deeper mathematical methods, such as homogenization techniques or fluid-dynamic and diffusion limits, as commented again in [317].

In the literature, there two types also of hybrid models [317]. The most common one, and historically the first, uses a discrete approach to reproduce some elements of the system and a continuous approach to describe the others. For instance, discrete variables are typically used to reproduce cells, while density fields to represent extracellular molecules, ions, and ECM components. Examples of this standard hybrid approach are IBMs of solid tumor growth, where the duplication/death of cells is related to the diffusion of environmental nutrients through their boundaries [111, 112], or of angiogenic progression, where the movement of tip cells depends on the uptake of several extracellular chemical factors [14]. In the second and recent type of hybrid models, the same elements are instead described by spatially extended objects in selected portions of the domain and by means of continuous variables in the others. The basic idea is to use the modeling environment like a microscope, which is able to analyze the phenomenon of interest at the same time from different levels and to compare the results obtained with different approaches. These methodologies allow in principle to support an upscaling process and to infer the phenomenological behavior of single individuals from their microscopic characteristic, i.e., to build rheology-like *in silico* experiments. The efficiency of the cross-talk between the different subdomains is defined by suitable mathematic procedures, such as hydrodynamic limits [31, 222], homogeneous and/or coarse-graining methods [115].

One of the most widespread hybrid approaches, that is particularly suitable for biological problems, is the *cellular Potts model* (CPM, also called the Glazier–Graner–Hogeweg model, which was developed in [165, 169] and reviewed in [20, 164, 252]). As a generalization of the grid-based Ising model, the CPM is a stochastic Monte Carlo method, whose energy minimization philosophy drives the evolution of the simulated system.

The CPM domains are regular lattices, with a defined list of objects, which can be both *discrete generalized cells* or *continuous fields* (giving rise to the hybrid characteristic of the theoretical environment). Generalized cells are spatially-extended, mesoscopic, cell-scale elements: they may correspond to unicellular organisms, cells, cell subcompartments, clusters of cells, ECM fibers, or portions of non-cellular materials. The fields instead represent the spatio-temporal evolution of microscopic quantities, such as diffusive ions and molecules. The attributes of the discrete individuals, and the rules for their dynamics and interactions with fields are described by an effective potential formalism, and are included in a system energy that is given by the *Hamiltonian.* This functional, which may include true energies (e.g., cell–cell adhesion) and terms that mimic energies (e.g., the response of a cell to a chemical gradient), describes the state of the system and drives all the lattice rearrangements with an algorithm of stochastic minimization (a modified Metropolis procedure which we will describe in detail in the next chapter). As long as a biological mechanism can be described with an energetic formalism, it can be included in the CPM framework. Thus, the CPM is not for a specific type of biological problems, but can be considered as a framework for model building. Such a flexibility, as well as a simplicity of implementation, makes the CPM unusually rewarding to work with.

The CPM method is becoming an increasingly common technique for the mathematical modeling of a wide range of biological phenomena, including: avascular and vascular tumor growth [25, 334, 365, 395], gastrulation [114], skin pigmentation [273], yeast colony growth [403], stem cell differentiation [424], fruiting body formation of *Dictyostelium discoideum* [253], epidermal formation [343], hydra regeneration [269], retinal patterning [266], wound healing [247], biofilms [313], chick limb-bud growth [75, 76, 314], differential adhesion-driven cell rearrangement [164], cellular differentiation and growth of tissues, blood flow and thrombus development [414, 415, 416], and angiogenesis [259, 260].

The scope of this book is to present some innovative extensions for the CPM approach, which are able to increase its biological realism, accuracy, and predictive power. In particular, we focus on ways to integrate and interface the standard method with detailed descriptions of microscopic dynamics located not only in the external space but also within the simulated elements. We aim therefore to introduce some *nested characteristics* in the basic hybrid environment, that reproduce the multiscale organization typical of biological development, where, as seen, the individual behavior is driven by the constant interplay between different levels of description.

Indeed, the main body of this work is organized as follows. In order to have an exhaustive overlook of the CPM approach, Chapter 1 explains in details its basic version, with comments on its historical journey, from the original Ising method to its last developments, the Glazier–Graner–Hogeweg model (GGH model). In particular, along the text, we give some ideas on how the CPM, born for physical phenomena, has been then used to deal with biological problems.

In Chapters 2 and 3, we apply the basic CPM to the case of two medical problems, stemming by the collaboration with experimental centers. The first one is a joint work with Roeland Merks of the NCSB-NISB Science Park and with Enzo Medico at the Department of Oncological Sciences and Laboratory of Oncogenomics of the Institute for Cancer Research and Treatment (who originally stimulated the study of the system from a mathematical view point, carrying out the experimental part) reproduces the behavior of two cell lines, ARO (thyroid carcinoma-derived cells) and MLP-29 (mouse liver progenitor cells), in response to stimulations with hepatocyte growth factor (HGF). In particular, we show how the agonist-induced increased motility and decreased adhesiveness are responsible for cell–cell dissociation and tissue invasion of ARO individuals. On the other hand, it is demonstrated that, in addition to the biological mechanisms above, it is necessary to include the directional persistence in cell motion and the HGF diffusion to describe the branching process characteristic of MLP-29 individuals.

The second model deals with the dynamical analysis of an *in vitro* trans-mesothelial migration assay of ovarian cancer cells, isolated or aggregated in multicellular spheroids. In particular, in good agreement with the experimental evidences provided by Ada Funaro and her collaborators in the Laboratory of Immunogenetics of the Azienda Ospedaliero-Universitaria S.Giovanni Battista-Molinette. It is shown that the overall process is regulated by both the activity of matrix metalloproteinases (MMPs) and by the adhesive properties of malignant individuals. Indeed, it is found that single cell invasion is more conservative than spheroid invasion, which is more disruptive and achieves monofocal or multifocal characteristics according to the relative adhesive affinity among cancer cells or between them and the mesothelial layer.

The first part of Chapter 4 concerns a critical review of the main advantages and limitations of the basic CPM. In particular, we comment both its most relevant features and the main drawbacks that limit its accuracy. Some improvements of the method are then presented. The first one consists in giving biological entities a more realistic representation, that can be achieved compartmentalizing cells into physiologically functional subunits, such as nucleus, cytosol, plasmamembrane or organelles in case of cells. The second one consists in introducing a procedure to nest in the standard CPM specific models of microscopic biochemical pathways, which realistically modulate selected biophysical and biomechanical properties of simulated individuals (and therefore the parameters related to their motility, adhesivity strength, elasticity, and so on), ultimately affecting their phenotypic behavior. Finally, alternative laws for the Boltzmann transition function, the core of the CPM algorithm, are proposed with a special emphasis on the description of the motility of individuals and on the actual possibility of performing energetically favorable steps.

The following Chapters 5–10 are devoted to test the flexibility and the improvements of the extended method by giving several sample applications, also in this case originated by the collaboration with different experimental teams looking for a computational support in their research. In particular, such examples range from unicellular processes to multicellular phenomena.

The first application, in Chapter 5, is developed to reproduce and analyze a *wound-healing assay*. The proposed approach quantifies the migratory capacity of the overall cell population, as well as it allows the evaluation of statistical descriptive parameters (such as average velocity, directional movement, and final displacement) characterizing the behavior of single cells, drawing the attention to specific correlations. The analysis of the phenomenon puts on evidence the possibility of differentiating and grouping the cell mass into three sub-populations, sited at increasing distances from the edge of the scratch. The computational findings also describe the motility of the culture in different conditions and, consistently with published experimental observations, provide that the maximal healing is seen when the more external cells can undergo a reorganization toward a motile phenotype.

The second proposed application, in Chapters 6 and 7, deals with the reproduction of selected features of a tumor-derived *tubulogenic assay*, focusing on the role of VEGF-induced calcium signals in regulating its driving mechanisms. At the cellular level, the migratory phenomenology of a single tumor-derived endothelial cell (TEC) is reproduced in various experimental conditions. In particular, our study describes the polarization process of the motile individual and investigates ways to inhibit its locomotion ability, fundamental for the overall malignant vascular progression, by interfering with different molecular processes, i.e., by virtually testing specific pharmacological interventions. At the multicellular level, the simulation results, in agreement with experimental analysis provided by the group headed by Luca Munaron at the Department of Life Sciences and Systems Biology of the Università degli Studi di Torino, instead allow to identify the key mechanisms of the overall network formation. In particular, we provide evidence that the nascent pattern is characterized by precise topological properties, regulated by the initial cell density in conjunction with the degree of the chemotactic response and the directional component of cell migration. Moreover, the analysis of the link between microscopic calcium-dependent subcellular dynamics and macroscopic cell behavior confirms the efficiency of some pharmacological anti-angiogenic interventions that are currently in use and in trial, more interestingly, proposes some new therapeutic approaches, that are counter-intuitive but potentially effective.

The third biological phenomenon analyzed with the extended CPM is the tissue invasion of a solid tumor (Chapter 8). In particular, if the population of cancer cells is described at the mesoscopic scale, a continuous approach of reaction-diffusion equations is employed to describe the evolution of microenvironmental variables, as the nutrients and the proteins present in the extracellular matrix and the proteolytic enzymes secreted by the tumor. The resulting composite model quantifies the influence of selected changes in the mechanisms involved in tumor invasion and, more interestingly, puts in evidence possible therapeutic approaches, potentially effective in controlling the morphology of the disease, which defines its malignancy, such as the alteration in homotypic and heterotypic adhesive properties of malignant cells, the inhibition in their ability to remodel and the disruption of the haptotactic movement. The simulation framework is also extended by including cell proliferation which, following experimental evidence, is regulated by the intracellular level of growth factors. Interestingly, in spice of the increment in cellular density, the depth of invasion does not significantly increase, as one could have expected.

The same model is then applied in the last part of Chapter 8 to reproduce the early growth of a solid tumor spheroid. Also in this case, the behavior of each cancer cell is determined by a balance of interaction forces, such as homotypic (cell–cell) and heterotypic (cell–matrix) adhesions and haptotaxis, and is mediated by its molecular state, which regulates the motility and proliferation rate. The resulting model captures the different phases of the development of the tumor mass, i.e., its exponential growth and the subsequent stabilization in a steady-state due to limitations in vital molecules. The proposed approach also predicts the influence on the cancer morphology of changes in specific intercellular adhesive mechanisms.

Chapter 9 is focuses on cell migratory phenotypes on planar and in three-dimensional matrices, composed of a fibrous collagenous mesh and a homogeneous fluid component, in a joint work with Katarina Wolf at the Department of Cell Biology, Nijmegen Center for Molecular Life Science, of the Radboud University Nijmegen Medical Center. In the 2D case, the model provides a biphasic dependence of cell migratory abilities on both cell-substratum adhesiveness and matrix density. Moreover, it is proven that the directional component of cell movement is strongly correlated with the topological distribution of the fibrous network of the ECM. In the 3D case, we investigate the effects of the matrix topological and mechanical microstructure, specifically pore size and elasticity. In particular, it is observed that cell motility has a subtle bimodal relation with matrix pore size. Indeed, at intermediate pore dimensions, cell dispersion is maximal at sufficiently high fiber stiffness. Finally, an in-depth multidimensional analysis reveals that cell locomotion within rigid matrices characterized by small pores can only be achieved through the deformation of cell nucleus or the activation of cell proteolytic machinery, able to degrade the extracellular space overcoming steric obstacles.

An analogous model is finally applied in Chapter 10 to simulate migration in a micro-fabricated migration chip, which tests the active invasive behavior of cancer cells into threedimensional narrow channels. Indeed, it is found that the migratory behavior of cells within subcellular and subnuclear structures strongly relies on the deformability of the cytosol and of the nucleus, respectively. Further, two migration dynamics can be identified: a stepwise way, characterized by fluctuations in cell length, within channels smaller than nucleus dimensions and a smooth sliding (i.e., maintaining constant cell length) behavior within channels larger than the nuclear cluster.

In the Appendix, we first give a brief overview of the most used CPM packages, with comments of their main issues and suggestions on ways to overcome them. A guide to build an own-property code for a CPM application employing the extensions presented in this book is also provided in the same section.

Always trying to keep in mind the interdisciplinary nature of the subject, we tried to use a language that can be understood by both Ph.D. students and researchers working in both life and exact sciences. We are aware that this might lead to a lack of deepness is some concepts, that can be found in the cited literature, but at the same time we hope in this way to close the gap and overcome the language barriers between such different research fields.

Symbol Description

Ω	CPM domain
$\mathbf{x}$	lattice site
$\partial\mathbf{x}$	border of $\mathbf{x}$
$\sigma(\mathbf{x})$	site identification number, i.e., spin
$\mathbf{r_x}$	position vector of $\mathbf{x}$
$\mathbf{x}'$	neighbor of site $\mathbf{x}$
$\Omega'_{\mathbf{x}}$	overall neighborhood of $\mathbf{x}$
Σ_σ	discrete object
$\partial\Sigma_\sigma$	border of Σ_σ
$\tau(\Sigma_\sigma)$	type of Σ_σ
$\Sigma_{\sigma'}$	neighbor of Σ_σ
δ	Kronecker delta
H	Hamiltonian functional
T	Boltzmann temperature
$J_{\Sigma_\sigma,\Sigma_{\sigma'}}$	binding force per unit of area
$A^i_{\Sigma_\sigma}$	target value of object attribute
$a^i_{\Sigma_\sigma}$	actual value of object attribute
$\lambda^i_{\Sigma_\sigma}$	elastic modulus of object attribute
$\mathbf{F}$	external force
η	compartmentalized individual
$\theta(\eta)$	type of η
$\partial\eta$	border of η
$\mathbf{s}_{\Sigma_\sigma}$	internal state vector of Σ_σ

Abbreviations

AA	arachidonic acid
ARO	thyroid carcinoma-derived cell
ATP	adenosine triphosphate
CA	cellular automata
CAI	carboxyamidotriazole
CFSE	carboxyfluorescein succinimidyl ester
CPM	cellular Potts model
CTM	cell type map
DAH	differential adhesion hypothesis
DMEM	Dulbecco's minimal essential medium
EC	endothelial cell
ECM	extracellular matrix
ER	endoplasmic reticulum
FCS	fetal calf serum
HGF	hepatocyte growth factor
HPMC	human pleural mesothelial cell
IBM	individual cell-based model
MAPK	mitogen-activated protein kinase
MCS	Monte Carlo Step
MLP-29	mouse liver progenitor cell
MMP	matrix metalloproteinase
NCCE	non-capacitive calcium entry
NO	nitric oxide
NSOCE	non-store-operated calcium entry
OvCa	ovarian cancer
PBS	phosphate buffered saline
PDE	partial differential equation
PI3K	phosphatidylinositol 3-kinases
PM	plasmamembrane
SF	scatter factor
RD	reaction-diffusion equation
TEC	tumor-derived endothelial cell
uPa	urinary-type plasminogen
VEGF	vascular endothelial growth factor

Part I

Basic Cellular Potts Model and Applications

Chapter 1

Basic CPM

1.1 The CPM Domain

All cellular Potts model approaches include a list of objects, a description of their interactions, and rules for their dynamics. The CPM domains are d-dimensional lattices $\Omega \subseteq \mathbb{R}^d$, where $d = 1, 2, 3$. The term *lattice* defines a regular repeated graph, formed by identical d-dimensional closed grid sites $\mathbf{x} \in \mathbb{R}^d$, and characterized by periodic or fixed boundary conditions in each direction. The volumetric extension of Ω is equal to the total number of its sites, that therefore represent the basic unit of length of the system. Each site $\mathbf{x} \in \Omega$ is uniquely identified by its location and is labeled by an integer, $\sigma(\mathbf{x}) \in \mathbb{N}$, where σ can be interpreted as a degenerate *spin* value coming from the original Ising approach [85, 199, 315]; see Figure 1.1(A). With an abuse of notation, $\mathbf{x}$ also usually denotes the closed elementary spatial region (e.g., the *voxel*) centered in $\mathbf{x}$. As classically adopted in CPM models, the border of a lattice site $\mathbf{x}$ is denoted by $\partial\mathbf{x}$, one of its neighbors by $\mathbf{x}'$, while its overall neighborhood by $\Omega_{\mathbf{x}}^{'}$, i.e., $\Omega_{\mathbf{x}}^{'} = \{\mathbf{x}' \in \Omega : \mathbf{x}' \text{ is a neighbor of } \mathbf{x}\}$.

Objects in the CPM are either *discrete* or *continuous*: this coexistence provides to the method its *hybrid* characteristic. Discrete objects Σ_σ are finite, spatially extended lattice subdomains of contiguous sites with the same spin σ, i.e.,

$$\Sigma_\sigma = \{\mathbf{x} \in \Omega : \sigma(\mathbf{x}) = \sigma\}.$$

They are therefore undifferentiated *functional units* that, in basic CPMs, represent single biological elements with the typical mesoscopic length scale of a cell. Examples are bacteria, unicellular organisms, single cells, ECM fibers, or other substrates; see Figure 1.1(B-C). Trivially, a collection of N discrete individuals Σ_σ is defined by N integer spin $\sigma = 1, 2, \ldots, N$. The borders between sites with different degenerated spins that are thus shared between a couple of objects define their membranes, i.e.,

$$\partial\Sigma_\sigma = \bigcup_{\mathbf{x}\in\Sigma_\sigma, \mathbf{x}'\in\Omega_{\mathbf{x}}^{'} : \mathbf{x}'\notin\Sigma_\sigma} (\partial\mathbf{x} \cap \partial\mathbf{x}').$$

Each unit Σ_σ has a set of attributes (both geometrical, such as volume and surface, and biophysical, such as velocity and elasticity), and an associated type

$\tau(\Sigma_\sigma)$ (e.g., endothelial cell, ECM fiber, fibroblast, ...). Mesoscopic, cell-level objects rearrange their boundaries to realistically reproduce shape changes and motion. Moreover, they can grow, die, duplicate (and the daughter objects typically inherit their parent's properties [186, 270, 314]), and carry a set of possible rules for transitions between types (a Cell Type Map, CTM [84]). The notation $\Sigma_{\sigma'}$ is usually used to identify a discrete object neighbor of Σ_σ.

Continuous objects, or *fields*, represent the spatiotemporal evolution of microscopic entities, that may reside within the discrete objects (as DNA, RNA, cytosolic ions, and proteins), or in the external environment (as nutrients, growth factors, matrix proteins, matrix metalloproteinases). They are described as variable concentrations with standard *reaction-diffusion* (RD) equations, whose general form is

$$\frac{\partial c}{\partial t}(\mathbf{x},t) = \underbrace{\nabla \cdot [D_c(\mathbf{x},t)\nabla c(\mathbf{x},t)]}_{diffusion} + \underbrace{F(c)}_{reaction\ term} , \tag{1.1}$$

where $c(\mathbf{x},t)$ denotes the local concentration (i.e., at site $\mathbf{x}$) of the chemical substance, D_c its diffusion coefficient, and $F : \mathbb{R}_+ \to \mathbb{R}$ is the reaction term. In classical CPM applications, D_c is homogeneous in space and constant in time; however, it is possible to drop these restrictions. Equations of type (1.1) may apply to the entire domain Ω or to selected subregions, with fixed or moving boundaries (as in the case of intracellular chemicals). Indeed, as we will explain more in detail in Appendix A, these continuous equations are numerically solved using *finite difference schemes* on grids that exactly match the CPM domain, and that are discretized at the same resolution.

The specific interactions between discrete cell-level objects and continuous molecular-level objects can be characterized either by the reaction term in Equation (1.1), as in the case of cell absorption and secretion of chemical diffusants [25, 183, 254], or by constitutive laws relating phenotypic behavior of discrete individuals to the evolution of specific microscopic variables, as we will see in Chapter 4. The coupling between the dynamics of microscopic and mesoscopic objects provides to the CPM its hybrid characteristic. Obviously, each application of the method needs the specification of the initial condition of the lattice, i.e of both the initial condition of the continuous fields and the initial spatial configuration of the discrete objects.

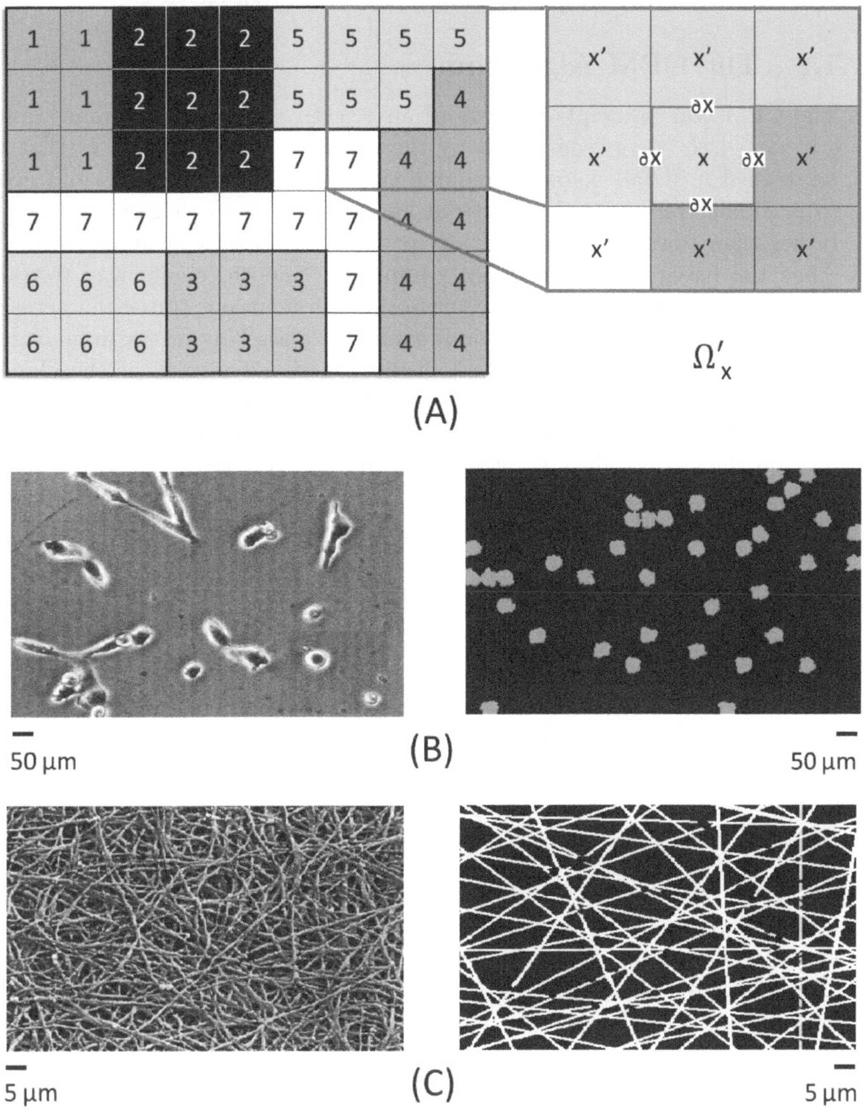

FIGURE 1.1: (A) Typical 2D rectangular CPM lattice. The integers (σ) identify single discrete objects Σ_σ. The couples of individuals Σ_1 and Σ_4, and Σ_5 and Σ_6 are of the same type τ, and are identified by the same gray scale. In the zoom view, it is possible to see a lattice site, $\mathbf{x}$, its border $\partial\mathbf{x}$, and its first-nearest neighbors, $\Omega'_{\mathbf{x}}$. (B) Population of experimental and virtual cells. (C) Network of experimental and virtual anisotropic fibers of collagen I. Experimental images courtesy of the Department of Life Sciences and Systems Biology, Universitá degli Studi di Torino.

1.2 The CPM Algorithm

The CPM core principle consists of an iterative stochastic minimization of a system free energy, described by the *Hamiltonian functional* H that will be defined in detail below. Simulated objects, in fact, rearrange and evolve to gradually reduce such a pattern energy looking for a global minimum, rather than toward a configuration in which multiple local minima coexist. Since the energy gradient is not completely smooth, on its way to global minimization, the system has to move through transient states characterized by higher energies than the previous configurations. This energy minimization philosophy is implemented by adopting a modified version of the classical Metropolis algorithm for Monte Carlo–Boltzmann thermodynamics [169, 263].

It evolves in time using repeated probabilistic updates of the site identification spins. Procedurally, at each simulation time step, t, a lattice site $\mathbf{x}$, belonging to an object Σ_σ, is randomly selected (*source voxel*), and proposed to copy its spin $\sigma(\mathbf{x})$ into an arbitrary unlike neighbor $\mathbf{x}' \notin \Sigma_\sigma$ (*target voxel*). The proposed change in the lattice configuration (also called *spin flip*) is accepted with a classical *Boltzmann transitional probability*, which is a relic of the CPM descent from statistical physics [315]:

$$P(\sigma(\mathbf{x}) \to \sigma(\mathbf{x}'))(t) = \begin{cases} e^{-\Delta H|_{\sigma(\mathbf{x})\to\sigma(\mathbf{x}')}/T} & \Delta H|_{\sigma(\mathbf{x})\to\sigma(\mathbf{x}')} > 0\,; \\ 1 & \Delta H|_{\sigma(\mathbf{x})\to\sigma(\mathbf{x}')} \leq 0\,, \end{cases} \tag{1.2}$$

where

$$\Delta H|_{\sigma(\mathbf{x})\to\sigma(\mathbf{x}')} = H_{after\ spin\ flip} - H_{before\ spin\ flip}, \tag{1.3}$$

is the net variation in the total energy of the system as a consequence of the spin update and $T \in \mathbb{R}_+$ is a Boltzmann temperature measured in units of energy. T does not reflect any conventional thermal temperature, but in basic CPMs broadly correlates to an overall *system motility*. The nomenclature of T originates from the fact that membrane agitation rates in biological individuals play a role analogous to real temperatures in ordinary thermodynamics [268]. From a statistical point of view, T represents the likelihood of the energetic unfavorable changes in lattice configurations, since it determines the rate of their acceptance. For very small values, the system evolution is almost deterministic, and it can be trapped in local minima. For very large T all moves are accepted, and the simulated objects are characterized by a biased random walk in the absence of potential barriers. Their motility, in fact, overcomes the constraints set by the local environment, since $\Delta H|_{\sigma(\mathbf{x})\to\sigma(\mathbf{x}')}/T \to 0$ for all the proposed displacements of their sites. It is useful to underline that in [162] the motility T is allowed to depend on the object type ($T = T_{\tau(\Sigma_\sigma)}$); however, it is also possible to relate it to each single unit, $T = T_{\Sigma_\sigma}$. This aspect is described in more detail in Sections and 4.3 and 4.4.

After the discrete object Σ_σ has evolved through a spin flip, both equations that describe the variation of continuous fields, and the attributes of all the objects are rederived on the basis of new lattice configuration.

The basic step of the Metropolis algorithm is then iterated until the end of the simulation time or until the whole system reaches an energetic global minimum, if it exists. The unit of time of all CPM approaches is the *Monte Carlo Step* (MCS). An MCS corresponds to a fixed number of trial lattice updates, which is usually a multiple of the total number of sites of Ω (i.e., 1 MCS $= k \cdot \text{Volume}_\Omega$) and which has to be translated into the actual unit of time (i.e., seconds, hours, days). A direct correspondence between the model and the actual time scale may therefore not be straightforward, giving rise to one of the main criticisms of the method. However, a realistic correspondence is usually set by fitting *a posteriori* the temporal dynamics of the simulated phenomenon with the relative experimental counterparts.

1.3 The Hamiltonian

The discrete *effective energy* of the system, given by the *Hamiltonian* H, may contain a variable number of terms, that can be grouped as:

$$H(t) = H_{adhesion}(t) + H_{constraint}(t) + H_{force}(t). \tag{1.4}$$

$H_{adhesion}$ describes the adhesive/repulsive interfacial energy between all the pairs of discrete objects that interact across their common membrane. $H_{adhesion}$ is based on Steinberg's *Differential Adhesion Hypothesis* (DAH) [169, 377, 378]. The DAH proposes that individuals in the same aggregate adhere to each other with different strengths, according to their type. Such a hierarchy of contact forces is one of the main driving mechanisms behind the evolution of biological systems, whose final organization maximizes the overall strength of interface interactions (or, in other words, minimizes the overall adhesion energy). However, the DAH says nothing about the dynamics of moving objects: differential adhesion itself, in fact, only helps to select the most favorable configuration among the different possibilities that have been explored. Evidence supporting DAH has been observed in a wide array of biological systems, especially in the embryonic stage of life. For example, it successfully explains how cellular adhesive properties can operate to determine tissue reorganization during cell sorting [16, 164, 169]. The typical formulation of DAH-derived $H_{adhesion}$ is

$$H_{adhesion}(t) = \sum_{\substack{\mathbf{x},\mathbf{x}' \in \Omega'_{\mathbf{x}}: \\ (\partial\mathbf{x} \in \partial\Sigma_\sigma) \cap (\partial\mathbf{x}' \in \partial\Sigma_{\sigma'}) \neq \emptyset}} J_{\tau(\Sigma_{\sigma(\mathbf{x})}),\tau(\Sigma_{\sigma'(\mathbf{x}')})}(t), \tag{1.5}$$

where, as seen, $\mathbf{x}$ and $\mathbf{x}'$ are two neighboring sites and Σ_σ and $\Sigma_{\sigma'}$ two neighboring objects. The coefficients $J_{\tau(\Sigma_\sigma),\tau(\Sigma_{\sigma'})} \in \mathbb{R}$ are the binding forces per unit area, the first type of the so-called Potts parameters, and are obviously symmetric w.r.t. the indices. In basic CPMs, they only depend on the type of discrete objects that are in contact (i.e., $\tau(\Sigma_\sigma)$ and $\tau(\Sigma_{\sigma'})$ in Equation (1.5)), as they are not a characteristic of each single unit; see, for example, [261, 268, 334, 342]. Moreover, they are uniformly distributed over the whole surface of the discrete objects, neglecting microscopic inhomogenities, such as a clusterization or a different strength of adhesion in well-localized parts of their membranes. This issue will be discussed in more detail in Section 4.3, as the model will be extended in this respect.

In the case of cells, such contact strengths give a qualitative measure of the expression of adhesion molecules in the individuals on either side of the common border, whose activity defines their binding properties. At least two classes of J's can be identified: those relative to the adhesion between cells and extracellular material (and thus modeling the activity of cell–matrix adhesion molecules, such as integrins), and those that mediate the adhesion between cells of either the same or of different populations (and thus related to the expression of cell–cell adhesion molecules, such as cadherins). A surface contact force can also be defined with an external undifferentiated medium (for example, culture medium, air, generic substrate), but it is biologically meaningless and therefore negligible.

The term $H_{constraint}$, whose use also comes from the physics of classical mechanics, sums the energetic components that describe the object attributes. They are written as energetic penalties that increase as the objects deviate from a designed state. In a characteristically elastic form, it is classically written as

$$H_{constraint}(t) = \sum_{\Sigma_\sigma} \sum_{i-constraint} \lambda^i_{\Sigma_\sigma}(t) \left[a^i_{\Sigma_\sigma}(t) - A^i_{\Sigma_\sigma}(t)\right]^2, \qquad (1.6)$$

where $a^i_{\Sigma_\sigma}(t)$ is the actual value of the i-attribute of individual Σ_σ, and $A^i_{\Sigma_\sigma}(t)$ is its target value that usually characterizes an object type and that can also vary in time. The Potts parameters $\lambda^i_{\Sigma_\sigma} \in \mathbb{R}_+$ take the role of elastic moduli, which determine the weight of the relative energetic constraint, and thus the importance of the relative attribute. Low values of $\lambda^i_{\Sigma_\sigma}$, in fact, allow the discrete unit Σ_σ to deviate more from the configuration that satisfies the constraint. Indeed, since the energetic contributions given in Equation (1.6) smoothly decrease to a minimum when the attributes are satisfied, the modified Metropolis algorithm automatically drives any configuration toward one that satisfies the constraints. Obviously, the simulated system is not able to exactly satisfy all the constraints of all the objects Σ_σ at any given time t, since multiple attributes may be in conflict: this leads to lattice configurations characterized by local energetic minima.

Among others, the energetic components relative to geometrical attributes of discrete objects, such as their volume and surface, are of particular relevance. These components depend on the actual measures of each mesoscopic element, $a_{\Sigma_\sigma}^{volume}(t)$ and $a_{\Sigma_\sigma}^{surface}(t)$, as well as on the same quantities in the relaxed/undeformed state $A_{\Sigma_\sigma}^{volume}(t)$ and $A_{\Sigma_\sigma}^{surface}(t)$. In particular, $\lambda_{\Sigma_\sigma}^{volume}$ regulates the conservation of mass of the discrete objects, and encodes all the bulk effects: moreover, their growth can be realistically included by assuming that $A_{\Sigma_\sigma}^{volume}(t)$ increases during the simulation [75, 314]. $\lambda_{\Sigma_\sigma}^{surface}$ instead represents the inverse compressibility of Σ_σ, the ease with which it can change its shape. If $\lambda_{\Sigma_\sigma}^{surface}$ is very large, Σ_σ has negligible elasticity, and its membrane is tight. In particular, for $\lambda_{\Sigma_\sigma}^{volume}, \lambda_{\Sigma_\sigma}^{surface} \to \infty$, Σ_σ behaves as a rigid body. In the case of cells, $\lambda_{\Sigma_\sigma}^{volume}$ and $\lambda_{\Sigma_\sigma}^{surface}$ regulate, respectively, the growth and the change of shape due to active reorganizations of the actin cytoskeleton triggered by both internal (such as small G-protein activity), or external stimuli (such as ECM contact guidance). Obviously, in a bidimensional model the volume of an object relates to the surface of its 2D CPM unit and the surface to its perimeter.

Attribute constraints, and the relative energetic penalties, can also regulate interactions between objects. Their form is analogous to Equation (1.6):

$$H_{constraint}(t) = \sum_{\Sigma_\sigma, \Sigma_{\sigma'}} \sum_{j-constraint} \lambda_{\Sigma_\sigma, \Sigma_{\sigma'}}^{j}(t) \left[a_{\Sigma_\sigma, \Sigma_{\sigma'}}^{j}(t) - A_{\Sigma_\sigma, \Sigma_{\sigma'}}^{j}(t) \right]^2, \tag{1.7}$$

where, in this case, Σ_σ and $\Sigma_{\sigma'}$ may or may not be neighboring objects. A typical example is a linear spring connecting the center of mass of a pair of cells. In this case, $A_{\Sigma_\sigma, \Sigma_{\sigma'}}^{j}$ represents the equilibrium length of the connection, and $a_{\Sigma_\sigma, \Sigma_{\sigma'}}^{j}$ is the actual distance between the center of mass of the two neighboring cells. Such a constraint is useful, for instance, to represent tight junctions between endothelial cells in a mature capillary that maintain the integrity of the vessel [365].

It is useful to underline that, in the basic CPM, both the target attributes of objects and the relative Potts parameters are usually the same for all individuals of the same type, i.e.,

$$\begin{cases} \lambda_{\Sigma_\sigma}^{i} = \lambda_{\tau(\Sigma_\sigma)}^{i}; \\ \lambda_{\Sigma_\sigma, \Sigma_{\sigma'}}^{j} = \lambda_{\tau(\Sigma_\sigma), \tau(\Sigma_{\sigma'})}^{j}; \\ A_{\Sigma_\sigma}^{i} = A_{\tau(\Sigma_\sigma)}^{i}; \\ A_{\Sigma_\sigma, \Sigma_{\sigma'}}^{j} = A_{\tau(\Sigma_\sigma), \tau(\Sigma_{\sigma'})}^{j}. \end{cases}$$

Relations (1.6) and (1.7) give the classical formulations of the energetic constraints describing object attributes adopted in CPM applications. However, more suitable types of potentials can be used, as we will see in Chapter 4.

The last term in Equation (1.4) includes the energetic counterparts of the forces (both effective and generalized) that act on the simulated individuals. All these contributions are described with the same architecture [99]:

$$H_{force}(t) = -\sum_{\mathbf{x}\in\Sigma_\sigma}\sum_{k-force} \mu^k_{\Sigma_{\sigma(\mathbf{x})}}(t)\,\mathbf{F}^k(t)\cdot\mathbf{r_x}, \tag{1.8}$$

where $\mathbf{r_x} = (i_\mathbf{x}, j_\mathbf{x}, k_\mathbf{x})^T$ is the position vector of lattice site $\mathbf{x}$, which is the application point of force $\mathbf{F}^k$, and $\mu^k_{\Sigma_\sigma}$ is the relative Potts parameter, which measures the effective strength of the force on object Σ_σ (i.e., also in this case it can be common for all the objects of the same type: $\mu^k_{\Sigma_\sigma} = \mu^k_{\tau(\Sigma_\sigma)}$) .

The most diffused example in CPM applications is the action due to the presence of extracellular chemical substances (which are described as continuous CPM objects) on a population of cells (which are typical discrete objects):

$$H^{chemical}_{force}(t) = -\sum_{\Sigma_\sigma}\sum_{\mathbf{x}\in\Sigma_\sigma} \mu^{chem}_{\Sigma_{\sigma(\mathbf{x})}}(t)c(\mathbf{x},t), \tag{1.9}$$

where $c(\mathbf{x},t)$, defined with Equation (1.1), is the concentration of the chemical sensed by the cell in $\mathbf{x}$ (which can be modeled as the local chemical concentration in site $\mathbf{x}$ itself [203, 259], or in its neighborhood [350, 351]), and the Potts coefficient $\mu^{chem}_{\Sigma_\sigma}$ is, in this case, an effective chemical potential of cell Σ_σ. Moreover, the net energy difference caused by such a chemical force is

$$\Delta H^{chemical}_{force}\Big|_{\sigma(\mathbf{x})\to\sigma(\mathbf{x}')} = \mu^{chem}_{\Sigma_{\sigma(\mathbf{x})}}[c(\mathbf{x},t) - c(\mathbf{x}',t)], \tag{1.10}$$

where $\mathbf{x}\in\Sigma_\sigma$ and $\mathbf{x}'\notin\Sigma_\sigma$ are the two neighboring lattice sites randomly selected during the trial update at time t [342].

If $\mu^{chem}_{\Sigma_\sigma}$ is a constant, Σ_σ has a linear chemical sensitivity. In particular, $\mu^{chem}_{\Sigma_\sigma} > 0$ yields to its motion up the gradient of c (which is thus a *chemoattractant*, and the relative force is called *chemotaxis*), while $\mu^{chem}_{\Sigma_\sigma} < 0$ yields to its motion in the opposite direction (and c is a *chemorepellent*). Moreover, if c is a nondiffusive fixed substrate, Equation (1.9) is a representation of a *haptotactic* force, as in [246, 395].

The importance of each term in the *Hamiltonian* (i.e., of each simulated biological mechanism) is defined by the magnitude of the relative Potts parameters, which act as a sort of *penalty coefficient* (whose unit of measure is obviously established by the type of the relative energetic contribution). In this respect, a crucial role in determining the evolution of the system is therefore played by the hierarchy of the Potts coefficients, and not by their exact values. This consideration allows one to overcome one of the main limitations of the CPM: given its energetic nature, a direct one-to-one correspondence between Potts parameters and experimental quantities is not straightforward, it being only possible to infer empirical relationships. This issue will be approached more in details in Chapter 4. However, it is useful to underline that, in the following, we will use the terms *low* and/or *high* to set the place of a single parameter in the overall hierarchy of Potts coefficients.

1.4 Evaluation of Some Kinematic Parameters

The *position* of an object Σ_σ is established by the position of its center of mass

$$\mathbf{x}_{\Sigma_\sigma}^{CM}(t) = \frac{1}{a_{\Sigma_\sigma}^{volume}(t)} \sum_{\mathbf{x}\in\Sigma_\sigma} \mathbf{x}, \tag{1.11}$$

where $a_{\Sigma_\sigma}^{volume}(t)$ is its actual volume.

The *instantaneous velocity* of Σ_σ is calculated as the velocity of its center of mass:

$$\mathbf{v}_{\Sigma_\sigma}(t) = \frac{\mathbf{x}_{\Sigma_\sigma}^{CM}(t) - \mathbf{x}_{\Sigma_\sigma}^{CM}(t-\Delta t)}{\Delta t}, \tag{1.12}$$

over $\Delta t = 1$ MCS, as done in [261, 309, 350].

Indeed, if $\Delta t > 1$ MCS, we measure a *mean velocity* of the individual. In particular, the *average velocity* of Σ_σ over an entire simulation is given by

$$\overline{\mathbf{v}}_{\Sigma_\sigma}(t) = \frac{\mathbf{x}_{\Sigma_\sigma}^{CM}(t_{final}) - \mathbf{x}_{\Sigma_\sigma}^{CM}(0)}{t_{final}}, \tag{1.13}$$

where $\mathbf{x}_{\Sigma_\sigma}^{CM}(0)$ is the initial position of its center of mass and t_{final} corresponds to the final time of the observation period of interest.

The *mean square displacement* (MSD) at time t of an individual Σ_σ, $< d_{\Sigma_\sigma}^2(t) >$, is calculated as

$$< d_{\Sigma_\sigma}^2(t) >=< (\mathbf{x}_{\Sigma_\sigma}^{CM}(t) - \mathbf{x}_{\Sigma_\sigma}^{CM}(0))^2 >, \tag{1.14}$$

where $\mathbf{x}_{\Sigma_\sigma}^{CM}(0)$ is defined in Equation (1.13). Following [106, 421], the square displacements are averaged over all previous time steps in order to take into account the back and forth motions exhibited by the moving individual.

It is useful to emphasize that the above definition of kinematic parameters can be easily extended to the compartmentalization approach that will be introduced in Chapter 4, by noting that the center of mass of the individual will have to be calculated for the entire element and not for each single objects.

1.5 Some Illustrative Simulations

In order to clarify the applicative significance of the specific terms in the *Hamiltonian* introduced in the previous section, we now illustrate some biologically relevant test simulations.

The elastic moduli $\lambda_{\Sigma_\sigma}^i$ and $\lambda_{\Sigma_\sigma,\Sigma_{\sigma'}}^j$ in Equations (1.6) and (1.7) essentially ensure that the relative individuals maintain their biophysical attributes close

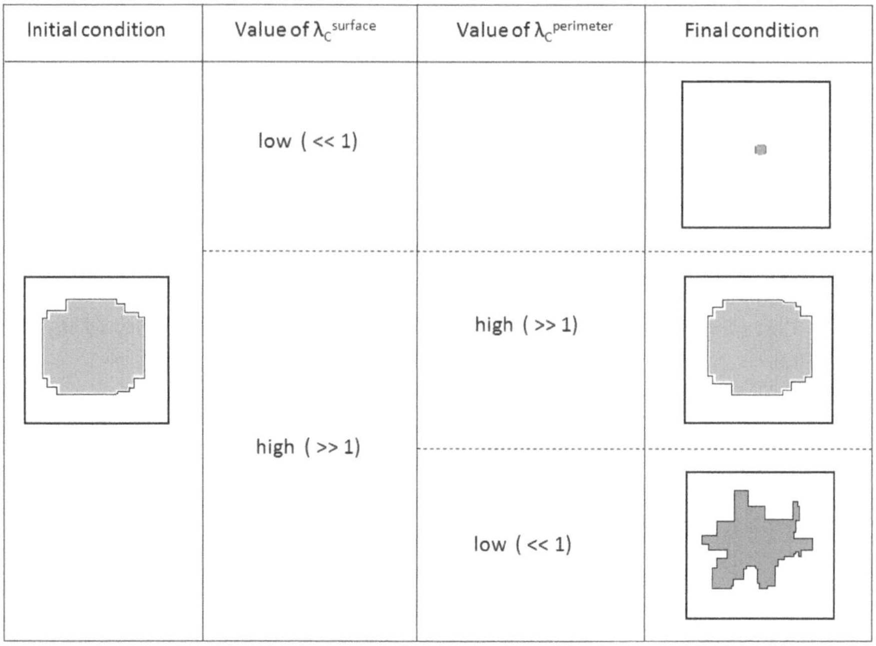

FIGURE 1.2: Illustrative simulations of the biological role of the geometrical constraints. In a 100×100-site domain Ω, an initially round cell of 75 sites of diameter (i.e., initial area of 4415 sites^2 and initial perimeter of 235.5 sites) is placed. The cell target measures $A_C^{surface}$ and $A_C^{perimeter}$ coincide with the initial dimensions. Top line: If the surface constraint $\lambda_C^{surface}$ is low (i.e., 0.2 site^{-2}), the cell shrinks. Middle line: If both the surface and the perimeter elastic moduli are high (i.e., $\lambda_C^{surface} = \lambda_C^{perimeter} = 50$), the cell freezes as soon as it reaches its target measure. Bottom line: If the cell surface constraint is high (i.e., $\lambda_C^{surface} = 50$ site^{-2}), whereas the perimeter constraint is low (i.e., $\lambda_C^{perimeter} = 0.2$ site^{-1}), the cell continuously remodels while preserving its overall area.

to their target states/values. As already sketched, the most relevant parameters of this family are the geometrical constraint. To provide a simple description of their role, we reproduce the random movement of a single cell $\tau = C$ placed in a homogeneous medium $\tau = M$, which may represent the fluid component of the extracellular matrix; see Figure 1.2. We do not consider other terms in the *Hamiltonian*. If $\lambda_C^{surface}$ is low (i.e., $\ll 1$), the cell shrinks, regardless of its initial dimension and of the value of $\lambda_C^{perimeter}$; Figure 1.2 (top line). On the contrary, if $\lambda_C^{surface}$ is high (i.e., $\gg 1$), the cell rapidly reaches its target measures. After this transient, the cell freezes if also $\lambda_C^{perimeter} \gg 1$, while it undergoes random fluctuations that, however, preserve its area, if $\lambda_C^{surface} \ll 1$; Figure 1.2 (middle and bottom lines). It is straightforward to

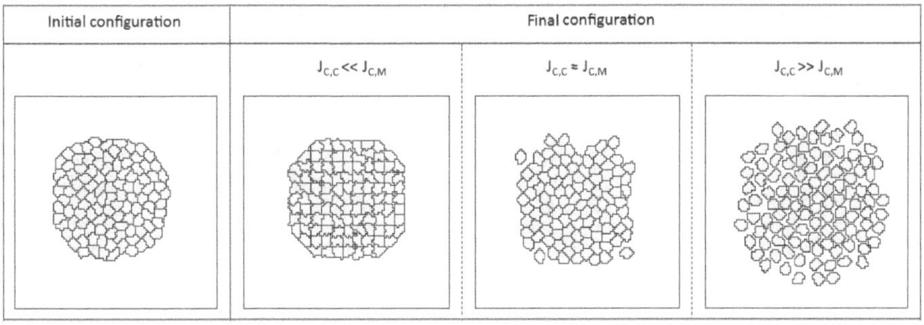

FIGURE 1.3: Illustrative simulations of the biological role of the adhesive energies. In a 200 × 200-site domain Ω, a round cluster of cells is placed. The cells are all of the same type, with an initial diameter of 8 sites, target measures $A_C^{surface}$ and $A_C^{perimeter}$ equal to the initial dimensions, $\lambda_C^{surface}$ = 50 site^{-2}, and $\lambda_C^{perimeter} = 0.2$ site^{-1}. The final configurations correspond to $J_{C,C} = 0.4 \ll J_{C,M} = 25$ (left panel); $J_{C,C} = J_{C,M} = 4$ (middle panel); and $J_{C,C} = 25 \gg J_{C,M} = 0.4$ (right panel).

notice that similar dynamics can he obtained in the three-dimensional case by substituting, respectively, $\lambda^{surface}$ with λ^{volume} and $\lambda^{perimeter}$ with $\lambda^{surface}$.

As stated above, the values of the binding energies Js in Equation (1.5) describe the relative preference of simulated individuals for creating adhesive *homotypic* (i.e., with individuals of the same type) or *heterotypic* (with individuals of other types or with extracellular components) boundaries. To analyze how specific variations of such adhesion strengths affect the overall dynamic, as well as the final configuration, of a biological system, we first examine the evolution of a round cluster of cells, all of the same type $\tau(\Sigma_\sigma) = C$, plated in the center of the standard extracellular medium, $\tau = M$; see Figure 1.3. For the sake of simplicity, the cell initial and target measures coincide, with a high surface constraint $\lambda_C^{surface} \gg 1$ that preserve their dimensions and a low perimeter constraint $\lambda_C^{perimeter} \ll 1$ that allows their remodeling and fluctuations.

If $J_{C,C} \ll J_{C,M}$, which means an intercellular adhesion significantly higher than the cell–medium adhesion, and hence either an overexpression of cadherin molecules or a downregulation of integrin ligands, the cellular spheroid remains compact and strictly packed, trying to minimize its external surface, because for the cells it costs more, from an energetic viewpoint, to float in the medium rather than stay attached to each other (Figure 1.3 (left panel)).

If the two contact energies are instead comparable (i.e., $J_{C,C} \approx J_{C,M}$), a small friction with external cells is able to randomly detach and reattach from the rest of the cluster, whose overall border continuously remodels (Figure 1.3 (middle panel)).

Finally, for larger values of cell–cell adhesion (i.e., $J_{C,C} \gg J_{C,M}$, which represent the model counterpart of downregulation in cadherin activity), the

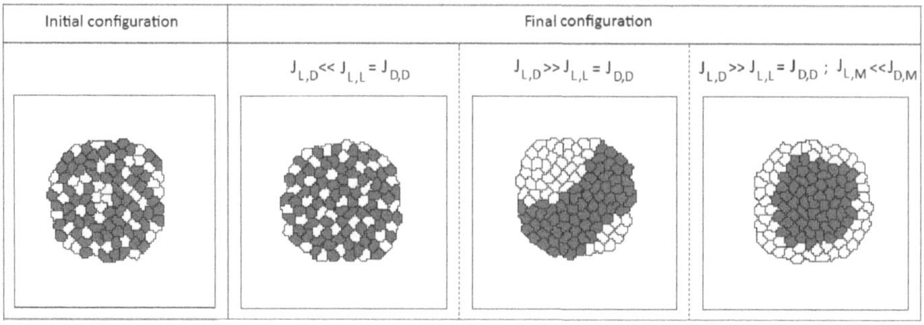

FIGURE 1.4: Illustrative simulations of the biological role of the adhesive energies. In a 200 $\times$ 200-site domain Ω, a round cluster of cells is placed. The cells are of two different types (i.e., light and dark), with initial and target measures as in the previous case, as well as the values of $\lambda_L^{surface} = \lambda_D^{surface} = 50$ site^{-2} and $\lambda_L^{perimeter} = \lambda_D^{perimeter} = 0.2$ site^{-1}. The final configurations correspond to $J_{L,D} = 0.4 \ll J_{L,L} = J_{D,D} = 20$ and $J_{L,M} = J_{D,M} = 0$ (left panel); $J_{L,D} = 20 \gg J_{L,L} = J_{D,D} = 0.2$ and $J_{L,M} = J_{D,M} = 0$ (middle panel); $J_{L,D} = 20 \gg J_{L,L} = J_{D,D} = 0.2$ and $J_{L,M} = 0.8 \ll J_{D,M} = 12$ (right panel).

aggregate quickly dissociates into isolated individuals, displaying a scattered phenotype. In fact, not only do external cells lose contact and spread away from the main spheroid, but a repulsion occurs also among individuals within its core, because for the cells it costs more to stay attached to each other than to float in the medium (Figure 1.3 (right panel)).

As we will described more in detail in the following chapters, the dispersion of single cells upon low intercellular adhesion is of particular relevance in the early stages of solid cancer growth and development. The malignant individuals able to escape from the main tumor mass have in fact the greatest invasive potential, as they can invade the host tissue and further metastasize.

In multicellular organisms, the relative adhesion of various cell types to each other or to noncellular components surrounding them is also fundamental. From the late 1950s, it has been widely noticed that during embryonic development the behavior of cell aggregates resembles that of viscous fluid. A random mixture of two types of embryonic cells, in fact, spontaneously reorganizes to reestablish coherent homogeneous tissues. A similar process is a key step also in the regeneration of normal animal from aggregates of dissociated cells of adult hydra. It also explains the layered structure of the embryonic retina. These phenomena, commonly called cell *sorting*, involve neither cell division nor differentiation, but are entirely caused by spatial rearrangements of cell positions due to differences in the specific adhesivities; see [165, 169] and references therein. Indeed, specific hierarchies of adhesive strengths lead to specific configurations of the cellular aggregate.

The most famous simple and intuitive CPM simulation reproducing biological cell sorting consists of a cellular aggregate formed by two types of randomly positioned individuals, namely, light $\tau = L$ and dark $\tau = D$, and placed in the standard extracellular medium $\tau = M$; see again [165, 169]. The target measures of all cells are the same as in the previous simulations, as well as their geometrical constraints (i.e., we use high values of $\lambda_L^{surface} = \lambda_D^{surface}$ and low values for $\lambda_L^{perimeter} = \lambda_D^{perimeter}$). If the adhesive energy between the two cell types is higher than the two self–contact interactions (i.e., $J_{L,D} < J_{L,L} = J_{D,D}$), cells heterogeneously mix to form an experimentally observed checkerboard (Figure 1.4 (left panel)). Again, this is due to the fact that making heterotypic bonds is more convenient from the energetic point of view than making homotypic bonds with cells of the same type.

If the homotypic adhesions are stronger than the heterotypic ones (i.e., $J_{L,D} > J_{L,L} = J_{D,D}$), we find a spontaneous cell sorting, with the formation of small clusters of cells of the same type within the spheroid; Figure 1.4 (middle panel). In particular, if further, the edge adhesion between a cell type and the medium is significantly low (i.e., $J_{L,D} > J_{L,L} = J_{D,D}$ and $J_{L,M} < J_{D,M}$), a single cluster of this cell type forms in the center of the spheroid, surrounded by a crew of individuals of the other type (Figure 1.4 (bottom right panel)): this phenomenon is called *engulfment* and was first investigated in a CPM in [169]. This is due to the fact that after clustering, the cell type that is able to build stronger bonds with the medium (i.e., lower energetic costs) will tend to stay in the exterior layers of the aggregate.

As previously seen, the chemical responses are the most important and commonly used forces described by the term (1.8) of the *Hamiltonian*. To provide a useful analysis of how it works, we simulate the chemotactic movement of a single cell, $\tau = C$, influenced by the chemical concentration gradients created in the extracellular medium, $\tau = M$, by a punctual source; see Figure 1.5. The chemical substance evolves as the following:

$$\frac{\partial c}{\partial t} = \underbrace{D_c \nabla^2 c}_{diffusion} - \lambda_c c + \underbrace{S}_{source}, \tag{1.15}$$

where, as seen, $c = c(\mathbf{x}, t)$ denotes the actual concentration of the peptide at the medium site $\mathbf{x}$. The coefficients of diffusivity, D_c, and of degradation, λ_c, are assumed to be constant throughout the domain. S describes the production of the chemical at a constant rate per unit of time by the discrete source. If the cell chemical strength μ_C^{chem} is positive, the cell undergoes a gradual transition from the initial symmetric stationary state to a polarized migratory state characterized by clearly distinguishable leading and trailing edges, and moves in the direction of the source up the concentration gradients. On the contrary, if μ_C^{chem} is negative, the cell again somehow elongates, but starts to move in the opposite direction, down the concentration gradients. Finally, if $\mu_C^{chem} = 0$, the cell remains more or less hemispheric and starts a slow random

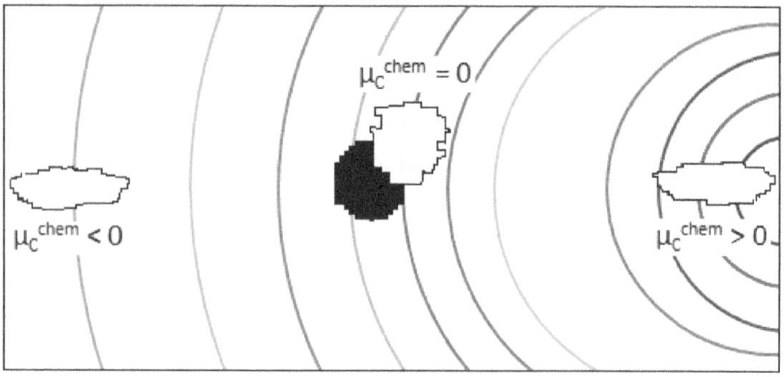

FIGURE 1.5: Illustrative simulations of the biological role of chemotaxis. In a 300 × 100-site domain Ω, an initially round cell of 30 sites of diameter is placed; see black individual. The target measures of the simulated cell, $A_C^{surface}$ and $A_C^{perimeter}$, coincide to its initial dimensions. $\lambda_C^{surface} = 50$ site^{-2} and $\lambda_C^{perimeter} = 0.2$ site^{-1}. A point source of a chemical, that evolves satisfying Equation (1.15), is applied in $\mathbf{x} = (300, 150)$. If $\mu_C^{chem} = 10\ \mu\text{M}^{-1}$, the cell moves in the directions of higher morphogen concentrations, displaying a persistent directional migration. If $\mu_C^{chem} = 0\ \mu\text{M}^{-1}$, the motion of cells is uncorrelated from the chemical field. Finally, if $\mu_C^{chem} = -10\ \mu\text{M}^{-1}$, the cell moves in the opposite direction w.r.t. the chemical gradient. Isolines indicate chemical levels.

motion, almost remaining in the initial position, since it is not affected by the chemical field.

Chapter 2

HGF-Induced Cell Scatter

2.1 Biological Introduction

Hepatocyte growth factor (HGF), originally identified as a mitogen for hepatocytes [271, 288], was subsequently shown to be identical to scatter factor (SF) [45, 92, 97], a ligand able to induce epithelial cell dissociation [91]. HGF, produced primarily by mesenchymal cells, is therefore a unique growth factor that elicits multiple cellular responses via its receptor, a tyrosine kinase encoded by the proto-oncogene Met [257, 288]. The main biological processes regulated by HGF are mitogenesis, motility and morphogenesis [389, 394], cell dissociation, migration through the extracellular matrix, acquisition of polarity, and tubule formation [338, 380]. This combination of events, also known as *invasive growth*, is fundamental during the embryonic development of most epithelial tissues. When inappropriately activated, this genetic program confers invasive ability to normal and neoplastic epithelial cells [271, 407].

Overexpressions of the HGF receptor Met have been observed in primary thyroid cells [223, 261], suggesting that Met may mediate some of the phenotypic alterations observed in these cells: these data allow one to precisely define the role of the HGF and its receptor in the evolution of thyroid tumors, finding which of the many cellular events known to follow Met activation effectively take place in thyroid cancer cells. Indeed, an experimental model of a poorly differentiated thyroid carcinoma-derived cell line, ARO, provides evidence that HGF stimulation does not increase the growth rate, probably due to the high basal level of proliferation, but it increases motility [101, 104]. In particular, HGF induces a mesenchymal transition with a dramatic change from a virtually nonmotile phenotype to a phenotype with high short-range motility: cells loosen contacts, dissociate, move from their original site and start wandering in close proximity, displaying an evident ability to invade the open spaces and modifying the interaction with the extracellular matrix; see Figure 2.1 (the full time-lapse movie is available at calvino.polito.it/~preziosi/AROex.avi). It is noteworthy that in response to HGF, ARO cells do not show a dramatic cytoskeletal reorganization or polarization but lose their tendency to maximize the contact area with the surrounding cells and the physiological regulation of contact inhibition mechanism. The change in the above-mentioned adhesive properties is a consequence of the change in the transmission of a number

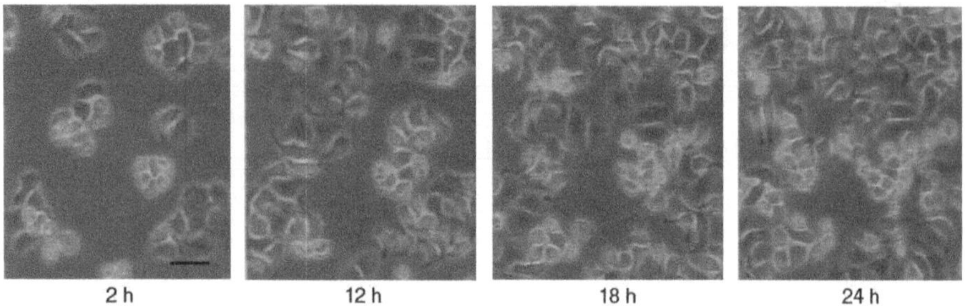

FIGURE 2.1: Phase contrast time-lapse microscopy of in vitro dissociation and migration of ARO cells, grown in 10% fetal calf serum and incubated in the presence of 100 U/mL of hepatocyte growth factor (HGF) for 2 h, 12 h, 18 h, and 24 h. The scale bar is 25 μm long.

of local, contact-dependent signals (juxtracrine). In fact, we can hypothesize that the loss of intercellular adhesions does not cause by itself a dissociation phenotype, since E-cadherin -/- cells may still form strong adhesive junctions. Instead, the loss of E-cadherin signaling, related to HGF downstream signaling pathways (e.g., via BCL9-2, [43]), may be responsible for this behavior [260, 324]. The above assumptions are validated also by the fact that in the absence of HGF the transmission of the juxtracrine is recovered so that the dissociated and disconnected cells have a tendency to get back together and form new compact colonies (not shown).

Another scattering example is given by cultures of MLP-29, a strongly HGF-responsive epithelial cell line established from mouse embryonic liver. MLP-29 are small cells with ovoidal nuclei, forming tightly packed colonies. After the addition of nanomolar concentration of HGF, the cells start to elongate, migrate, and differentiate, leading the islands to assume the characteristic stellar configuration within 1 day, as shown in Figure 2.2 (the full time-lapse movie is available at calvino.polito.it/~preziosi/MLPex.avi). This morphogenic response is dose dependent and consists of three steps:

1. Within the first 6 h, the colony borders start roughening and most of cellular junctions (but not all) start breaking or reorganizing.

2. In the following 6 h there is a cytoskeletal reorganization followed by an asymmetric polarization and an active migration of cells, which lead to the formation of some sprouts.

3. Within 24 h, at the end of the branching process, the colonies assume the characteristic stellar shape, with sprouts formed by one or few elongated cells that keep a thin adhesion contact with the rest of the island.

None of these events is observed in control medium or after stimulation with other growth factors: in those cases, MLP-29 cells tend to form spherical and

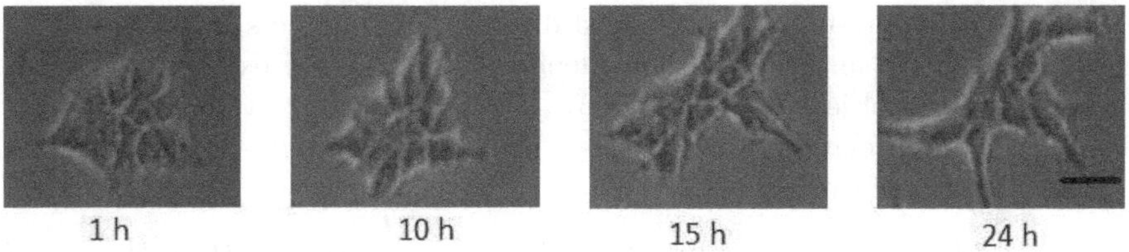

FIGURE 2.2: Typical time sequence of in vitro branching morphogenesis of a colony of MLP-29 cells, incubated for 24 h in DMEM - 1% FCS, containing recombinant hepatocyte growth factor (HGF), diluted to the final concentration of 100 ng/mL, at 1 h, 6 h, 10 h, 15 h, 20 h, and 24 h after incubation. The scale bar is 25 μm long.

compact aggregates of undifferentiated, almost round cells, with the interstitial spaces almost filled with microvilli protruding from the cell surface along the entire perimeter of each cell [271].

In HGF-driven scattering, the cells in the middle of an island do not undergo a dramatic change of shape, keeping their original elliptical shape, and passively move only when dragged by the nearest sprout. On the other hand, the cells at the margin acquire a polar configuration driven by a cytoskeletal remodeling: they become highly asymmetric and have distinct apical, lateral, and basal surfaces that correspond to a leading and trailing surface during their migration. In particular, they migrate faster along their long axes than along their short axes: such an anisotropic, cell-shape-dependent motility results in strong directional persistence because the cells must reorient before moving toward a new direction and also introduces two timescales into scattering: fast migration along the cells' long axes produces rapid branching and polarization, whereas the slower sideways migration coarsens the pattern as the cells align and close small lacunae. The polarized cells, in the part of their surface in contact with the ECM, extend filopodia toward the open space and keep a thin adhesion point with the main corpus of the colony.

The branches, generally formed by couples of parallel elongated cells, are radially disposed with angles ranging between 15 and 100 degrees. If the HGF is no longer added to the culture, it can be observed as a sort of involution of the sprouts, and the colonies tend to become round and compact again (not shown).

2.2 Mathematical Model for ARO Aggregates

The scattering process is modeled on a bidimensional domain Ω, which is consistent with the planar substrates used in the experiments. The ARO individuals are represented as standard discrete CPM objects Σ_σ of type $\tau = C$. They reside in an isotropic homogeneously distributed extracellular matrix, a generalized object of type $\tau = M$. The ARO colony evolves following the rules given in relation (1.2), where T is an overall motility of the culture and the *Hamiltonian* reads as

$$H(t) = H_{adhesion}(t) + H_{shape}(t) = H_{adhesion}(t) + H_{volume}(t) + H_{surface}(t), \tag{2.1}$$

which, expanding all terms for readers' convenience, becomes:

$$\begin{aligned} H(t) \quad = \quad & \sum_{\substack{\mathbf{x},\mathbf{x}' \in \Omega'_{\mathbf{x}}: \\ (\partial \mathbf{x} \in \partial \Sigma_\sigma) \cap (\partial \mathbf{x}' \in \partial \Sigma_{\sigma'}) \neq \emptyset \\ \tau(\Sigma_{\sigma(\mathbf{x})}) = \tau(\Sigma_{\sigma'(\mathbf{x}')}) = C}} J_{C,C} + \sum_{\substack{\mathbf{x},\mathbf{x}' \in \Omega'_{\mathbf{x}}: \\ (\partial \mathbf{x} \in \partial \Sigma_\sigma) \cap (\partial \mathbf{x}' \in \partial \Sigma_{\sigma'}) \neq \emptyset \\ \tau(\Sigma_{\sigma(\mathbf{x})}) = C \wedge \tau(\Sigma_{\sigma'(\mathbf{x}')}) = M}} J_{C,M} + \\ & + \sum_{\Sigma_\sigma} \lambda_{\Sigma_\sigma}^{surface} (a_{\Sigma_\sigma}^{surface}(t) - A_C^{surface})^2 + \\ & + \sum_{\Sigma_\sigma} \lambda_{\Sigma_\sigma}^{perimeter} (a_{\Sigma_\sigma}^{perimeter}(t) - A_C^{perimeter})^2. \end{aligned} \tag{2.2}$$

$H_{adhesion}(t)$ has the same form as in Equation (1.5): in particular, since there is only one type of cell placed in an extracellular medium, the Js reduce to cell–matrix $J_{C,M}$ and cell–cell $J_{C,C}$ adhesive strengths. The terms in $H_{shape}(t)$ are instead the energetic constraints regulating the changes in cell dimensions and shape and are given as in Equation (1.6). Indeed, the target dimensions, corresponding to the initial measures, and the elastic moduli are the same for all ARO individuals, i.e., for all Σ_σ such as $\tau(\Sigma_\sigma) = C$:

$$\begin{cases} A_{\Sigma_\sigma}^{surface} = A_C^{surface}; \\ A_{\Sigma_\sigma}^{perimeter} = A_C^{perimeter}; \\ \lambda_{\Sigma_\sigma}^{surface} = \lambda_C^{surface}; \\ \lambda_{\Sigma_\sigma}^{perimeter} = \lambda_C^{perimeter}. \end{cases}$$

In particular, following the experimental observations provided in the previous section, both the growth rate and the elasticity of the AROs are here assumed to be negligible, and therefore we set high values of $\lambda_C^{surface} = \lambda_C^{perimeter}$.

In the case of ARO aggregates, HGF stimulation is modeled only through its effects both on the motility of individuals and on their adhesive interactions. In particular, we preliminarily hypothesize a directed proportionality between the amount of HGF added in the culture and the model parameters T and $J_{C,C}$: increasing doses of growth factor therefore correspond to increasing values of the cell agitation rate T and of the intercellular adhesion coefficient, which results in weaker cell–cell contact junctions. $J_{C,M}$ remains instead fixed in each simulation, as HGF stimulation has not been demonstrated to affect the expression of the activity of integrin molecules.

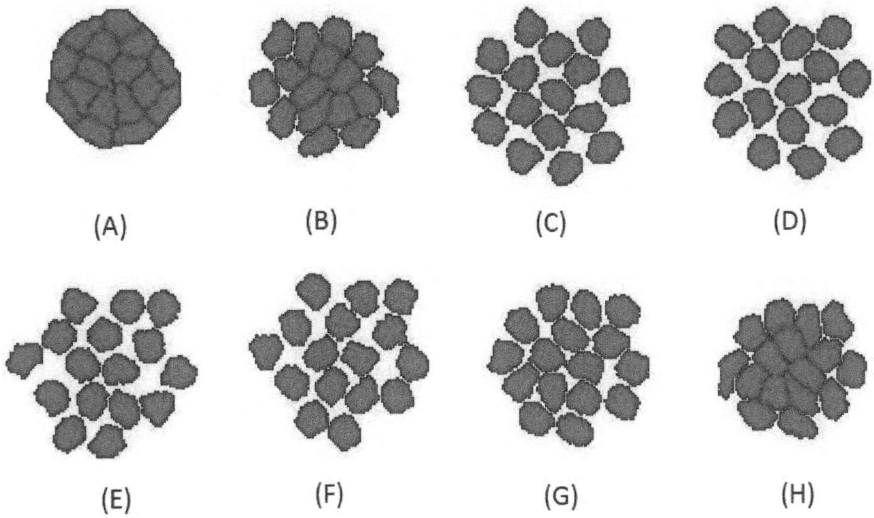

FIGURE 2.3: Cell-cell dissociation process in a simulation initiated with a clump of 16 ARO cells: (A) 6 h, (B) 12 h, (C) 18 h, (D) 24 h, (E) 30 h, (F) 36 h, (G) 42 h, and (H) 48 hours. After 3000 MCS ($\approx$ 24 hours), the HGF is no longer added in the culture.

2.3 Scattering of ARO Aggregates

Each simulation is initiated with a mass of 16 virtual cells over an area of 300×300 lattice sites, which we positioned in a larger domain Ω of 500×500 sites to minimize boundary effects. Each CPM lattice site corresponds to a square of size 2 μm $\times$ 2 μm. A Monte Carlo Step (MCS) corresponds to 30 s: for this choice of timescale, the mean of the cell velocities agrees well with *in vivo* observations. The initial area A_C^{volume} of the simulated ARO cells is set to 50 lattice sites, corresponding to 200 μm^2, while their perimeter $A_C^{perimeter}$ to 48, i.e., 96 μm. Assuming that the HGF has already been added in the culture, and following the hypothesis made in the previous section, we first use high values for the motility parameter T and for $J_{C,C} \gg J_{C,M}$; see Table C.1 in Appendix C.

Figure 2.3(A-D) shows a typical simulation of the time-sequence evolution of a culture of ARO cells in the presence of HGF. After an initial roughening, the cells start to detach and to radially spread in close proximity, keeping an almost elliptical shape. In Figure 2.3(E-H), we also simulate the loss of HGF in the culture medium by decreasing $J_{C,C}$ while maintaining the other parameters fixed: as in the experiments, the cells aggregate again in an closely adjoined island, confirming the hypothesis of the recovered E-cadherin signaling.

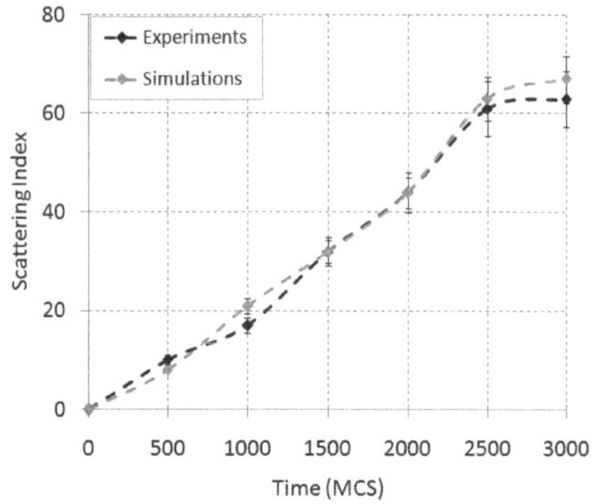

FIGURE 2.4: Comparison of the evolution in time of the Scattering Index (as defined in Equation (2.3) of ARO cells for experimental (dark line) and simulation (light line) evidences. Error bars show standard deviations over ten representative colonies for the experiments and over ten simulations.

The evolution of the simulated colony compares well with the experimental evidences, having also the same temporal dynamics, as shown in Figure 2.4, where we plot the *scattering index*

$$S_I(t) = \frac{A(t)}{A(0)} - 1, \tag{2.3}$$

with $A(t)$ defined as the area of the minimal convex polygon enclosing the aggregate at time t and $A(0)$ the one at $t = 0$. The experimental scatter begins slightly slower than the simulated; however, after the initial phase ($t > 1500$ MCS), S_I increases at comparable rates in both cases.

To further study the robustness of the model and to identify the critical features of the output, we analyze how the results of the simulations depend on the parameters by measuring the scattering index after 3000 MCS (i.e., 1 day). Figure 2.5 (left panel) shows that, at suitable T, our simulated aggregates do not scatter for low values of intercellular adhesion parameter $J_{C,C}$, corresponding to strong intercellular adhesion, whereas they dissociate when cells adhere less strongly: this observation well corresponds with the experiments where ARO aggregates scatter at high concentrations of HGF (see [101]).

In the right panel of Figure 2.5, we also look for the motilities required for the process: for $T < 15$, there is no detectable scattering within the time limit (3000 MCS), independently of the adhesion energies $J_{C,C}$ (the cells dissociate, but do not spread away), while for $T > 60$ they break up into small pieces or fragments, a well-characterized nonbiological artifact of the CPM due to the

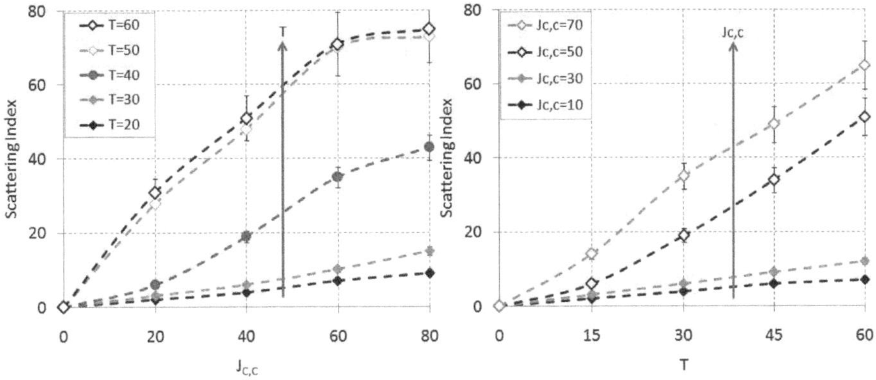

FIGURE 2.5: Dissociation process in response of HGF in a culture of ARO cells. The scattering index at t = 3000 MCS, defined as (2.3), as a function of (left panel) cell–cell adhesion energy $J_{C,C}$ at $T = 20$, $T = 30$, $T = 40$, $T = 50$ and $T = 60$ and of (right panel) Boltzmann temperature T at $J_{C,C} = 10$, $J_{C,C} = 30$, $J_{C,C} = 50$ and $J_{C,C} = 70$. Error bars show s.d. over 10 simulations.

fact that too high agitation rates result in a chaotic sequence of spin flips [260] (not shown). For the in-range values (i.e., $30 < T < 60$), the process becomes more significant and quicker for every T-increment.

These results validate our hypothesis that HGF concentration can be modeled by increasing simultaneously T and $J_{C,C}$, since both an enhancement in cell motility and the breakage in E-cadherin bonds are needed for ARO cell scattering, as seen in experiments. In particular, such distinctive effects of HGF on ARO cells are clearly detected at a certain range of doses (i.e., in a certain range of model parameters), since they are not elicited for lower quantities. Therefore, it can be concluded that the ARO cells acquire not only an HGF-dependent phenotype but an HGF dose-dependent one.

Figure 2.6 shows a definitive confirmation of the HGF-induced increase of cell motility, comparing the single-cell trajectories during a 24 h (= 3000 MCS) time-lapse, both in the control and in the stimulated case resulting from the experiment and the numerical simulation: the value of the mean effective distance covered by an ARO cell is about 5 μm in the absence and 12 μm in the presence of the growth factor (the mean is over ten representative cells, with a standard deviation of ≈ 1 μm in both cases), which confirms that, under normal conditions, ARO cells are characterized by a virtual nonmotile phenotype.

The enhancement of cell motility and dissociation can be further captured by a *wound healing* assay. This experimental model tests, in fact, the ability of a cell line to fill gaps created in cell monolayers and is generally considered a simple and reliable test for quantitative evaluation of cell migratory phenotype. Indeed, a confluent monolayer of ARO cells, grown in fetal calf serum,

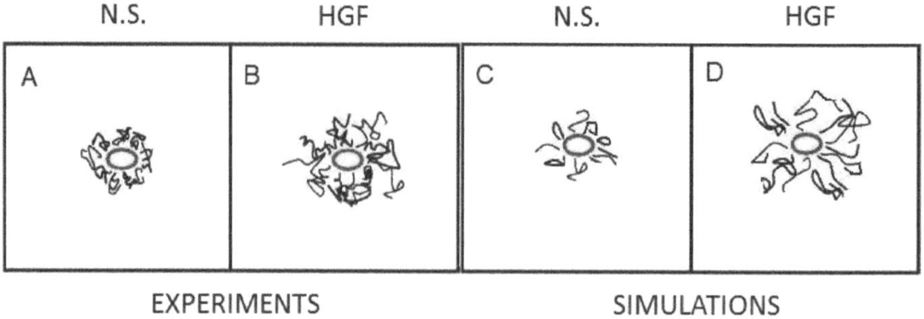

FIGURE 2.6: HGF-mediated increase of motility: comparison of single-cell motion tracks obtained by 24-h time-lapse video-recordings (A-B, see [101]) and 3000 MCS simulations (C-D) of ARO cells cultured in the absence (N.S.) and in the presence of HGF, as indicated. In the experiments, the trajectories are obtained by tracking continuously the motion of the nucleus of a cell, while in the simulations, they are formed by the position of the center of mass of the single cells at each time step. Experimentally, this would correspond to track the position of the nucleus every 30 seconds (1 MCS is set equal to 30 s). Ten representative tracks were chosen for each case and optimally arrayed for picture presentation. The oval positioned at the center of each group of tracks represents the size of a single cell.

is wounded with a pipette tip and incubated for 24 h in the presence HGF; see [101].

In the simulation shown in Figure 2.7, the culture is formed by two masses of 120 virtual cells distributed over an area of about 700 μm $\times$ 700 μm (350 $\times$ 350 lattice sites), while the other parameters are the same as in the simulations of Fig 2.3. The gap is of about 120 μm (60 lattice sites): the maximum value, deduced from the above results, for which the two colonies encounter within the time limit (24 h); for the full simulation refer to calvino.polito.it/~preziosi/AROsim2.avi. The time needed to invade the space between the colonies, in the cases of $T = 40$ and $T = 50$, is evaluated in Figure 2.8 as a function of different cell–cell adhesion parameter values. For strong intercellular bonds, corresponding to $J_{C,C} < 30$, ARO cells display barely no detectable healing after 3000 MCS (i.e., nearly 24 h), which is indicative of an inability to invade the tissue, while for an intermediate range (corresponding to $30 < J_{C,C} < 70$) the time necessary to fill the gap decreases until a sort of limit threshold characteristic of the process: about 800 MCS at $T = 40$ and 600 MCS at $T = 50$.

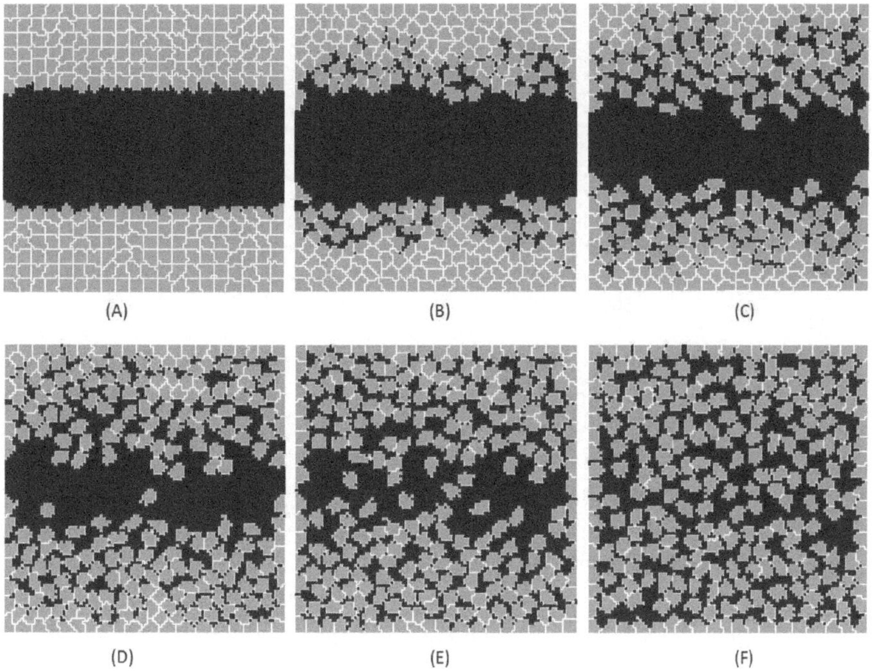

FIGURE 2.7: *Wound-healing* process: evolution in time of a simulation initiated with two masses of 120 ARO cells: (A) 1 h,(B) 8 h, (C) 12 h, (D) 16 h, (E) 20 h, and (F) 24 h.

2.4 Mathematical Model for MLP-29 Aggregates

As in the case of AROs, MLP-29 cells are represented as standard discrete CPM objects Σ_σ of type $\tau = C$, which are cultured in an isotropic extracellular medium of type $\tau = M$. In MLP-29 colonies, the stimulation with HGF cause not only an enhancement of cell motility and dispersion, but also a chemotactic stimulus. The MLP-29 individuals therefore move toward directions of higher morphogen concentrations and, still trying to keep contact with the rest of the aggregate, they elongate. In addition, such a polarization process triggers a significant persistent component of motion of MLP29 cells. A change in the direction of movement, in fact, take time, as it requires the reorganization of the cell cytoskeletal machinery.

These ingredients, not contained in the basic model for AROs but essential for the branching morphogenesis and the subsequent formation of the stellar shape aggregates experimentally found, need to be implemented by the addition of appropriate terms in the *Hamiltonian* in Equation (2.1).

In particular, the chemical response is given by relation (1.9), which results

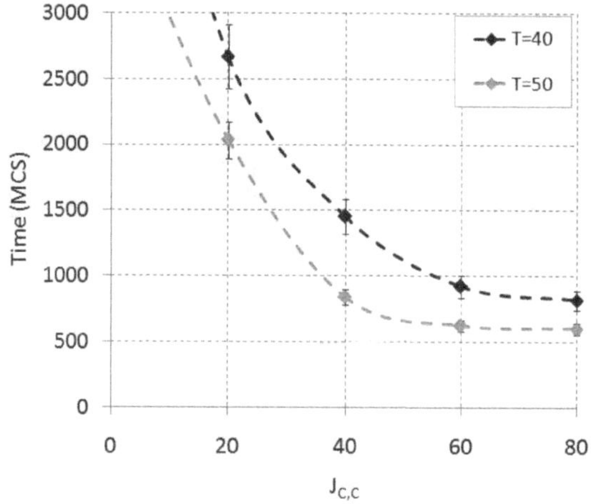

FIGURE 2.8: Time for wound healing as function of the cell–cell adhesion energy, for $T = 40$ (dark line) and $T = 50$ (light line). The error bars show standard deviations over ten simulations.

in a net energy difference at each copy attempt equal to (1.10). In this case $c(\mathbf{x}, t)$ is the local concentration of HGF, which is assumed to be present everywhere in the layer under the cells, while the positive parameter μ_C^{chem} is the chemotactic sensitivity, constant in time and homogeneous for all individuals. Indeed, the chemotaxis coefficient is set equal to 0 at cell–cell interfaces, ensuring that chemotactic extensions occur only at cell-ECM surfaces, mimicking contact inhibition chemotaxis.

The persistent, shape-dependent motion of MLP-29 is modeled with a further energy term, which coherently is a running mean over the cell past movements [20, 349]:

$$H_{persistence} = \mu_C^{pers} |\mathbf{v}(t) - \mathbf{v}(t - \Delta t)|^2, \tag{2.4}$$

where $\mathbf{v}(t)$, defined in Equation (1.12), is an average velocity of the cell, as $\Delta t = 60$ MCS. μ_{pers} controls the persistence: if $\mu_C^{pers} = 0$ the cells undergoes uncorrelated Brownian motion, while if μ_C^{pers} is very large their motion is almost ballistic.

With respect to the model of ARO colonies, the HGF is now explicitly represented as a continuous object (i.e., a field), whose addition, diffusion, and absorption are described from a macroscopic point of view, with a standard diffusion-reaction equation of type (1.1):

$$\begin{aligned} \frac{\partial c}{\partial t} &= \underbrace{D_n \nabla^2 c}_{diffusion} - \underbrace{\varepsilon_c c \delta(\tau(\Sigma_{\sigma(\mathbf{x})}), C)}_{uptake} + \\ &- \underbrace{\lambda_c c(1 - \delta(\tau(\Sigma_{\sigma(\mathbf{x})}), C))}_{decay} + \underbrace{\alpha_c}_{addition}, \end{aligned} \tag{2.5}$$

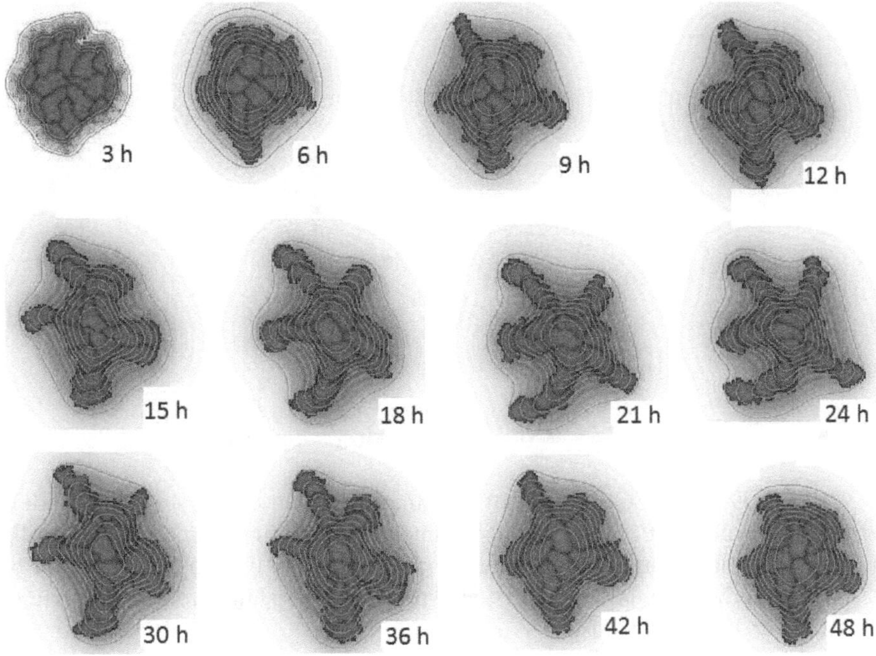

FIGURE 2.9: (See color insert.) Branching morphogenesis in a simulation initiated with a clump of 16 MLP-29 cells. After 3000 MCS $\approx$ 24 h the HGF is no longer added in the culture and regression of cell scattering is observed. Isolines (green) indicate 10 HGF levels relative to the maximum concentration in the simulation.

where $\delta(\tau(\Sigma_{\sigma(\mathbf{x})}), C)) = 1$ within cells and 0 in the extracellular environment. D_c is the characteristic diffusion coefficient and λ_c the decay rate, which are assumed constant throughout ECM. ε_c is the uptake rate of the HGF, while its input is set at a constant rate α_c over the entire domain.

2.5 Scattering of MLP-29 Aggregates

As in the case of the ARO model, an MCS corresponds to 30 s, while the size of a lattice site is set equal to 2 μm. In order to create a hybrid simulation environment, the PDE for the HGF evolution is solved using a finite difference scheme on a lattice that matches the CPM lattice, using a sufficient number of diffusion steps per MCS to guarantee the stability of the numerical method. The reader is referred to Appendix A for more details of the computational implementation.

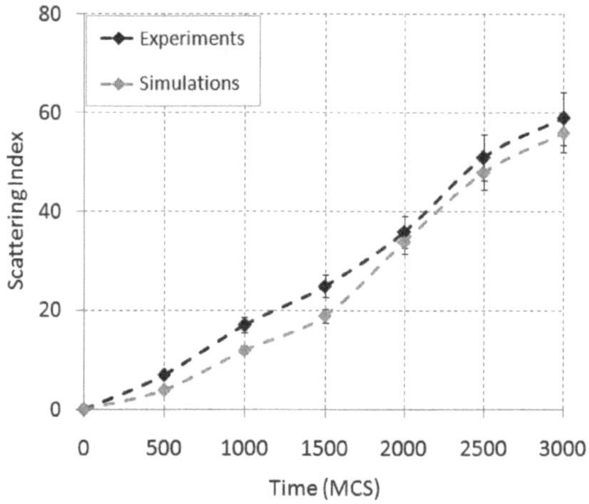

FIGURE 2.10: Comparison of the time-evolution of the scattering index of a round colony of 16 cells for experimental (dark line) and simulation (light line). The parameters of the simulations are those of Figure 2.9. The error bars show standard deviations over ten representative islands for the experiments and over 10 simulations.

We initiate the simulation with a mass of 16 virtual cells, each of them with an area of $\approx 160\ \mu\text{m}^2$ (40 lattice sites), over an area of $\approx 700\ \mu\text{m} \times 700\ \mu\text{m}$ (350×350 sites), which is positioned inside a larger domain Ω of $1000\ \mu\text{m} \times 1000\ \mu\text{m}$ (500×500 sites) to minimize the boundary effects. The HGF-driven enhancement of cell motility is modeled by increasing the value of T, e.g., taking it proportional to the concentration of the growth factor. Assuming that the cells do not grow during the process, we set high $\lambda_C^{surface}$. The active cell cytoskeletal remodeling, which results in cell elongation, is instead permitted by a low $\lambda_C^{perimeter}$. The effective cell–cell adhesion is the same as the cell–ECM adhesion, since experimental analysis suggests a reorganization of intercellular bonds rather than a completely breakage of them as in the case of ARO individuals. Finally, to model significant chemotaxis and persistent component in cell movement, we set high $\mu_C^{chem} = \mu_C^{pers}$. In accordance with culture measurements provided in [101], the HGF addition and uptake rates are set equal (i.e., $5{\cdot}10^{-4}\ \text{s}^{-1}$), while $\lambda_c \ll \varepsilon_c$. All the values of the parameters used are summarized in Table C.2 in the Appendix C.

Figure 2.9 shows a typical simulation of a time-sequence evolution of a colony of MLP-29 cells in the presence of HGF. After an initial stage in which the surface of the colony roughens, with some cells radially protruding from the surface, some branches develop still keeping contact with the neighboring cell and the center of the island: this gives the colony the star-shaped configuration experimentally found. As shown in the same image, as soon as the HGF is no longer added (after 24 h $\approx$ 3000 MCS), the sprouts start a sort of

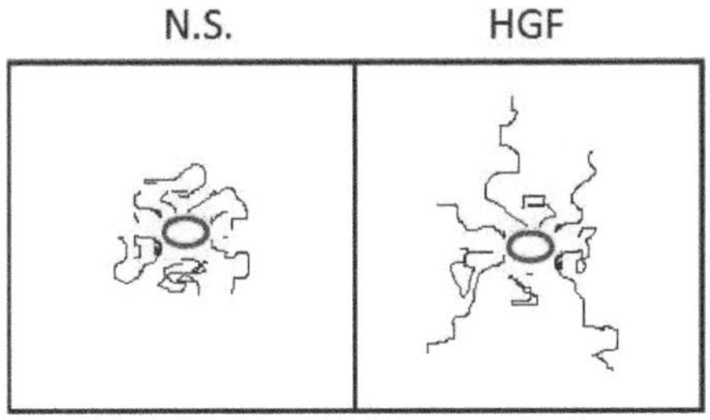

FIGURE 2.11: Hepatocyte growth factor-mediated increase of motility as shown by manual time lapse analysis. Single cell motion tracks obtained from 24-hour lapse simulations of MLP-29 cells culture in the absence (N.S.) and in the presence of HGF. For reference, the oval positioned at the center of the group of tracks represents the size of a single cell.

involution and the island becomes compact again, a biological behavior that it is possible to see also experimentally (see [271]). This second phase is modeled by imposing $\alpha_c = 0$ in Equation (2.5), while the other parameters are kept unchanged.

The evolution of the simulated colony agrees well with the experimental evidence from a quantitative point of view, as shown by Figure 2.10, which plots the time evolution of the scattering index of the island, defined in (2.3) in the two cases. It is noteworthy also that the simulated branching develops through the three temporal phases as pointed out by the experiments in the introduction of this chapter. The scattering index does not significantly increase after ≈ 2500 MCS since it is possible to hypothesize an equilibrium between the different forces: the tendency of the cells to elongate and to respond to HGF can no more overcome the adhesion with the rest of the colony and the area constraint. A breakage of this equilibrium would create the scattered island fragment. An analogous behavior is experimentally seen with the scattered colonies that reach and keep a limit conformation before the detachment of some branches.

From a mechanical viewpoint, the branching process is due to the fact that the presence of the growth factor produces a continuous, inward, normal force at the boundary of the cell island, creating a buckling-like instability, as the chemotactic force compresses small initial bumps laterally, producing sprouts. Indeed, MLP-29 cells in the bumps are more likely to extend pseudopods than cells in the valley between the bumps, since filopodia at growing tips are more frequent because they have a lower energy cost, making the rate of pseudopod extension critical to pattern evolution. In particular, for the instability to persist, it is required that the cells in the valley must retract while the others protrude, leading to the development of the branch they belong to. At equilibrium, the tracks of the zones of HGF-action have therefore

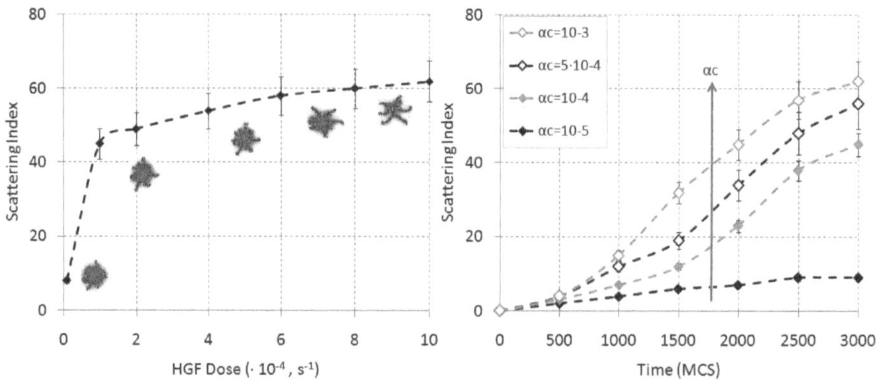

FIGURE 2.12: Left panel: scattering index of 16 MLP-29 cell clusters after 3000 MCS ($\approx$ 24 h) as a function of the HGF secretion rate α. Insets are representative configurations after 3000 MCS. Right panel: comparison of time evolution of S_I for $\alpha_c = 10^{-5}$ s^{-1}, $\alpha_c = 10^{-4}$ s^{-1}, $5 \cdot 10^{-4}$ s^{-1}, and 10^{-3} s^{-1}. All standard deviations are evaluated over over 10 simulations.

a quasi-Gaussian profile, as it levels off toward the clusters boundary, while its inflection point is at the cell–cluster boundary.

The analysis of the dynamics of single individuals permits one to characterize in greater detail the differentiation of cells in the aggregate: those in the sprouts, after the polarization, have a persistence motion in the direction of the branch they belong to, whereas the others (in the valley or in the center) keep a round shape and have a more isotropic dynamic, as shown in Figure 2.11. In particular, it is useful to emphasize that the anisotropic migration requires no extra assumptions in the CPM and produces long persistence time in the presence of HGF cues, because, as seen, cells change direction slowly. The long persistence time introduces two time scales into the branching morphogenesis: the faster time scale along the longer axes produces the growth of the spikes, and the slower time scale, radially directed, coarsens the main corpus of the island.

To have a deeper idea of the quantity of HGF needed for sprouting and to test whether its dose affects the results, the simulation model is run with increasing values of α_c (i.e., $= \varepsilon_c$), while the other parameters are unchanged. The initial island does not sprout for $\alpha_c < 10^{-4}$ s^{-1}, requiring a minimal level of HGF to start the process, while above this value a sort of phase transition is observed (see Figure 2.12), and the branching process regularly develops.

The tendency of cells to polarize, elongate, and have a persistent dynamics affects the geometry of the branches. This is studied running a set of simulations keeping the same initial configuration as that in Figure 2.9 and gradually increasing μ_C^{pers} (the other parameters are unchanged, with $\alpha_c = \varepsilon_c = 5 \cdot 10^{-4}$ s^{-1}). For $1 < \mu_C^{pers} < 45$ the sprouts become longer and thinner with an almost linear trend respect to the μ_C^{pers} increments. In fact, for very low values

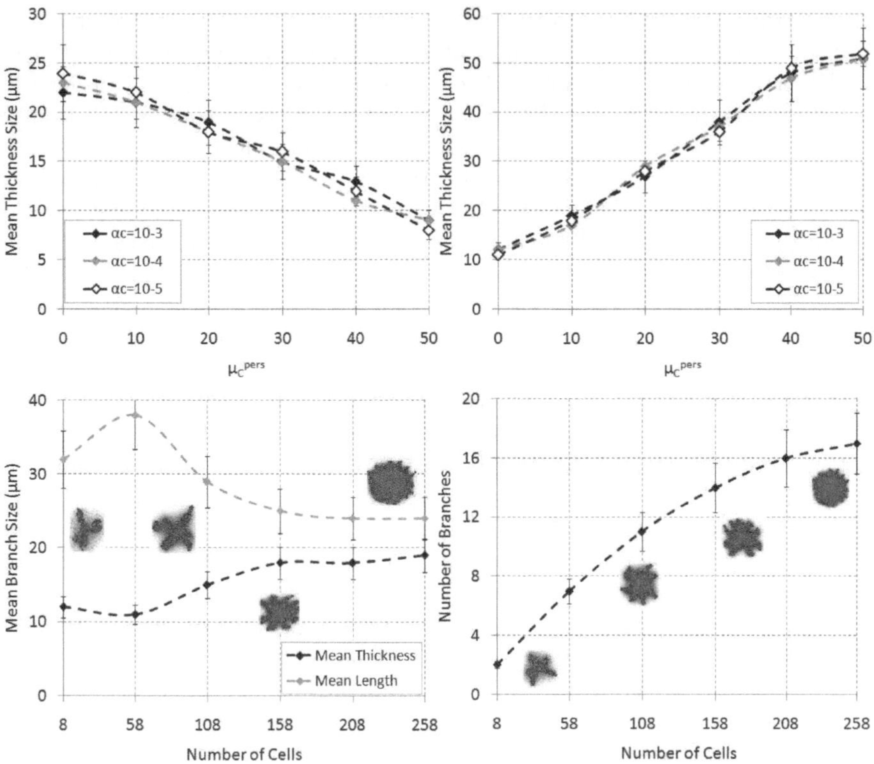

FIGURE 2.13: Mean length (top line, left panel) and mean thickness (top line, right panel) of the branches formed in a 16 MLP-29 cells round island after 3000 MCS ($\approx$ 24 h) as a function of the persistence penalty coefficient μ_C^{pers} in the cases of $\alpha_c = \varepsilon_c = 10^{-5}$ s^{-1}, $\alpha_c = \varepsilon_c = 10^{-4}$ s^{-1}, and $\alpha_c = \varepsilon_c = 10^{-3}$ s^{-1}. (Bottom line, left panel) Mean length (light line) and mean thickness (dark line) as functions of the initial number of cells forming the colony for $\alpha_c = \varepsilon_c = 5 \cdot 10^{-4}$ s^{-1}. (Bottom line, right pane) Number of the branches formed in a time-lapse of 3000 MCS ($\approx$ 24 h) as a function of the initial cell density of the island. Insets are representative configurations after 3000 MCS. The error bars indicate standard deviations over 10 simulations.

of the parameter they are almost formed by four or five parallel cells that are not completely polarized; see Figure 2.13 (top line). For $\mu_C^{pers} > 45$ the branches tend to become only one cell wide and comprise of two or three elongated cells in length. For still higher values, there is a sort of limit threshold. Obviously the branches cannot be formed by less than a single cell and those that are longer than ≈ 52 μm tend to detach from the main corpus of the colony and, as isolated cells, to migrate in the direction of their longer axes, a behavior also observed in the experiments. It is also worth noticing that the geometry of the branches is not dramatically influenced by the dose of HGF, since the curves related to different quantities of the growth factor have almost the same slope and do not differ much in the particular values, but may also be related to the number of cells forming the studied island (as shown in Figure 2.13 (bottom line, left panel)). For this reason, some simulations are performed starting with colonies with different densities.

As the number of cells increases, the lattice is taken wider, to avoid overlapping problems. For low densities (say, $n < 60$) the branches are quite long and thin, so that it can be hypothesized that the colony assumes a star-shaped configuration, with a little main corpus. For high densities (say, $n > 60$) the sprouts tend to become shorter and thicker, while at still higher densities (say, $n > 130$) there are limit thresholds for both length and thickness. The decrement of the length and the relative increment of the thickness are almost specular and has the consequence that the area of a virtual branch remains almost constant, independently of the cell density, which can only lead to a bigger main corpus of the island. The cell density affects also the number of the branches developing in a time-lapse of 24 h: for very low values (say, $n < 8$) there is no development of sprouts, and for high densities (say, $n > 108$) there is a constant value for their quantity, as shown in Figure 2.13 (bottom line, right panel). In the medium range, the number of branches increases with the number of cells forming the island. The higher quantity of sprouts appearing in bigger colonies may have two explanations: first, bigger colonies have more cells available for branching, and then they have wider external surfaces and so more space for the growth of the sprouts.

In the model it is assumed that the MLP-29 decision to whether extend or retract a pseudopod depends on the concentration difference between the retracted and extended position of the pseudopods, regardless the absolute concentration of the HGF. However, at higher concentrations, we can hypothesize that most HGF receptors will saturate, thus becoming insensitive to HGF levels, a situation that occurs experimentally and that is possible to find also in the sprouting process of blood vessel [260]. It is possible to hypothesize that the viscoelastic properties of the matrix substrate, as well as its chemical composition, may affect the diffusive behavior of the HGF, ultimately resulting in delayed scattering phenomena with unstable branches.

Chapter 3

Mesothelial Invasion of Ovarian Cancer

3.1 Biological Introduction

Ovarian cancer (OvCa) is the fifth leading cause of tumor-related deaths in the Western world (with little change in its incidence in recent decades) [4], the second most common gynecological carcinoma, and the leading cause of death from gynecological malignancies. In fact, its mortality rate is high, compared to other cancers [52]. Significant contributors to its high mortality rate are the vague or absent symptoms in the early stages of the disease, the lack of reliable tumor markers, and the recurrences, generally fatal, despite good initial responses to chemotherapy. For these reasons, long-term survival is rare since 70% of the patients present at diagnosis extensive and widespread intraperitoneal dissemination throughout the abdomen and the pelvis [62]. The majority ($\approx$ 90%) of ovarian cancer stems from surface epithelium that overlies the ovary [126].

Clinically, ovarian cancer progression is divided into four stages by the Federation Internationale de Ginecologie et d'Obstretique (F.I.G.O.) based on tumor spread. In the early stage (stage I), the disease is confined to one or both the ovaries, while in stage II it has begun spreading, with localized extensions into the adjacent pelvic tissues and organs. As the disease progresses into stage III, the tumor has spread to the upper abdominal cavity until at the final stage IV it metastatizes, proliferating and reaching distant, extra-peritoneal sites. In this last phase, after detaching from the ovarian capsule, the original malignant cells exfoliate, as single individuals, aggregates, or multicellular spheroids, from the primary tumor into the abdominal cavity, where they disseminate transported by peritoneal fluids and ascites. In particular, the ascites, present in almost one third of ovarian cancer patients, is a generally voluminous exudative fluid, containing selected cell populations mainly consisting of tumor cells, lymphocytes, and mesothelial cells, which facilitates tumor cell transport throughout the abdominal cavity [320].

The migration of neoplastic cells is also regulated by environmental factors such as ECM components, cytokines, growth factors, or chemotactic factors, mainly secreted by the mesothelium, an epithelial-like monolayer [287], that lines the organs of the abdominal cavity and is the elective site of ovarian car-

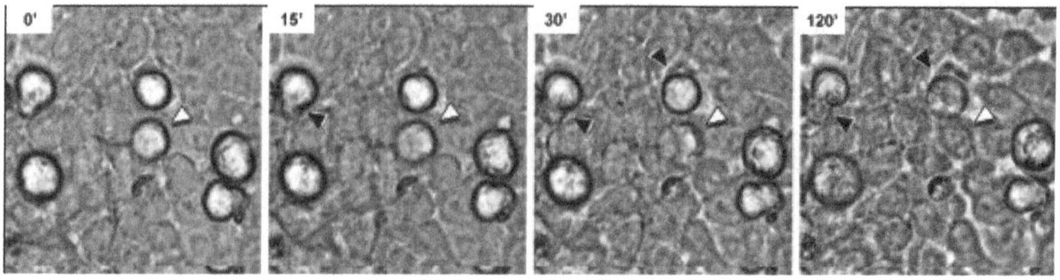

FIGURE 3.1: Ovarian cancer (OvCa) transmesothelial migration was visualized by real-time microscopy using nonmalignant Met-5A mesothelial cells grown on fibronectin. Samples were placed on a heated stage set at 37°C, and images were taken using a 40x objective. OvCa cells were added on mesothelial monolayer, and frames were captured every 15 s for 2 h. Frames corresponding to 0, 15, 30, or 120 min of the movie are shown. The migration of a single cell (white arrows) through a mesothelial cell junction is appreciable without damage to the mesothelial layer. At the end of tumor cell transmigration (120 min) the junction is restored. After 30 min, a number of cells in the same field started to migrate (black arrows).

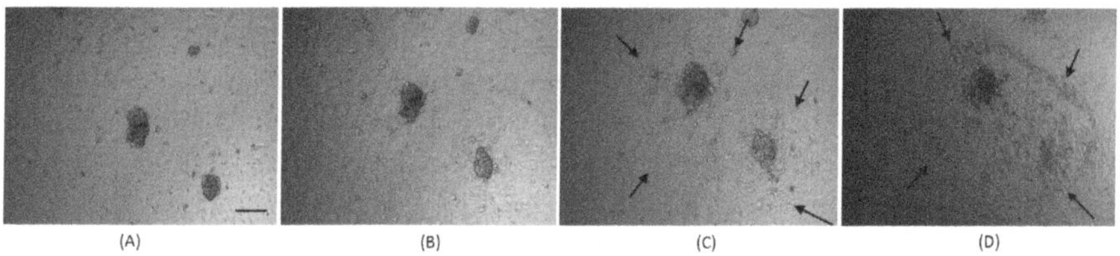

FIGURE 3.2: Spheroid dissemination and mesothelial monolayer invasion assay: spheroids (5-10/well) were seeded onto the HPMC and digitally photographed at (A) 1 h after plating ($t = 0$), (B) 1 and (C) 3 and (D) 7 days.

cinoma disaggregation, dissemination, and metastatic outgrowth [3, 290, 419]. In fact, malignant cells able to survive finally implant on and invade through the mesothelial lining of the peritoneum, establishing secondary tumors, often without the need to enter the vasculature. The successful metastatic process, governed by the biophysical properties of cancer cells combined with the remodeling of intra- and intercellular proteins, that regulate cell–cell adhesion (for example, cadherins) and cell–ECM interactions (for example, integrins), can be therefore divided into two main steps: i) adhesion of the tumor cells to the mesothelial layer and ii) invasion of the mesothelial layer.

3.1.1 Single Cell Transmigration

The transmesothelial migration of ovarian cancer cells has been reproduced and analyzed *in vitro* at the Laboratory of Immunogenetics of the Azienda

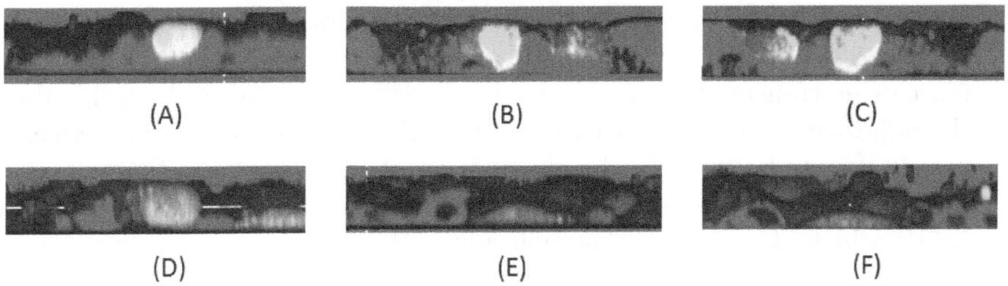

FIGURE 3.3: (See color insert.) Transmesothelial migration assay: nonmalignant transformed human pleural mesothelial cells (HPMCs), Met-5A cells were labeled with the PKH26GL cell linker kit (colored in red), seeded on coverslips coated with fibronectin (10 μg/mL), and grown until confluence at 37°C. The human ovarian carcinoma cell line NIH:OVCAR-3 was stained with 5 μM CFSE 20 min at 37°C (colored in green). CFSE-labeled tumor cells (1.5×10^5) were plated on HPMCs and incubated for 3 h at 37°C. Nonadherent OvCa were carefully removed. Samples were analyzed by sequential scanning of the XY planes recorded along the Z-axis (step size: 1.5 μm) and then processed using the 3-dimensional reconstruction software bioView3D and visualized as orthogonal views. In (A) a cell is on the top of the mesothelial layer. In (B) and (C) cells are inducing opening in the layer and passing through. In (D) a cell is in the middle of the layer. In (E) and (F) cells crossed the mesothelial layer, and the cells rejoined closing the gaps previously created over the malignant cells.

Ospedaliero-Universitaria S.Giovanni Battista-Molinette di Torino. The experimental model consists of plating an ovarian cancer cell line (NIH: OVCAR-3) on a nonmalignant transformed human mesothelial cell layer (Met-5A), grown on an extracellular matrix protein, as reproduced in Figure 3.3. Experimental evidence then shows that single ovarian tumor cells' adhesion to the monolayer is mainly mediated by the interactions between β1-integrin and some of its epitopes [368] and selected ECM proteins (such as laminin, fibronectin, vitronectin, and type I and IV collagen), which are secreted by the mesothelial cells and form a sort of pericellular matrix around the layer. CD44-hyaluronan and E-cadherins also interact with the mesothelial cells [234, 291].

The subsequent transmigration across the mesothelium requires instead the activity of selected matrix metalloproteinases (MMPs), endopeptidases that predominantly degrade any structural components of the ECM, along with a variety of cell adhesion molecules [406]. They also affect the relative cellular signaling pathways and functions and control cell migration. Moreover, the MMPs are involved in the release of cell-membrane-bound precursors of many growth factors, whose receptors are MMP substrates, and are able to cleave and activate their own zymogen forms [122]. In pathological conditions, the MMPs actively contribute to cancer progression: clinical data suggest that benign tumor cells acquire malignant properties following upregulation of MMP expression and that, conversely, highly malignant cells become less aggressive when MMP activity is reduced. As provided by ovarian cancer transmigration assays, the overexpression of β1-integrins during the adhesive interactions between neoplastic individuals and the pericellular matrix results both in the downregulation of the E-cadherin function and in the activation and co-localization of MMP-2, MMP-9, MMP-14, and MT-1 MMP, which also cleave CD44 molecules, releasing their extracellular domain [136]. In particular, the activity of the secreted MMPs promotes cancer invasion by cleaving and regulating the normal function of the N-cadherins, which are the predominant cell–cell adhesion molecules holding the mesothelial cells together: the breakage of these bonds causes the retraction of the mesothelial layer at the attachment site of malignant cells, opening the way for the subsequent invasion [305], as shown in Figure 3.3(B-D). In this regard, some data suggest that the mesothelial cells themselves may produce low levels of MMPs and induce MMP expression by ovarian cancer cells [405].

Other data have instead shown that ovarian tumor cells secrete the urinary-type plasminogen (uPA), which participates in the conversion of plasminogen to plasmin and allows further amplification of submesothelial ECM degradation, and other serine proteases. However, uPa and serineproteases may play a less important role than MMPs in facilitating cancer invasion. Upon complete migration of cancer cells across the mesothelium, the normal function and signaling of N- and E-cadherins are recovered [345], as the mesothelial cells rejoin closing the gaps previously created over the migrated malignant cells [385], as shown in Figures 3.3(E-F). In this respect, we can say that single cell invasion is *conservative* [346].

3.1.2 Multicellular Spheroid Invasion

In the ascites of ovarian cancer patients, tumor cells also exfoliate as aggregates or multicellular spheroids, whose most important function is to create an anchorage-independent *in vivo* tumor microenvironment, supporting mechanisms of cell survival and growth through homotypic cell–cell adhesion [364]. The initial formation of ovarian spheroids is strongly regulated by β1-integrin subunits [68], but it is also mediated by compensatory mechanisms, through other cell–cell interactions, such as gap junctions, tight junctions, and desmosomes [425] and the activity of cadherins and different ECM components (including fibronectin, laminin, and type I and IV collagens), glycosaminoglycans, and proteoglycans incorporated into the spheroid from the ascites fluid [51]. Because of their low proliferative rate, spheroids are resistant to therapies directed against fast-growing tumor cells, to some forms of chemotherapy, and to apoptosis induced by radiation or by conventional therapeutic drugs [52, 207]. They have also altered kinetics of drug absorption. This evidence suggests that spheroids may represent a tenacious and important intermediate for secondary tumor growth and an accurate model system of ovarian metastasis [127, 363].

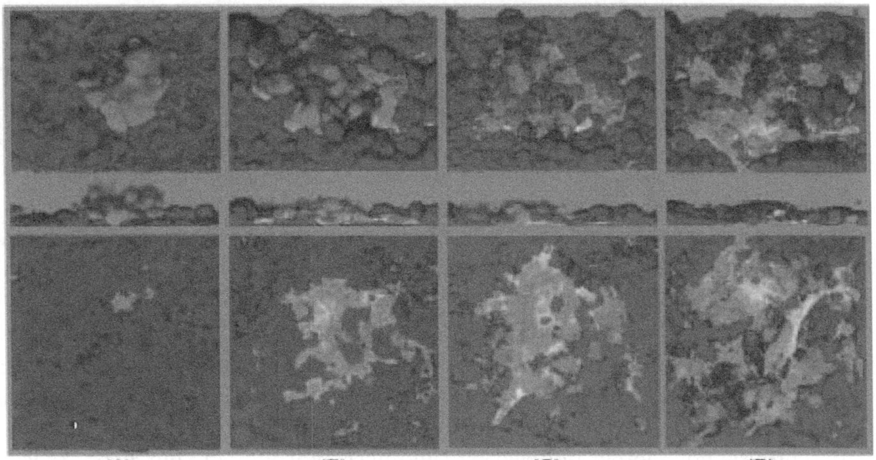

FIGURE 3.4: (See color insert.) Human peritoneal mesothelial cells (HPMCs) were labeled with PKH26GL (red), seeded on fibronectin-coated coverslips and grown until confluence at 37°C (as described in Figure 3.3). Spheroids were generated using a modification of the hanging droplet method [206]. Briefly, cells were stained in green and then placed (50 cells/15 μL of RPMI with 15% FCS) on the cover of a tissue culture dish, and the cover was placed over a dish containing PBS to prevent dehydration of the hanging droplets. After 4 days, spheroids were seeded on HPMCs layer and allowed to adhere for 3 h before imaging. Images show the 3D reconstruction of the same field visualized from the upper, transverse, and bottom sides at time zero (A), 24 h (B), 48 h (C), and 72 h (D).

To experimentally test the ability of a cancer multicellular spheroid to disseminate and to characterize its metastatic potential, the Laboratory of Immunogenetics of the Azienda Ospedaliero-Universitaria S.Giovanni Battista-Molinette di Torino provides specific transmesothelial assays. Indeed, neoplastic spheroids are generated, and seeded onto a mesothelial monolayer, generally anchored to an ECM-substrate; see Figure 3.4. The spheroid dissemination, defined as the tumor mass spreading on top of the monolayer without forming invasive foci, is shown to require the interaction of β1-integrins with the ECM proteins secreted by the mesothelial cells, with the contribution of other adhesion molecules. The quantified overall adhesion levels of ascites spheroids are somewhat lower than those reported for isolated carcinoma cells, which possibly reflects a change in cell adhesive ability upon acquisition of the spheroid morphology.

The process of invasion, determined by the establishment of proliferating foci of ovarian tumor cells within the same focal plane as the layer, starts with the retraction of some mesothelial cells (Figure 3.4(A)). Suddenly, the extracellular matrix molecules underlying the monolayer become exposed, allowing the spheroids to grow and to form large foci of invasion that appear to grow laterally, almost within the same plane of the layer (Figure 3.4(B)). The area occupied by the original spheroids progressively increases in size, as the mesothelial cells recede and are overtaken and replaced by the malignant mass (*disruptive invasion*; see Figure 3.4(C-D)). The invasion is regulated by the proteases secreted and activated by the tumor cells. Indeed, some experimental data indicate that ovarian cancer spheroids secrete much greater amounts of both pro-MMP-2 and MMP-9 compared to cells grown as a monolayer, and that both MMPs are present in the active form [53, 127, 136].

Quantitative studies evaluating the percentage of invasion demonstrate that tumor spheroids plated in the same well do not exhibit the same invasive properties despite sharing an identical environment, since only a proportion of them establishes foci of invasion. These data imply that the metastatic potential is not merely induced by the microenvironment, but likely relies on physiological differences in the invasive characteristics of every single spheroid.

3.2 Mathematical Model

The model aims to reproduce a typical *in vitro* experiment of ovarian cancer transmesothelial migration, characterizing the principal biological aspects. In particular, we focus on the differences between the invasion of a single cell and of cells aggregated in a spheroid. The simulation environment involves more entities with respect to the previous application: two cell populations, tumor ($\tau = C$) and mesothelial cells ($\tau = M$), and two types of substrate objects, matrix fibers ($\tau = E$) and the experimental dish ($\tau = S$), used only to indicate the low boundary of the domain, in addition to the interstitial/peritoneal fluid, isotropically distributed ($\tau = F$). Continuous fields are instead used to represent the evolution of chemotactic factors and tumor MMPs.

The system evolves following the usual rule given in relation (1.2). However, with respect to the previous example, we no longer use a single overall motility of the entire pattern, but T is specific for each type of individual, i.e., $T = T_\tau$.

The *Hamiltonian* includes surface and perimeter constraints for cells, terms for the adhesion between individuals, and the energetic counterpart of the cancer chemotactic movement:

$$H(t) = H_{adhesion}(t) + H_{shape}(t) + H_{chemotaxis}(t). \tag{3.1}$$

The form of $H_{adhesion}$ is the same as in Equation (1.5): the Js, as usually measuring the level of the expression of specific adhesion molecules. They are spatially homogeneous, meaning that the coupling strengths are uniformly distributed over the objects' surfaces and therefore the localization of adhesive proteins is homogenous. However, as will be described in detail in the next section, the Js may vary during the simulation, i.e., $J_{\tau(\Sigma_\sigma),\tau(\Sigma_{\sigma'})} = J_{\tau(\Sigma_\sigma),\tau(\Sigma_{\sigma'})}(t)$.

H_{shape} models the geometrical properties of the simulated individuals: given as in (1.6), the initial dimensions of discrete objects are assumed to be their target measures. In specific cases, the growth (respectively, death) of individuals is modeled by allowing their target measures to increase (respectively, decrease) with time, as we will explain below.

Finally, the chemotactic motion of malignant individuals along concentration gradients of signal molecules is implemented by relation (1.9), where μ_C^{chem} is the effective potential related to tumor cells and c represents the local level of the chemical substances.

The chemotactic factors are released in the interstitial/peritoneal fluid by both the mesothelial cells and the ECM molecules in order to direct the motion of tumor cells. Indeed, their spatiotemporal evolution is controlled by the following PDE:

$$\frac{\partial c}{\partial t} = \underbrace{D_c \nabla^2 c}_{diffusion} - \underbrace{\lambda_c c}_{decay} + \underbrace{\alpha_c}_{secretion} , \tag{3.2}$$

where the coefficients of diffusion D_c and decay λ_c are assumed constant in time and homogeneous throughout the extracellular fluid. $\alpha_c = \alpha_c(\mathbf{x}, t)$ describes the secretion of the chemoattractant:

$$\alpha_c(\mathbf{x}, t) = \begin{cases} \alpha_c & \text{if } \tau(\Sigma_{\sigma(\mathbf{x})}) \in \{M, E\} \text{ and } \exists \mathbf{x}' \in \Omega_{\mathbf{x}}' : \tau(\Sigma_{\sigma(\mathbf{x}')}) = F; \\ 0 & \text{else.} \end{cases} \tag{3.3}$$

As stated in Section 3.1, the tumor cells are able to activate and release different types of MMPs, m, whose evolution is described by

$$\frac{\partial m}{\partial t} = \underbrace{D_m \nabla^2 m}_{diffusion} - \underbrace{\lambda_m m}_{decay} + \underbrace{\alpha_m}_{release} , \tag{3.4}$$

where, as before, D_m and λ_m are, respectively, the diffusion and degradation rate, constant in time and homogeneous in space. $\alpha_m = \alpha_m(\mathbf{x}, t)$ models instead the secretion of proteinases over the external surface of the cancer in the interstitial fluid when meeting an ECM fiber, i.e.,

$$\alpha_m(\mathbf{x}, t) = \begin{cases} \alpha_m & \text{if } \tau(\Sigma_{\sigma(\mathbf{x})}) = C \text{ and } \exists \mathbf{x}' \in \Omega_{\mathbf{x}}' : \tau(\Sigma_{\sigma(\mathbf{x}')}) = E; \\ 0 & \text{else.} \end{cases} \tag{3.5}$$

During the invasion process, tumor MMPs are capable of degrading both the pericellular and the submesothelial matrix components: to model this

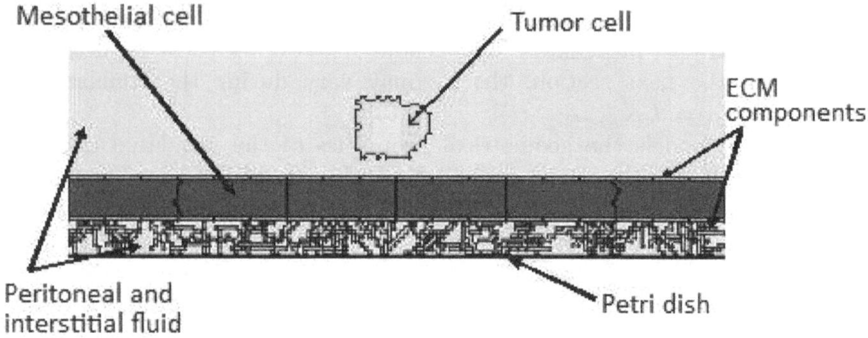

FIGURE 3.5: The initial condition of simulated single tumor cell mesothelial invasion and description of the symbols used in the model.

biological effect, a lattice site $\mathbf{x}$ within an ECM molecule becomes a generalized pixel of medium when the local level of MMPs ($m(\mathbf{x}, t)$) is sufficiently high (i.e., > 2.5). This change is implemented by changing $\tau(\Sigma_{\sigma(\mathbf{x})})$ from E to F.

3.3 Single Cell Transmigration

The *in vitro* transmesothelial migration of a single ovarian cancer cell is modeled in a two-dimensional 162 px × 50 px section illustrated in Figure 3.5 ($\approx$ 162 μm × 50 μm, since in our lattice 1 px corresponds to ≈ 1 μm), where the border at the bottom is the virtual Petri dish, as seen labeled with $\tau = S$. The mesothelial layer is formed by 27 px × 9 px rectangular cells ($\tau = M$), and surrounded by a sort of pericellular matrix made up for ECM fibers ($\tau = E$). The area between the simulated mesothelium and the dish is a mixture of interstitial fluid ($\tau = F$) and ECM molecules ($\tau = E$ again), that mimics the experimental sub-mesothelial-type matrigel used to anchor the monolayer. The virtual ECM components are arranged in random lines, representing fiber bundles. The round cancer cell ($\tau = C$), with a diameter of 14 px, is placed in proximity of the layer, within the peritoneal fluid (again $\tau = F$).

The overall simulation lasts 1200 MCS, which, fitting the experimental times, we set to correspond to about 5 h, defining 1 MCS $\approx$ 15 s. As outlined in the experiments, the mesothelial cells, anchored to fixed and rigid ECM fibers, form a continuous and stable pavement and barely move from their original position, maintaining their characteristic shape. Conversely, the tumor cell is subjected to dramatic surface and volume alterations and remodeling, since it has the ability to spread on and to suddenly cross the monolayer through gaps created by the retraction of mesothelial cells. These considerations lead

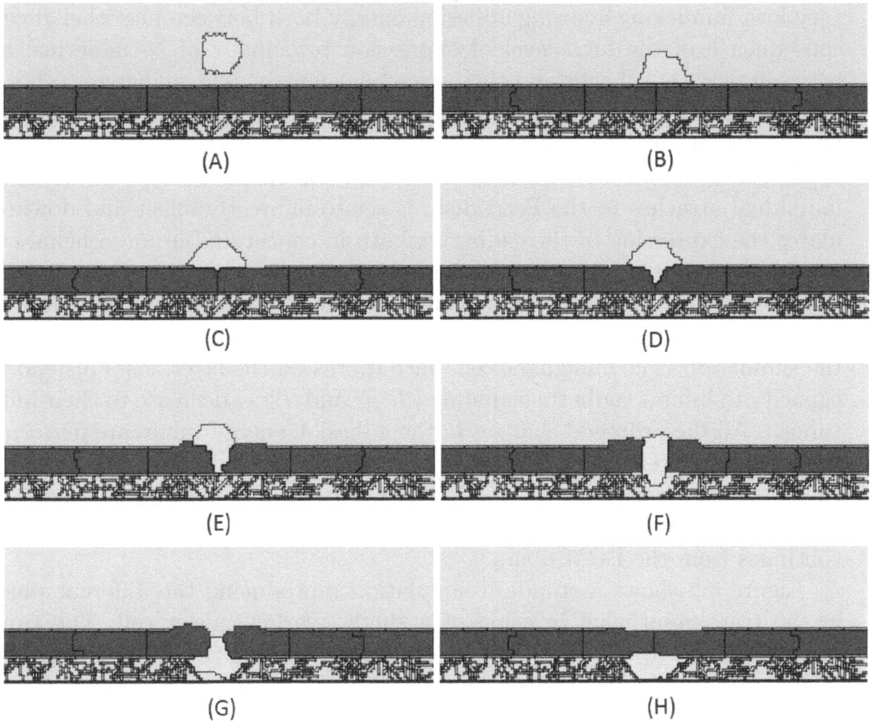

FIGURE 3.6: Simulated single cell transmesothelial migration at 0, 200, 400, 700, 800, 950, 1100, and 1200 MCS. In (B) the cell touches the mesothelial layer covered by the pericellular matrix and starts spreading over it. In (C) the production of MMPs starts degrading the pericellular layer. In (D) the cancer cell induces the loosening of the adhesion bonds between the epithelial cells that detach so that the cell progressively penetrates the layer. After the cell has migrated to the opposite side of the layer (G), the simulated mesothelium has closed back (H), and the tumor cell is considered completely infiltrated.

to set

$$\begin{cases} T_C \gg T_M; \\ T_E = 0; \\ \lambda_C^{surface} \ll \lambda_M^{surface} \ll \lambda_E^{surface}; \\ \lambda_C^{perimeter} \ll \lambda_M^{perimeter} \ll \lambda_E^{perimeter}. \end{cases}$$

The initial hierarchy of the Js is obviously important to maintain the structure of the mesothelial layer rigid, surrounded by the pericellular matrix and fixed on the submesothelial ECM fibers. In particular $J_{M,M}$ and $J_{M,E}$ are very low, mimicking a strong adhesion energy both between mesothelial cells, consequence of the high level of expression principally of N-cadherins, and between mesothelial cells and the pericellular matrix. The malignant cell's attachment via β1-integrin and CD44 molecule to the layer is instead modeled by decreasing both $J_{C,E}$ and $J_{C,M}$.

The activity of tumor MMPs, whose secretion is inhibited as the malignant individual attaches to the Petri dish, is set to indirectly affect and downregulates the expression of the cadherins both in cancer and in mesothelial cells (in this regard, it seems very promising to study the functions of the signaling molecule CD157 [150]), breaking cell–cell adhesion and making the mesothelial cells retract: mathematically $J_{C,M}$ and $J_{M,M}$ therefore increase during the simulation. The malignant cell, once it crossed the layer, loses instead the capacity to inhibit cadherin signaling ($J_{C,M}$ and $J_{M,M}$ decrease to their initial values). All the proposed changes in the adhesion energy values are performed with a unit increment (or decrement) every 10 MCS. Finally, the release of chemotactic signals from the mesothelial cells is inhibited after the adhesion of the tumor to the layer ($\alpha_c(\mathbf{x}, t) = 0$ in Equations (3.3) and (3.4)), while it continues from the ECM fibers.

Figure 3.6 shows a standard simulation reproducing the different phases of the transmesothelial invasion of a single ovarian cancer cell. The tumor cell takes time to adhere to the layer and to degrade the ECM components (see Figure 3.6(A-C)). The released MMPs break the intramesothelial adhesive contacts, making the layer retract and generating the penetration of the cell in the mesothelial pavement, as shown in Figure 3.6(D-G). As the malignant cell has passed, it loses the capacity to inhibit cadherin signaling, as the mesothelial junctions are recovered and the continuity of the layer restored, as shown in Figure 3.6(H). Indeed, the effective mesothelial infiltration period is 500 MCS long ($\approx$ 2 h).

To test our model and to underline how the single component mechanisms affect the biological outputs, we run sets of simulations changing one parameter at a time, while keeping the others fixed as in Table C.3 of Appendix C. The results are summarized in Figure 3.7. We have the confirmation that the cancer cell acquires malignant properties when MMP expression is strongly upregulated (with a strong decrement in the transmigration time) and that, in contrast, the malignant individual becomes less aggressive when the proteolytic secretion is reduced, resulting in the inability to degrade matrix components. Mathematically, low and high MMP releases are referred to their ratio with the threshold value (2.5) characteristic of the transition between an ECM fiber and the medium. With constant low values of $J_{M,M}$ (i.e., $J_{M,M} \ll J_{C,M}$), we instead simulate an overexpression of the N-cadherins connecting the mesothelial cells: this prevents the invasion of the tumor, which can only adhere at the top of the layer. On the contrary, high values of $J_{M,M}$ (i.e., $J_{M,M} \gg J_{C,M}$) mimic an unstable layer, which facilitates and accelerates the transmigration of the disease. Low values of $J_{C,M}$ (i.e., $J_{C,M} \ll J_{C,E}, J_{M,M}$) model instead an enhanced adhesive affinity between tumor and mesothelial

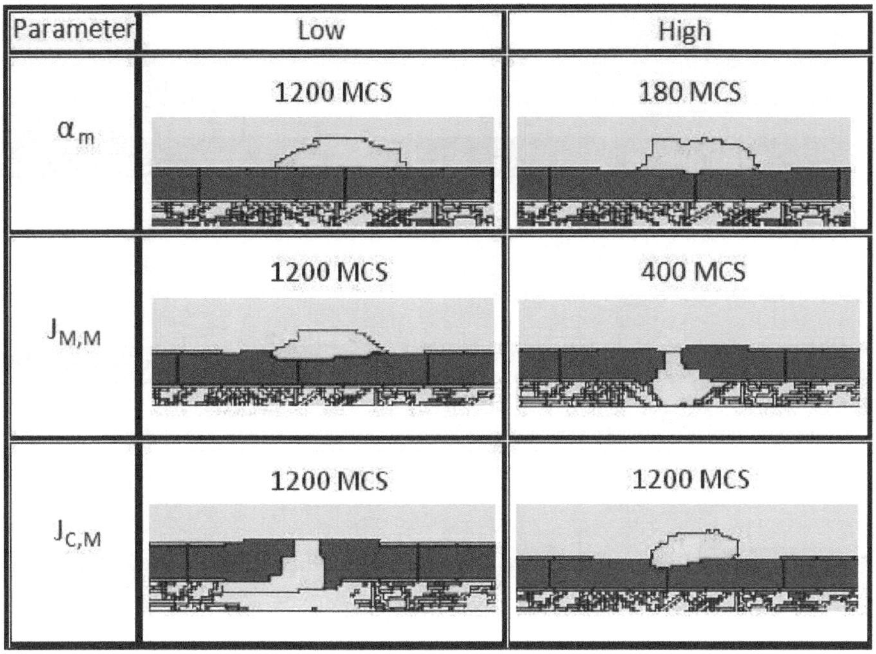

FIGURE 3.7: Summary of the simulations of single cell transmesothelial migration run with different parameters. All the other parameters not explicitly cited are the same as in Table C.3 given in Appendix C.

cells. This situation is biologically plausible, since ovarian carcinoma individuals are in origin epithelial-type cells, and therefore able to express the same cell adhesion molecules as the mesothelial cells. In this case, tumor cell forms close junctions with the layer, becoming part of it and losing its invasive capability. In contrast, for high values of $J_{C,M}$ (i.e., $J_{C,M} \gg J_{C,E}, J_{M,M}$), the tumor cell has few adhesion points with the virtual mesothelium and therefore it has not been able to activate the downstream cadherin-signals needed for the transmigration program.

3.4 Multicellular Spheroid Invasion

The initial conditions of the simulation of the spheroid invasion are the same as in Figure 3.5, but the tumor mass is formed by 10 virtual cells. The features of the model are structurally analogous to those of the previous section and the simulations last 12000 MCS (≈ 2 days), which allows one to neglect cancer cell proliferation. The only differences are the parameters explicitly cited in

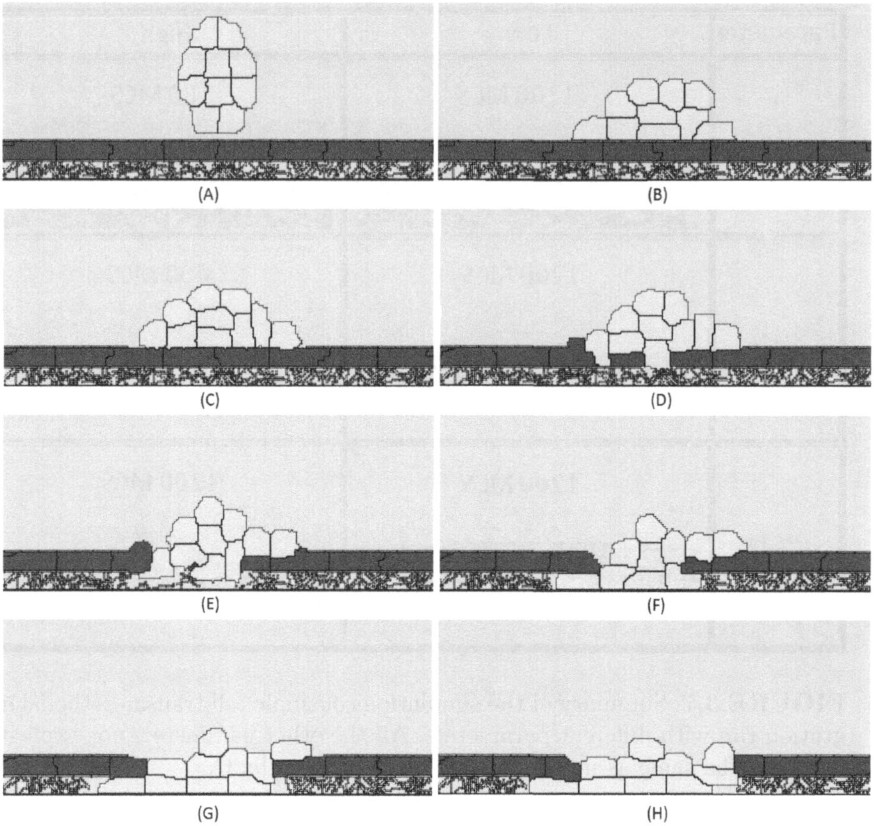

FIGURE 3.8: Simulated multicellular spheroid transmesothelial migration at 0, 2000, 4000, 6000, 7500, 9000, 10500, 12000 MCS. The tumor mass adheres to the layer (B), and the subsequent expression of MMPs degrades the ECM components and breaks the junctions between the mesothelial cells (C) and (D). The disseminated spheroid invades the virtual mesothelium through different foci of adhesion, causing the apoptosis of the mesothelial cells nearby (E), (F), and (G). At the end of the simulation (H), the cancer mass has overtaken a large area of the monolayer and starts the metastatic outgrowth.

Table C.4, and the fact that we also take into account the sedimentation of the spheroid due to gravity, which directs, along with the classical chemotactic cues, the motion of the tumor mass toward the layer. We use lower $A_C^{surface}$ and $A_C^{perimeter}$ and higher $\lambda_C^{surface}$ and $\lambda_C^{perimeter}$ since the tumor cells, acquiring a spheroid morphology, are tightly packed and resistant to change their shape, thereby remaining almost round with a constant volume/surface ratio.

The strong homotypic interactions, regulated by the expression of E-

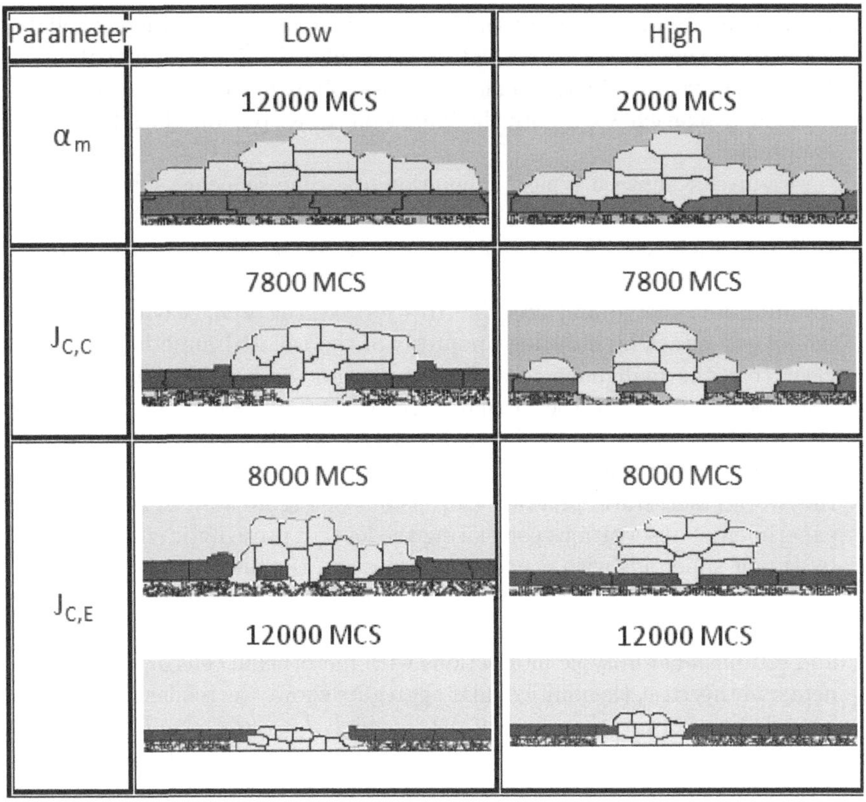

FIGURE 3.9: Summary of the simulations of multicellular spheroid invasion run with different parameters. All the parameters not explicitly cited are given in Tables C.3 and C.4 of Appendix C.

cadherins and by other intercellular mechanisms are instead modeled by an initial low value of $J_{C,C}$, which increases as the cancer aggregate starts disseminating over the monolayer. The higher value of $J_{C,M}$ with respect to the case of the single cell represents the lower overall spheroid–mesothelium adhesion reported in the experimental literature [234]. The low expression and avidity of cancer integrin subunits and the physical and geometrical constraints limiting its ability to spread on anchoring surfaces are modeled with a slow decrement (0.01 instead of 1 each 10 MCS) of the relative bond energy $J_{C,E}$.

The spatial and temporal evolution of the proteases' field are the same as in the case of the single cell, but the total amount of the secreted MMPs is higher (see again Table C.4) . This is due not only to the increased tumor mass but also to the fact that cancer cells, upon acquiring a spheroid morphology, overexpress the different types of proteases, thereby enhancing their invasive potential. Finally, since experimental evidence shows that the mesothelial layer generally disaggregates and retracts, and that isolated mesothelial cells beneath the cancer cells die, the target volume $A_M^{surface}$ of a virtual mesothelial cells surrounded by the malignant mass (30 px of common surface) is set to decrease 1 px each MCS until the limit value of 80 px, for which they undergo apoptosis.

Figure 3.8 shows a typical simulation of a time-sequence spheroid invasion of a mesothelial layer: w.r.t. a single individual, the tumor mass takes a longer time to attach to ECM components (see Figure 3.8(A-B)) and to invade the layer (see Figure 3.8(C-G)). At the end of the simulated process, multiple invading foci merge and are able to overtake the well, leaving little of the monolayer intact by inducing apoptosis of the isolated mesothelial cells. It is clear that the malignant cells disseminate not only in the same plane, but also above and below the virtual mesothelium, and degrade a large part of the ECM components, as shown in Figure 3.8(H). Also in the case of the multicellular aggregate, an increase in MMP release causes an acceleration in the overall metastatic program (top panels of Figure 3.9), in agreement with experimental observations correlating the level of proteolytic enzymes secreted by tumor spheroids with a poorer prognosis of the disease [122].

Low $J_{C,C}$ values ($J_{C,C} \ll J_{C,M}, J_{C,E}$) model the preference of the spheroid cells to maintain their strong homotypic connections, rather than disseminate and establish heterotypic interactions with mesothelial cells or matrix components: in this case, the multicellular aggregate shows the tendency to form only a single focus of invasion. In contrast, for high $J_{C,C}$ ($J_{C,C} \gg J_{C,M}, J_{C,E}$), the cancer cells quickly detach from the core of the spheroid and disseminate over a larger area of the mesothelial layer, invading in different sites (middle panels of Figure 3.9). A similar invasive phenotype can be also obtained by forcing an overexpression of β1-integrins in the tumor mass by strongly decreasing $J_{C,E}$ ($J_{C,E} \ll J_{C,M}, J_{C,C}$); bottom panels of Figure 3.9.

Part II

Extended Cellular Potts Model and Applications

Chapter 4

Extended Cellular Potts Model

4.1 Advantages and Limitations of the Basic CPM

Basic CPM approaches work surprisingly well in modeling a wide range of biological processes. In particular, they can provide important new insights into the principles of multicellular (tissutal) patterning in a number of phenomena, as they are able to analyze their driving mechanisms; see, for example, [183, 252, 314]. Moreover, CPM applications are a way of comparing the outcomes of different and equally plausible scenarios, providing a predictive value as well. It is possible in fact to analyze the system's response to a range of experimental perturbations, as shown in [246, 262, 351, 365].

The first advantage of the CPM compared to alternative cell-based modeling approaches that represent biological individuals as point particles or fixed-sized spheres or ellipsoids (for examples, see [11]) is that it differentiates between bound and unbound regions of their membranes. Moreover, morphologies and changes of shape of discrete elements are easily and realistically implemented. The key benefits of the CPM energetic formalism are its simplicity and extensibility. Almost any biological mechanism can, in fact, be included in the model, simply by adding an appropriate generalized potential term in the *Hamiltonian* H, as suggested in the main reviews of the method [20, 165]. It is therefore possible to easily comprehend the importance of each mechanism involved in the simulated phenomenon by only altering the relative Potts parameter, so that the other terms in the *Hamiltonian* scale accordingly. In particular, by equating all the other terms to zero, it is possible to understand whether such a mechanism is individually capable of producing the process of interest, or whether it requires cooperative processes.

However, most CPM approaches suffer from some limitations. First, the reproduction of biological entities is improbable, since they are usually represented by single discrete objects, which are isotropic and formed by equivalent and undifferentiated sites. This representation provides a useful level of abstraction, but also hides relevant inhomogeneous properties that characterize all biological individuals and become essential in several natural phenomena. For example, in the case of simulated cells reproduced by a single functional unit, on one hand the cytoskeleton and the plasmamembrane do not have an independent existence, while on the other, the nuclear envelope is not defined. Moreover, while certain model quantities, such as energy or temperature, have unambiguous meanings, most Potts parameters do not have a direct correspondence with biophysical measurable quantities, as also commented in [262]. This is a crucial drawback for a good quantitative comparison between *in silico* and *in vitro* results, and thus for a predictive value of CPM applications. Additionally, these constraint weights, which modulate the dynamic behavior of the simulated individuals, are generally static over the whole simulations, or have unrealistic variations. This situation in not plausible since real biological elements continuously change their biophysical and biomechanical properties as a consequence of continuous internal and external stimuli.

These considerations lead to one of the main criticisms of CPM approaches: most simulated phenomena emerge from quite strong *a priori* assumptions that are derived from experimental observations (see again [262]). In particular, the behavior of the simulated individuals is constrained by qualitative rules (such as the energetic constraints regulated by fixed parameters) that do not easily adapt during the evolution of the system. Furthermore, all the objects belonging to a given type (τ) are usually constrained to feature the same biophysical properties, such as the target states of most attributes or the adhesive strength, despite their individuality.

Finally, by treating simulated individuals with only a cell-level phenomenological approach, most CPM applications do not consider (or, in some cases, only approximately describe) the molecular scale of the biological organisms, as carefully explained in [165]. In fact, basic CPMs neglect the continuous flow of information between the microscopic and the mesoscopic scales, which, as seen in the introduction, is fundamental for developmental biology.

In the next sections, we propose some extensions for CPM applications in order to overcome the above-cited limitations and to improve the biological realism of the method. These developments are then applied in the following chapters to several biological problems, which range from unicellular processes to multicellular phenomena.

4.2 Compartmentalization Approach

As seen, most existing CPMs generally treat biological individuals as undifferentiated discrete objects, i.e., single functional units Σ_σ identified by a common spin σ. Different spin states therefore represent different simulated entities. However, biological elements are composed of different parts (such as the nucleus or the cytosol in an eukaryotic cell, or the microcompartments in bacteria), which play a fundamental and unique role in the development of the organism. Moreover, each of these parts is characterized by particular and well-defined biophysical, biochemical, and biomechanical properties.

The simplest and most realistic way of reproducing such complex mor-

Symbol	Basic CPM	Example	Compartmentalized CPM	Example
σ	object → single individual	σ= 1 , 2	object → individual compartment	σ= 1 , 2 , 3 , 4 , 5 , 6 , 7 , 8 , 9 , 10
η			cluster → single individual	η= 1 , 2 , 0 , 0
τ	object type → individual type	$\tau(\Sigma_{\sigma=1})$=endothelial cell $\tau(\Sigma_{\sigma=2})$=fibroblast	object type → compartment type	$\tau(\Sigma_{\sigma=1,5})$=nucleus $\tau(\Sigma_{\sigma=2,6})$=cytosol $\tau(\Sigma_{\sigma=3,7})$=PM $\tau(\Sigma_{\sigma=4,8})$=Golgi $\tau(\Sigma_{\sigma=9,10})$=medium
θ			cluster type → individual type	$\theta(\eta$= 1)=endothelial cell $\theta(\eta$= 2)=fibroblast

(A)

(1,3)	(1,3)	(1,3)	(1,3)	(1,3)	(1,3)	10	10	10	10	10
(1,3)	(1,2)	(1,2)	(1,2)	(1,2)	(1,3)	(2,7)	(2,7)	(2,7)	(2,7)	(2,7)
(1,3)	(1,2)	(1,2)	(1,1)	(1,2)	(1,3)	(2,7)	(2,6)	(2,6)	(2,6)	(2,7)
(1,3)	(1,2)	(1,2)	(1,2)	(1,2)	(1,3)	(2,7)	(2,8)	(2,6)	(2,6)	(2,7)
(1,3)	(1,4)	(1,2)	(1,2)	(1,2)	(1,3)	(2,7)	(2,6)	(2,5)	(2,6)	(2,7)
(1,3)	(1,3)	(1,3)	(1,3)	(1,3)	(1,3)	(2,7)	(2,6)	(2,5)	(2,6)	(2,7)
9	9	9	9	9	9	(2,7)	(2,6)	(2,6)	(2,6)	(2,7)
9	9	9	9	9	9	(2,7)	(2,7)	(2,7)	(2,7)	(2,7)

(B)

FIGURE 4.1: (A) Table of the symbolism used in both the standard CPM and in the compartmentalized extension, with appropriate examples. (B) Bidimensional square CPM domain representing the elements defined in the compartmentalized example in the above table. The bold borders indicate their external membranes $\partial\eta$, the light borders the internal ones. $\eta = 1, 2$ are compartmentalized individuals, each of which is formed by four units, while $\sigma = 9, 10$ are standard, noncompartmentalized individuals.

phologies is to introduce a *compartmentalization* technique. According to this approach, a collection of standard CPM objects can be clustered to form a compartmentalized element, which can better reproduce a real biological individual. In other words, if in the basic CPM a single discrete object represented an entire individual, it now represents one of its compartments. Technically, with the new procedure, the discrete units Σ_σ, identified as usual with their spin σ, share an additional attribute, a cluster id $\eta(\Sigma_\sigma) \in \mathbb{N}$, which defines the compartmentalized individual they belong to. Obviously, discrete units without η are not part of a compartmentalized entity, but represent, on their own, an entire element (as in the basic CPM). Apart from the type already defined for the discrete units, $\tau(\Sigma_\sigma)$, we can now introduce a type for the entire clusters, $\theta(\eta)$; see Figure 4.1(A). Compartmentalized individuals η of the same family θ are formed by the same number and type of units Σ_σ.

The borders between subunits belonging to the same individual represent its internal membranes, while the external membrane is defined as

$$\partial\eta = \bigcup_{\substack{\mathbf{x}\in\Sigma_\sigma\subseteq\eta \\ \mathbf{x}'\in\Omega'_{\mathbf{x}}:\mathbf{x}'\in\Sigma_{\sigma'}\nsubseteq\eta}} (\partial\mathbf{x}\cap\partial\mathbf{x}').$$

Indeed, here and in the following, we will use the simplified notations $\mathbf{x}\in\eta$ to identify a site $\mathbf{x}$ belonging to a compartmentalized individual η (i.e., it would write $\mathbf{x}\in\Sigma_\sigma\subseteq\eta$) and $\partial\mathbf{x}\in\partial\eta$ to identify that $\partial\mathbf{x}$ belongs to the external membrane of η. Referring to Figure 4.1 as an example, the discrete objects Σ_σ where $\sigma = 1,2,3,4$ are clustered in individual $\eta = 1$, which represents an endothelial cell, while objects Σ_σ such that $\sigma = 5,6,7,8$ form individual $\eta = 2$, which represents a fibroblast. Moreover, units $\Sigma_{9,10}$ represent entire standard noncompartmentalized entities.

This new representation of individuals requires a redefinition of the characteristic terms of the CPM. First, it is necessary to differentiate the contributions of $H_{adhesion}$ due either to the contact between the membranes of discrete units belonging to the same element, namely, $H^{int}_{adhesion}$, or to the contact between the membranes of discrete units belonging to different elements, namely, $H^{ext}_{adhesion}$:

$$H_{adhesion}(t) = H^{int}_{adhesion}(t) + H^{ext}_{adhesion}(t). \tag{4.1}$$

External and internal adhesive interactions, from a biological point of view, are in fact of a different nature. In particular, $H^{int}_{adhesion}$ models generalized contact forces between couples of objects that belong to the same individual, e.g., the nucleus and the cytosol in a cell:

$$\begin{aligned} H^{int}_{adhesion}(t) &= \sum_{\substack{\mathbf{x},\mathbf{x}'\in\Omega'_{\mathbf{x}}: \\ (\partial\mathbf{x}\in\partial\Sigma_\sigma)\cap(\partial\mathbf{x}'\in\partial\Sigma_{\sigma'})\neq\emptyset}} J^{int}_{\tau(\Sigma_{\sigma(\mathbf{x})}),\tau(\Sigma_{\sigma'(\mathbf{x}')})}(t)\,\delta_{\eta(\Sigma_{\sigma(\mathbf{x})}),\eta(\Sigma_{\sigma'(\mathbf{x}')})}(t) \\ &= \sum_{\substack{\mathbf{x},\mathbf{x}'\in\Omega'_{\mathbf{x}}: \\ (\partial\mathbf{x}\in\partial\Sigma_\sigma)\cap(\partial\mathbf{x}'\in\partial\Sigma_{\sigma'})\neq\emptyset \\ \eta(\Sigma_\sigma)=\eta(\Sigma_{\sigma'})}} J^{int}_{\tau(\Sigma_{\sigma(\mathbf{x})}),\tau(\Sigma_{\sigma'(\mathbf{x}')})}(t). \end{aligned} \tag{4.2}$$

The form of Equation (4.2) is analogous to that of Equation (1.5), and the Kronecker delta is

$$\delta_{m,n} = \{1, m = n; 0, m \neq n\}.$$

$J^{int}_{\tau(\Sigma_\sigma),\tau(\Sigma_{\sigma'})} \in \mathbb{R}_-$ account for high contact tensions, which prevent single individuals from splitting as the single compartments tend to fall apart to create the maximum possible common surface areas.

$H^{ext}_{adhesion}$ is formed instead by the effective adhesion energies between different compartmentalized individuals that interact with their external membranes:

$$H^{ext}_{adhesion}(t) = \sum_{\substack{\mathbf{x},\mathbf{x}' \in \Omega'_{\mathbf{x}}: \\ (\partial\mathbf{x} \in \partial\eta) \cap (\partial\mathbf{x}' \in \partial\eta') \neq \emptyset}} J^{ext}_{\theta(\eta),\theta(\eta')}(t). \tag{4.3}$$

The strengths $J^{ext}_{\theta(\eta),\theta(\eta')} \in \mathbb{R}_+$ depend, as a simple extension of the basic CPM, on the types of the respective interacting clusters.

To further increase the realism of the model, the bond energies J^{ext}s may be spatially inhomogeneous throughout the object membranes and may specifically depend on the single units in contact, and not only on their types, i.e.,

$$J^{ext}_{\theta(\eta),\theta(\eta')} = J^{ext}_{\theta(\eta),\theta(\eta')}(\partial\mathbf{x} \cap \partial\mathbf{x}', t).$$

These issues will be addressed in the following section. It is noteworthy that, if the objects in contact represent standard noncompartmentalized individuals, the relative energetic contributions are in the classical form of Equation (1.5).

The compartmentalized approach requires then to define the attributes, and the relative energetic contributions, of all the units and those that regulate their mutual interactions. A straightforward generalization of Equations (1.6) and (1.7) results in

$$H_{constraint}(t) = \sum_{\substack{\eta,\Sigma_\sigma \\ i-constraint}} \lambda^i_{\eta,\Sigma_\sigma}(t) \left[a^i_{\eta,\Sigma_\sigma}(t) - A^i_{\eta,\Sigma_\sigma}(t)\right]^2, \tag{4.4}$$

$$H_{constraint}(t) = = \sum_{\substack{(\eta,\eta'),(\Sigma_\sigma,\Sigma_{\sigma'}) \\ j-constraint}} \lambda^j_{(\eta,\eta'),(\Sigma_\sigma,\Sigma_{\sigma'})}(t) \left[a^j_{(\eta,\eta'),(\Sigma_\sigma,\Sigma_{\sigma'})}(t) - A^j_{(\eta,\eta'),(\sigma,\Sigma_{\sigma'})}(t)\right]^2, \tag{4.5}$$

where a^i_{η,Σ_σ} and $a^j_{(\eta,\eta'),(\sigma,\Sigma_{\sigma'})}$ are the actual values of the attributes, A^i_{η,Σ_σ} and $A^j_{(\eta,\eta'),(\Sigma_\sigma,\Sigma_{\sigma'})}$ are their target values, and $\lambda^i_{\eta,\Sigma_\sigma}$ and $\lambda^j_{(\eta,\eta'),(\sigma,\Sigma_{\sigma'})} \in \mathbb{R}_+$ are the relative Potts parameters .

The form of (4.4) and (4.5) is the simplest quadratic potential one can think of. However, the quantities $\left[a^i_{\eta,\Sigma_\sigma}(t) - A^i_{\eta,\Sigma_\sigma}(t)\right]^2$ and $\left[a^j_{(\eta,\eta'),(\Sigma_\sigma,\Sigma_{\sigma'})}(t) - A^j_{(\eta,\eta'),(\sigma,\Sigma_{\sigma'})}(t)\right]^2$ can be easily replaced by more sophisticated potentials $U(a^i_{\eta,\Sigma_\sigma}, A^i_{\eta,\Sigma_\sigma})$ and $U(a^j_{(\eta,\eta'),(\Sigma_\sigma,\Sigma_{\sigma'})}, A^j_{(\eta,\eta'),(\Sigma_\sigma,\Sigma_{\sigma'})})$ with the property

$$U(A^i_{\eta,\Sigma_\sigma}, A^i_{\eta,\Sigma_\sigma}) = U(A^j_{(\eta,\eta'),(\Sigma_\sigma,\Sigma_{\sigma'})}, A^j_{(\eta,\eta'),(\Sigma_\sigma,\Sigma_{\sigma'})}) = 0.$$

In particular, the form of (4.4) and (4.5) has the disadvantage that a finite energy is sufficient by a discrete unit to achieve a vanishing value of one of its constraints, for example, to shrink a cell to a point, a situation that should be avoided and that would in principle require infinite energy. For this reason, we prefer to use potentials that blow up in the case of $a^i_{\eta,\Sigma_\sigma}, a^j_{(\eta,\eta'),(\Sigma_\sigma,\Sigma_{\sigma'})} \to 0$. In particular, among the possible different forms, we define the constraint contributions to the *Hamiltonian* as

$$H_{constraint}(t) = \sum_{\substack{\eta,\Sigma_\sigma \\ i-constraint}} \lambda^i_{\eta,\Sigma_\sigma}(t) \left| 1 - \frac{A^i_{\eta,\Sigma_\sigma}(t)}{a^i_{\eta,\Sigma_\sigma}(t)} \right|^p, \tag{4.6}$$

$$H_{constraint}(t) = \sum_{\substack{(\eta,\eta'),(\Sigma_\sigma,\Sigma_{\sigma'}) \\ j-constraint}} \lambda^j_{(\eta,\eta'),(\Sigma_\sigma,\Sigma_{\sigma'})}(t) \left| 1 - \frac{A^j_{(\eta,\eta'),(\Sigma_\sigma,\Sigma_{\sigma'})}(t)}{a^j_{(\eta,\eta'),(\Sigma_\sigma,\Sigma_{\sigma'})}(t)} \right|^p, \tag{4.7}$$

with $p \in \mathbb{R}_+$. In this way, in addition to the just-stated advantages, all the components of $H_{constraint}$ are nondimensional, and thus all the relative Potts coefficients are coherently scaled to units of energy.

Finally, we characterize the forces acting on each object with

$$H_{force}(t) = -\sum_{\mathbf{x}\in\Sigma_\sigma} \sum_{k-force} \mu^k_{\eta,\Sigma_{\sigma(\mathbf{x})}}(t)\, \mathbf{F}^k(t) \cdot \mathbf{r}_\mathbf{x}(t), \tag{4.8}$$

where $\mu^k_{\eta,\Sigma_\sigma}$ is the Potts coefficient that measures the effective strength of the force $\mathbf{F}^k$ sensed by the unit Σ_σ of the individual η.

The compartmentalization technique is not entirely new in the CPM: it was in fact first introduced in [374], where the authors have subdivided a *Myxococcus xanthus* into strings of subcellular domains, in order to give the bacterium a particular geometry and to control its overall length. Another example is reported in [254], where a keratocyte has been represented with a set of undifferentiated hexagonal subunits, which have facilitated the reproduction of its polarization during motion.

Although these approaches are correct, the fact that the proposed subcellular compartments do not have an immediate or direct correspondence with real subcellular elements has limited the realism and the usefulness of the related models. Indeed, although each simulated element can in principle be compartmentalized in a variety of ways (for example, along symmetry planes, or in a fixed number of equivalent and undifferentiated subunits), a biologically plausible compartmentalization (which agrees with the compartmentalization suggested in *nature*), is preferable. In this way, the specific subunits, and the relative attributes, assume, in fact, experimentally relevant meanings and allow a detailed description of the microscopic biochemical and biomechanical mechanisms, which typically are strongly localized within well-defined individual subcompartments (as we will comment in the next section).

For example, referring to the ideal compartmentalization of a cell reproduced in Figure 4.2, an explicit representation of the plasmamembrane (PM),

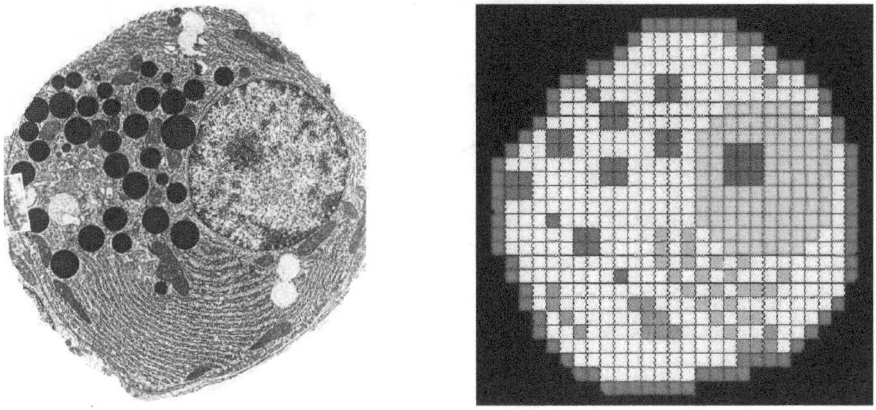

(A) (B)

FIGURE 4.2: Comparison between a cell section of (A) electron micrograph and (B) compartmentalized CPM cells. In both cases, it is straightforward to identify the nucleus, the nucleolus, the plasmamembrane, the Golgi apparatus, some mitochondria, and some secretory granules.

the cytosolic region, the nucleus, and other intracellular organelles (mitochondria, ribosomes, Golgi apparatus, secretory granules, ...) is mandatory to accurately model most of the main intracellular phenomena. For instance, the introduction of an explicit PM permits one to define the activity of surface chemical receptors, as well as to better describe the cell adhesive properties, i.e., to simulate the diffusion of adhesion molecules from the cell cytosol or their clusterization within particular regions. Moreover, it is possible to reproduce the active and continuous reorganization of the cytoskeleton, which provides the mechanical support for cells and mediates their coordinated and directed movements, in response to mechanical tensions and stresses from the local environment or to internal biochemical signals. The explicit representation of the cell nucleus would instead be mandatory to model genetic mechanisms, such as DNA duplication, RNA synthesis and diffusion, or transport of proteins through the nuclear membrane.

The compartmentalized approach is clearly flexible, since it allows the level of details to be tuned by only increasing or decreasing the number of units that form the clustered individual, or the number of lattice sites per functional unit. However, it is obviously computationally expensive, and often such a level of detail is neither required nor relevant for a good simulation of a number of biological processes. The optimal strategy depends on the phenomenon of interest, which therefore requires a preliminary analysis.

In particular, in the applications of the methods proposed in the next chapters, we will use a simplified representation of cells, which will not be

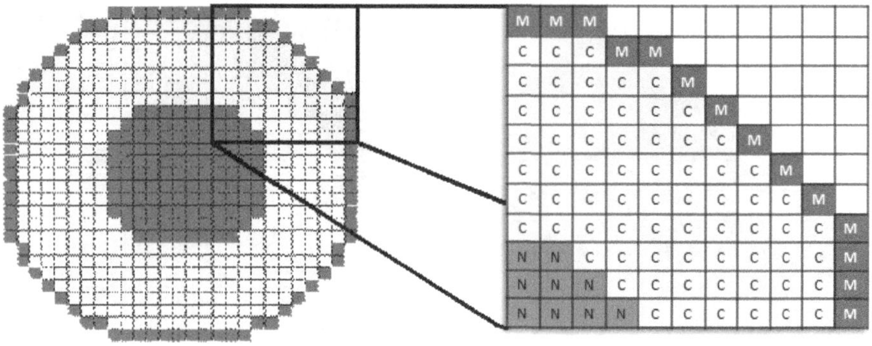

FIGURE 4.3: For representative purposes, simplified representation of a cell used in the applications of next chapters. The cell is sub-divided in the central nuclear cluster of type $\tau = N$, the surrounding cytosolic region of type $\tau = C$ and the plasma-membrane of type $\tau = M$. A further simplification made in some models will be an implicit representation of the PM, defined only as the interface between the cytosolic compartment and the extracellular environment.

described in full detail as in Figure 4.2 but will be reproduced only as bi- or three-compartmental individuals; see Figure 4.3. In the first case, they will be formed by the nucleus, a central almost spherical (or round in 2D) cluster of type $\tau = N$, and by the surrounding cytosol, of type $\tau = C$, while the plasma-membrane will be defined only as the border between the cytosolic region and the extracellular environment. In the second case, the plasma envelope will be instead explicitly represent as a further cell compartment, of type $\tau = M$, one site-width and enclosing the entire individual. In both cases, the nuclear membrane will be not explicitly model, being defined as the interface between the nucleus and the cytosolic units. This choice is justified from the fact that, with respect to the PM, no relevant processes will occur there during the simulated phenomena.

Obviously, the measures of the intracellular compartments will be clarified for each specific application, as they may vary according to the cell line of interest.

4.3 Nested Approach

In order to improve the realism of CPM applications, researchers have mainly been focused on defining more complex *Hamiltonians*, able to reproduce increasingly specific rules for object behavior and interactions [253].

However, as pointed out in [165, 262], little has been done to obtain an accurate description of the internal state of individuals whose evolution directly controls their phenomenology. In fact, microscopic mechanisms have only been modeled to influence the behavior of CPM objects via appropriate extra terms in the *Hamiltonian* [254], or with changes in cell mitotic rates [203] or types. In this regard, an intriguing approach has been used in a model for chick limb-bud development in [75, 76], where a threshold local concentration of activator TGF-β drove the differentiation of responsive cells in the active zone, eventually varying their properties (i.e., they became fibronectin-producing and upregulated the intercellular adhesion). In interesting models of thrombus formation [414, 415, 416], the activation of platelets was instead controlled by the level of chemical components, which derived from biochemical reactions of coagulation pathways in blood flow and on cell surface. Such approaches have given qualitatively correct results, and represent a useful starting point for further improvements of the method, aiming at accurately linking the cell-level phenomenology of simulated individuals to their microscopic level of organization, i.e., able to give to the CPM environment a *nested* characteristic.

Our main assumption is that the internal state of a biological individual (i.e., the microscopic level) regulates its biophysical properties (described by mesoscopic Potts coefficients) and not directly its dynamics (described by the terms in the *Hamiltonian*). An analogous idea was introduced in [186] for a specific case, but was barely developed. On the basis of this hypothesis, the comprehensive and general procedure to nest microscopic models for individual internal states within the mesoscopic CPM reads as follows.

Let Σ_σ denote a certain discrete object (which, as seen, can now represent a whole individual or one compartment): we define its internal state vector $\mathbf{s}_{\Sigma_\sigma} \in \mathbb{R}^n$. The length n of $\mathbf{s}_{\Sigma_\sigma}$ is defined by the number of internal factors (i.e., nutrients, proteins, growth factors, ..., all described by continuous objects) considered in the microscopic model, and represents a sort of internal degree of freedom of Σ_σ. Each component $s_{\Sigma_\sigma,l}$, where $l = 1, \ldots, n$, can be local (i.e., per site) and/or time-dependent (i.e., linked to a specific regulatory pathway, which needs to be modeled, as it will be explained hereafter). Hence, in general, $\mathbf{s}_{\Sigma_\sigma} = \mathbf{s}_{\Sigma_\sigma}(\mathbf{x}, t)$, where $\mathbf{x} \in \Sigma_\sigma$. The spatial localization of $\mathbf{s}_{\Sigma_\sigma}$ is mandatory to accurately represent internal inhomogeneities of Σ_σ, while its time dependence is necessary to reproduce its microscopic evolution.

For any Σ_σ, let us consider a generic Potts coefficient

$$\alpha \in \{\lambda^i_{\Sigma_\sigma}, T_{\Sigma_\sigma}, \mu^k_{\Sigma_\sigma}, \ldots\}.$$

We now define $\mathbf{s}^\alpha_{\Sigma_\sigma} \in \mathbb{R}^m$, where $m \le n$, the subvector of $\mathbf{s}_{\Sigma_\sigma}$ whose components influence the biophysical property of Σ_σ described by α. Therefore, the spatiotemporal evolution of α can be expressed as

$$\alpha(\mathbf{x} \in \Sigma_\sigma, t) = \alpha(\mathbf{s}^\alpha_{\Sigma_\sigma}(\mathbf{x}, t)) = f_\alpha(\mathbf{s}^\alpha_{\Sigma_\sigma}(\mathbf{x}, t)), \tag{4.9}$$

where $f_\alpha : \mathbb{R}^m \mapsto \mathbb{R}$ is a continuous function, which obviously needs to be appropriately defined in relation to the case of interest. In particular, if f_α is an increasing (respectively, decreasing) function of the component $s^{\alpha}_{\Sigma_\sigma,j}$, where $j = 1, \ldots, m$, we can therefore say that $s^{\alpha}_{\Sigma_\sigma,j}$ enhances (respectively, inhibits) α. In some cases it is useful to order the components of $\mathbf{s}^{\alpha}_{\Sigma_\sigma}$, to write it as $\mathbf{s}^{\alpha}_{\Sigma_\sigma} = (\mathbf{s}^{\alpha,A}_{\Sigma_\sigma}, \mathbf{s}^{\alpha,I}_{\Sigma_\sigma})$, where $\mathbf{s}^{\alpha,A}_{\Sigma_\sigma} \in \mathbb{R}^k$ (respectively, $\mathbf{s}^{\alpha,I}_{\Sigma_\sigma} \in \mathbb{R}^{m-k}$, with $k \leq m$) consists of the activators (respectively, the inhibitors) of the biophysical property defined by α.

According to the same notation, if

$$\gamma \in \{J^{int}_{\tau(\Sigma_\sigma),\tau(\Sigma_{\sigma'})}, J^{ext}_{\theta(\eta),\theta(\eta')}\},$$

for each local interface between neighboring objects (i.e., $(\partial\mathbf{x} \in \partial\Sigma_\sigma) \cap (\partial\mathbf{x}' \in \partial\Sigma_{\sigma'})$ or $(\partial\mathbf{x} \in \partial\eta) \cap (\partial\mathbf{x}' \in \partial\eta')$), we have

$$\gamma((\partial\mathbf{x} \cap \partial\mathbf{x}'), t) = \gamma(\mathbf{s}^{\gamma}_{\Sigma_\sigma}(\mathbf{x}, t), \mathbf{s}^{\gamma}_{\Sigma_{\sigma'}}(\mathbf{x}', t)) = g_\gamma(\mathbf{s}^{\gamma}_{\Sigma_\sigma}(\mathbf{x}, t), \mathbf{s}^{\gamma}_{\Sigma_{\sigma'}}(\mathbf{x}', t)), \quad (4.10)$$

where $g_\gamma : \mathbb{R}^m \times \mathbb{R}^m \mapsto \mathbb{R}$. In particular, the local adhesive strengths are determined by the (local) internal state of both elements, as they are not only a property of each single individual. Moreover, differentiating the components of the internal state vector that either enhance or downregulate the relative contact forces, we have that, if g_γ depends only on activators then $g_\gamma = g_\gamma(\mathbf{s}^{\gamma,A}_{\Sigma_\sigma}, \mathbf{s}^{\gamma,A}_{\Sigma_{\sigma'}})$ is a decreasing function of its components, while, if *vice versa* $g_\gamma = g_\gamma(\mathbf{s}^{\gamma,I}_{\Sigma_\sigma}, \mathbf{s}^{\gamma,I}_{\Sigma_{\sigma'}})$, then it is an increasing function.

Indeed, the Potts parameters that can locally vary (such as the adhesive interactions or the effective strengths of specific forces) require that the relative functions of the internal state vector $\mathbf{s}$ be local (i.e., they need to take into account the local concentration of the internal factors of interest), whereas the Potts parameters characterizing an entire individual (such as the motility or the elasticity) require that the relative functions of $\mathbf{s}$ be global (i.e., they need to take into account the overall level of the internal factor of interest).

Equations (4.9) and (4.10) state that the variation of the Potts coefficients of the element Σ_σ (as usual, either an entire individual or one of its compartments) is due to the evolution of its internal state: in this way, the mesoscopic biophysical properties of Σ_σ are no longer given *a priori* (or varied with prescribed rules), but are autonomously and continuously inherited from the flow of information coming from the microscopic molecular level, which in turn may be affected, for instance, by exchanges of signals with the external environment (i.e., as in a typical *hybrid* approach). They therefore assume a biologically more realistic and accurate characterization and, in principle, can be more easily compared with experimentally measurable quantities.

As explained in more detail in Appendix A, procedurally, at every simulation time step t, the microscopic model of each object Σ_σ is run. The outcome is then used to modulate the values of the relative Potts coefficients, which in turn rescale the pattern energy H. After the subsequent spin flip, the microscopic model is rederived, based on the new position of the object. The

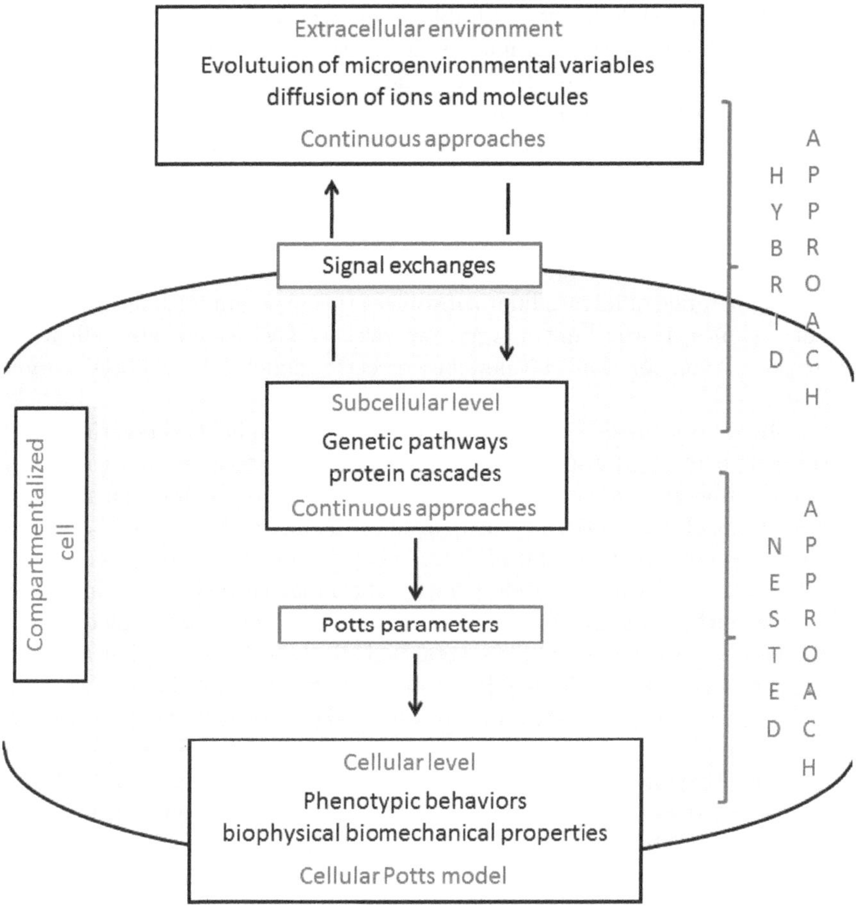

FIGURE 4.4: Hierarchy of scales and environments, the corresponding biological mechanisms, and our proposed modeling approach. Information flows from the microscopic biochemical processes to the cell-level phenomenology.

different levels of organization thus continuously give feedback and influence each other, as depicted in Figure 4.4.

The application of the new approach to biological cells is of particular interest. In this context, each component of the internal state vector $\mathbf{s}$ can represent the spatiotemporal variation of the concentration of intracellular ions and molecules (nutrients, nucleic acids, proteins, growth factors, ...), whose presence and quantity strictly regulate the cell activity and its biological properties. In particular, for any component $s_{\Sigma_\sigma,l}$ of $\mathbf{s}_{\Sigma_\sigma}$, given a well-characterized (although simplified) biochemical pathway

$$A + B \rightleftharpoons C + D \ldots \leftrightharpoons E + F \rightleftharpoons s_{\Sigma_\sigma,l}, \tag{4.11}$$

it is always possible to set a suitable model whose outcome is $s_{\Sigma_\sigma,l}$ itself. For the sake of simplicity, it is not restrictive to combine groups of reactions in (4.11) whose results have the same effect on the final result of the network. Biochemical kinetics can be accurately described by reaction-diffusion (RD) systems, which usually specialize in several coupled differential equations. As an example of the potentials of the proposed model development, the Boltzmann temperature T in Equation (1.2) is no longer a biologically meaningless agitation rate but becomes a variable property of each moving cell, assuming a well-defined value of *cell intrinsic motility*, which can be realistically mediated by a wide range of intracellular substances (i.e., calcium ions, fatty acids, ...). For instance, in the limit of very low levels of such chemicals, cells actually freeze, a behavior that can now be correctly modeled by a Potts parameter $T \approx 0$.

Microscopic models of intracellular dynamics can be used together with the compartmentalized approach described in the previous section (if Σ_σ represents a subregion of compartmentalized individual η, the internal state vector is $\mathbf{s}_{\eta,\Sigma_\sigma}$), to further increase the realism of the model. In fact, if Σ_σ represents a cell subunit, the biochemical processes defined in Equation (4.11) are accurately localized within a well-defined subcellular compartment, as occurs in reality. Such an integration between the two proposed extensions of the CPM therefore allows one to handle several biological mechanisms, that are difficult to reproduce with the basic CPM. For example, the explicit representation of the cell plasmamembrane permits one to straightforwardly model a wide range of surface-receptor-activated intracellular pathways, as well as specific protein cascades that mediate the activity of the cell adhesion molecules. Moreover, the geometrical properties of the cell cytosolic compartment, such as its elasticity, can now evolve according to a model of the active reorganization of the actin filaments, which is powered, for example, by ATP hydrolysis.

A more realistic representation of the mitotic process is also possible. In most existing CPMs, the cell cycle is not modeled explicitly, since cells usually undergo duplication when they reach a fixed volume [25, 334, 365]. With the presented CPM improvements, it is instead possible to incorporate appropriate intracellular signaling cascades regulating cell-cycle-dependent events.

The proposed method to interface the basic mesoscopic CPM with models of microscopic dynamics can be considered a general guide to reproduce complex interactions between the different levels of organization that are typical of biological phenomena. In fact, it can be applied to a number of situations, given the knowledge of

- The pathways of interest.
- The specific functional laws that describe their influence on the biophysical properties of individuals, i.e., the functions f and g.

Indeed, all these components are integrated and interfaced together, constituting a hybrid-nested simulation environment characterized by a constant flux of information from finer to coarser levels, where the kinetics of molecular subcellular networks strongly deliver the mesoscopic cell properties, described, as explained, by typical Potts parameters. The resulting modeling framework, utilizing the advantage of both discrete and continuous techniques, is therefore able to span all the different biological spatiotemporal scales involved, which range from 10^{-8} m and 10^{-9} s for the intracellular molecular processes to 10^{-6} m and hours for the cell-level phenomenology.

The sample applications given in the following chapters will be useful to demonstrate the efficacy of the above-defined procedure and to clarify the complex notation adopted in this section.

4.4 Motility of Individuals

The motility of individuals plays an important role in all biological phenomena, both physiological (such as embryo development and organogenesis, organism growth and survival, or wound healing) and pathological (such as inflammation and atherosclerosis, cancer invasion, or metastatization). An accurate description of the motility of individuals is therefore a fundamental issue for all computational approaches, and it is one of the most attractive features of the CPM. The Metropolis algorithm allows, in fact, CPMs to represent naturally the continuous, exploratory behavior of migrating organisms through biased extensions and retractions of their boundaries. Moreover, by updating only one spin at a time, the individuals move gradually, rather than in jumps, as in some other approaches (for instance, multispin dynamics such as Kawasaky dynamics are also possible, as commented in [165]). The CPM technique also allows one to differentiate between the isotropic intrinsic motility of each individual, which is described by its Boltzmann temperature T (which can be approximately compared to a diffusion coefficient from a continuous point of view), and the directional, force-based component of its motion. Since a difference in a potential energy might be related to a force, at any given time t, for any site $\mathbf{x}$ of the domain Ω, the local negative gradient of the functional H can be related to the local applied force.

Indeed, since a difference in a potential energy might be related to the work done by a force, we can compute the change of energy due to the spin flips related to an object Σ_σ as

$$\frac{\Delta H}{\Delta t} = -\Pi = -\sum_{\mathbf{x}\in\Sigma_\sigma} \mathbf{F}_{\mathbf{x}\in\Sigma_\sigma} \cdot \frac{\Delta \mathbf{x}}{\Delta t} = -\sum_{\mathbf{x}\in\Sigma_\sigma} \mathbf{F}_{\mathbf{x}\in\Sigma_\sigma} \cdot \mathbf{v}_{\mathbf{x}\in\Sigma_\sigma}, \tag{4.12}$$

where Π is the power of all forces $\mathbf{F}$ acting on site $\mathbf{x}$ of the object Σ_σ and $\mathbf{v}_{\mathbf{x}\in\Sigma_\sigma}$ the local velocity.

In CPMs, it is assumed that for any element Σ_σ (as usual, both a whole individual or a subcompartment), the local velocity, and not the acceleration, is proportional to the local force. This relation is called *overdamped force-velocity response*, and it is characteristic of other IBMs [111, 113]:

$$\mathbf{F}_{\mathbf{x}\in\Sigma_\sigma} = \alpha_{\mathbf{x}\in\Sigma_\sigma}\mathbf{v}_{\mathbf{x}\in\Sigma_\sigma} = \frac{1}{k_{\mathbf{x}\in\Sigma_\sigma}}\mathbf{v}_{\mathbf{x}\in\Sigma_\sigma}. \tag{4.13}$$

The coefficient $k_{\mathbf{x}\in\Sigma_\sigma}(t)$ is the *net rate of transition* of site $\mathbf{x}$, the difference between its probability of moving and staying still, $P(\sigma(\mathbf{x}) \to \sigma(\mathbf{x}')) - P(\sigma(\mathbf{x}) \nrightarrow \sigma(\mathbf{x}'))$, as in [165]. As analytically demonstrated again in [165] and in [254] for a particular case, $k_{\mathbf{x}\in\Sigma_\sigma}(t)$ is related to the Boltzmann temperature. We can therefore write

$$\mathbf{F}_{\mathbf{x}\in\Sigma_\sigma} \propto \frac{1}{T_{\Sigma_\sigma}}\mathbf{v}_{\mathbf{x}\in\Sigma_\sigma}. \tag{4.14}$$

As far as the above-mentioned published results are concerned, we here prefer to use a proportional dependence between $k_{\mathbf{x}\in\Sigma_\sigma}(t)$ and $T_{\Sigma_{\sigma(\mathbf{x})}}(t)$ and not an equation, since the exact relation between the Monte Carlo spin copy attempts and the continuous time, as well as the kinematics application of the Metropolis-like algorithm, are still debated and are a persistent source of criticism.

Using (4.14) and assuming, for the sake of simplicity, the proportionality coefficient constant throughout the object, (4.12) can be rewritten as

$$\frac{\Delta H}{\Delta t} \propto -\sum_{\mathbf{x}\in\Sigma_\sigma}\frac{1}{T_{\Sigma_\sigma}}\mathbf{v}^2_{\mathbf{x}\in\Sigma_\sigma} = -\frac{1}{T_{\Sigma_\sigma}}\sum_{\mathbf{x}\in\Sigma_\sigma}\mathbf{v}^2_{\mathbf{x}\in\Sigma_\sigma}, \tag{4.15}$$

given that T_{Σ_σ} is a global property of the entire object. Let us now decompose the velocity $\mathbf{v}_{\mathbf{x}\in\Sigma_\sigma}$ as

$$\mathbf{v}_{\mathbf{x}\in\Sigma_\sigma} = \mathbf{v}^{CM}_{\Sigma_\sigma} + \mathbf{w}_{\mathbf{x}\in\Sigma_\sigma}, \tag{4.16}$$

where $\mathbf{v}^{CM}_{\Sigma_\sigma}$ is the velocity of the object center of mass (defined in Equation (1.11)) and $\mathbf{w}_{\mathbf{x}\in\Sigma_\sigma}$ a local fluctuation. Simple calculations lead to

$$\sum_{\mathbf{x}\in\Sigma_\sigma}\mathbf{v}_{\mathbf{x}\in\Sigma_\sigma} = \sum_{\mathbf{x}\in\Sigma_\sigma}\mathbf{v}^{CM}_{\Sigma_\sigma} + \sum_{\mathbf{x}\in\Sigma_\sigma}\mathbf{w}_{\mathbf{x}\in\Sigma_\sigma} = \mathbf{v}^{CM}_{\Sigma_\sigma}a^{volume}_{\Sigma_\sigma} + \sum_{\mathbf{x}\in\Sigma_\sigma}\mathbf{w}_{\mathbf{x}\in\Sigma_\sigma}, \tag{4.17}$$

where the second term of the sum vanishes. Therefore, we obtain

$$\sum_{\mathbf{x}\in\Sigma_\sigma}\mathbf{v}^2_{\mathbf{x}\in\Sigma_\sigma} = (\mathbf{v}^{CM}_{\Sigma_\sigma})^2 a^{volume}_{\Sigma_\sigma} + \sum_{\mathbf{x}\in\Sigma_\sigma}\mathbf{w}^2_{\mathbf{x}\in\Sigma_\sigma}, \tag{4.18}$$

that, substituted in (4.12), leads to

$$\frac{\Delta H}{\Delta t} \propto -\frac{1}{T_{\Sigma_\sigma}}\left[(\mathbf{v}^{CM}_{\Sigma_\sigma})^2 a^{volume}_{\Sigma_\sigma} + \sum_{\mathbf{x}\in\Sigma_\sigma}\mathbf{w}^2_{\mathbf{x}\in\Sigma_\sigma}\right]. \tag{4.19}$$

Given that

$$\frac{\Delta H}{\Delta t} = \frac{\Delta H}{\Delta \mathbf{x}^{CM}_{\Sigma_\sigma}}\mathbf{v}^{CM}_{\Sigma_\sigma},$$

it is possible to finally conclude that

$$\frac{\Delta H}{\Delta \mathbf{x}^{CM}_{\Sigma_\sigma}} \propto -\frac{1}{T_{\Sigma_\sigma}} \left[\mathbf{v}^{CM}_{\Sigma_\sigma} a^{volume}_{\Sigma_\sigma} + \frac{1}{\mathbf{v}^{CM}_{\Sigma_\sigma}} \sum_{\mathbf{x}\in\Sigma_\sigma} \mathbf{w}^2_{\mathbf{x}\in\Sigma_\sigma} \right], \tag{4.20}$$

or, with another viewpoint,

$$a^{volume}_{\Sigma_\sigma} \mathbf{v}^{CM}_{\Sigma_\sigma} \propto -T_{\Sigma_\sigma} \frac{\Delta H}{\Delta \mathbf{x}^{CM}_{\Sigma_\sigma}} - \frac{1}{\mathbf{v}^{CM}_{\Sigma_\sigma}} \sum_{\mathbf{x}\in\Sigma_\sigma} \mathbf{w}^2_{\mathbf{x}\in\Sigma_\sigma}. \tag{4.21}$$

Some comments on the consequences of relation (4.13) are the following:

- It is characteristic of an extremely viscous regime, such as the biological environment where the individuals thus evolve under strong damping.

- It is a definitive confirmation that discrete CPM objects move in order to minimize the total energy. The modulus of the local velocity of Σ_σ, at any given time t, depends on the magnitude of the energy difference due to the proposed spin flip, as well as on its intrinsic motility $T_{\Sigma_\sigma}(t)$, which, in our extended approach, is coherently a variable property of each unit Σ_σ.

- Given that T_{Σ_σ} is now a property of each object Σ_σ, different individuals (or different individual subcompartments) may have different velocities even if they experience the same energy gradient. A consequence is that the one-to-one relationship between a MCS and the standard Boltzmann temperature T characteristic of the basic CPM is lost.

- Given that the energy functional H is the sum of the terms that represent multiple biological mechanisms with the same architecture, it is straightforward to evaluate the contribution of each of them to the local velocity of unit Σ_σ. In fact, for any mechanism i, by equating all the other terms to zero, we obtain

$$a^{volume}_{\Sigma_\sigma} \mathbf{v}^{CM}_{\Sigma_\sigma} \Big|_{i-mechanism} \propto -T_{\Sigma_\sigma} \frac{\Delta H^{i-mechanism}}{\Delta \mathbf{x}^{CM}_{\Sigma_\sigma}} - \frac{1}{\mathbf{v}^{CM}_{\Sigma_\sigma}} \sum_{\mathbf{x}\in\Sigma_\sigma} \mathbf{w}^2_{\mathbf{x}\in\Sigma_\sigma}. \tag{4.22}$$

  In particular, it is possible to differentiate the contributions of either short- or long-range mechanisms:

$$a^{volume}_{\Sigma_\sigma} \mathbf{v}^{CM}_{\Sigma_\sigma} \propto -T_{\Sigma_\sigma} \left[\frac{\Delta H^{short\ range} + \Delta H^{long\ range}}{\Delta \mathbf{x}^{CM}_{\Sigma_\sigma}} \right] - \frac{\sum_{\mathbf{x}\in\Sigma_\sigma} \mathbf{w}^2_{\mathbf{x}\in\Sigma_\sigma}}{\mathbf{v}^{CM}_{\Sigma_\sigma}} \tag{4.23}$$

  the former of which includes, for example, adhesion and haptotaxis, while the latter includes chemotaxis.

However, even with the new meaning and expression assumed by the Boltzmann temperature, the model in Equation (1.2) can be further improved with the use of alternative laws that are better able to describe experimental observations on the ruffling of individual membranes. In particular, the main weakness of Equation (1.2) is that, in the case of nonpositive net differences in the system energy caused by the proposed displacement ($\Delta H|_{\sigma(\mathbf{x})\to\sigma(\mathbf{x}')} \leq 0$), each element Σ_σ is certainly going to move, regardless of its T_{Σ_σ}. This lacks biological realism, since the effective motion of an organism is closely dependent on its motility. For example, a "frozen" cell (i.e., with negligible intrinsic motility) does not extend its pseudopods toward a chemical source even if it senses a high chemotactic gradient (which, in the absence of other external forces, results in $\Delta H \ll 0$). This situation is the model counterpart of experimental cells pretreated with cytochalasin B, or held at 4°C, as clearly shown in [268]. This issue can easily be addressed using a modified family of transition probabilities $P(\sigma(\mathbf{x}) \to \sigma(\mathbf{x}'))$, which takes into account the object motility $T_{\Sigma_{\sigma(\mathbf{x})}}$ also in the case of energetically favorable displacement attempts. A possible option is to use

$$P(\sigma(\mathbf{x}) \to \sigma(\mathbf{x}'))(t) = \begin{cases} p(T_{\Sigma_{\sigma(\mathbf{x})}}(t))e^{-\Delta H|_{\sigma(\mathbf{x})\to\sigma(\mathbf{x}')}/T_{\Sigma_{\sigma(\mathbf{x})}}(t)} & \\ & \Delta H|_{\sigma(\mathbf{x})\to\sigma(\mathbf{x}')} > 0\,; \\ p(T_{\Sigma_{\sigma(\mathbf{x})}}(t)) & \Delta H|_{\sigma(\mathbf{x})\to\sigma(\mathbf{x}')} \leq 0\,, \end{cases} \tag{4.24}$$

where $p(T_{\Sigma_\sigma}(t)) : \mathbb{R}_+ \mapsto [0,1]$, a sort of *maximum transition probability*, is a continuous and increasing function of T_{Σ_σ}, and is characterized by

$$\begin{cases} p(0) = 0; \\ \lim_{T_{\Sigma_\sigma}\to+\infty} p(T_{\Sigma_\sigma}) = 1. \end{cases} \tag{4.25}$$

With reference to the above-mentioned example, Equations (4.24) and (4.25) state that if $T_{\Sigma_\sigma} \to 0$, cell Σ_σ does not move, even if it senses a steep energy gradient, as it is really frozen. Again, according to the discussion in [165, 254], we obtain

$$\mathbf{F}_{\mathbf{x}\in\Sigma_\sigma} = \alpha_{\mathbf{x}\in\Sigma_\sigma}\mathbf{v}_{\mathbf{x}\in\Sigma_\sigma} = \frac{1}{k'_{\mathbf{x}\in\Sigma_\sigma}}\mathbf{v}_{\mathbf{x}\in\Sigma_\sigma} \propto \frac{1}{p(T_{\Sigma_\sigma})T_{\Sigma_\sigma}}\mathbf{v}_{\mathbf{x}\in\Sigma_\sigma}, \tag{4.26}$$

where $k'_{\mathbf{x}\in\Sigma_\sigma}(t)$ is the new *net rate of transition* coming from the proposed probability function in Equation (4.24). Furthermore, considering the expression of (1.2) itself and comparing it with (4.24) (i.e., the latter is equal to the former multiplied by the factor $p(T_{\Sigma_\sigma}(t))$), it is clear that

$$k'_{\mathbf{x}\in\Sigma_\sigma}(t) = p(T_{\Sigma_{\sigma(\mathbf{x})}}(t))k_{\mathbf{x}\in\Sigma_\sigma}(t),$$

and that therefore

$$k'_{\mathbf{x}\in\Sigma_\sigma}(t) \propto p(T_{\Sigma_{\sigma(\mathbf{x})}}(t))T_{\Sigma_{\sigma(\mathbf{x})}}(t).$$

Summing up, we finally obtain:

$$a_{\Sigma_\sigma}^{volume}\mathbf{v}_{\Sigma_\sigma}^{CM} \propto -p(T_{\Sigma_\sigma})T_{\Sigma_\sigma}\frac{\Delta H}{\Delta \mathbf{x}_{\Sigma_\sigma}^{CM}} - \frac{1}{\mathbf{v}_{\Sigma_\sigma}^{CM}}\sum_{\mathbf{x}\in\Sigma_\sigma}\mathbf{w}_{\mathbf{x}\in\Sigma_\sigma}^2, \tag{4.27}$$

where the local velocity of Σ_σ is again dependent on its motility $T_{\Sigma_\sigma}(t)$, which is now scaled by the value of $p(T_{\Sigma_\sigma}(t))$. In Figure 4.5, we plot an example of Boltzmann-like probability given in Equation (4.24), as a function of both temperature T_{Σ_σ} and the magnitude of the energy difference. In particular, in the following

$$p(T_{\Sigma_\sigma}(t)) = \tanh(T_{\Sigma_\sigma}(t))$$

is chosen.

However, alternative functions of p can also be used as well: obviously, each of them might have different degrees of success, when compared to experimental data.

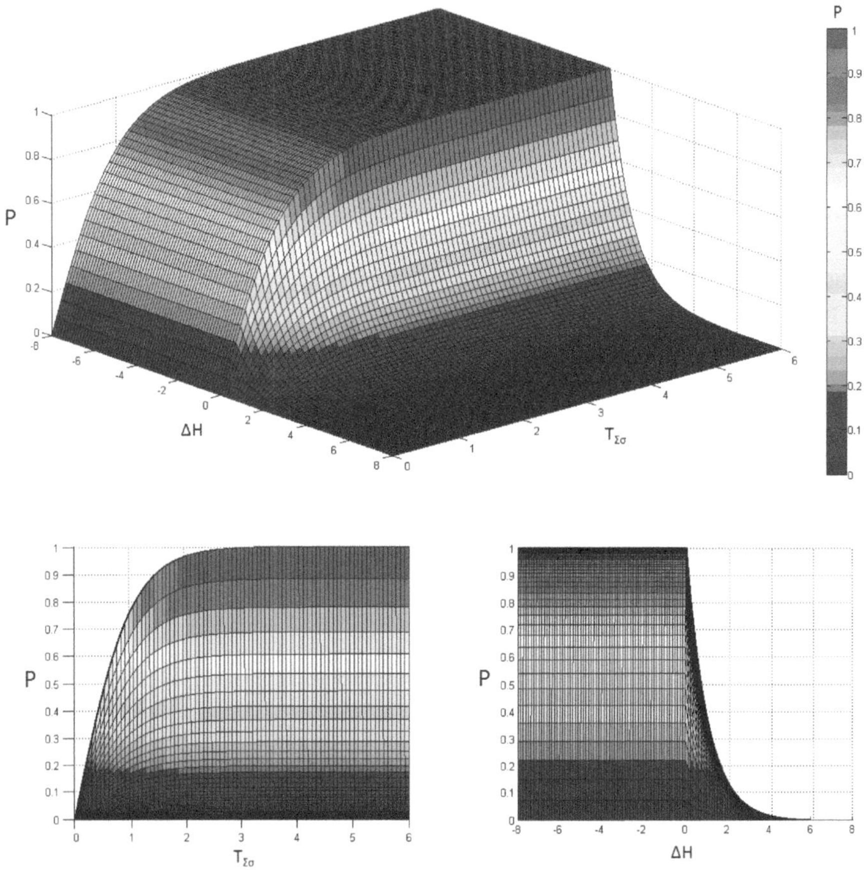

FIGURE 4.5: (See color insert.) Diagram showing the variation of the Boltzmann-like transition probability Equation (4.24) as a function of both the intrinsic motility of the individual, T_{Σ_σ}, and of the sensed net energy gradient due of the displacement attempt, $\Delta H|_{\sigma(\mathbf{x})\to\sigma(\mathbf{x}')}$. In particular, $p(T_{\Sigma_\sigma}(t)) = \tanh(T_{\Sigma_\sigma}(t))$. The small panels represent the Boltzmann-like sections in the case of fixed $\Delta H = -2$ (left), and fixed $T_{\Sigma_\sigma} = 4$ (right).

Chapter 5

Wound Healing Assay

5.1 Biological Introduction

Cell migration plays a critical role in a diverse array of processes both in physiological conditions and in pathological situations. In developing embryos, the coordinated movement of cells of different origin, over short and long distance paths, is fundamental for organogenesis, as defects at all stages of such a cooperative migration lead to severe embryonic malformations, resulting in drastic overall consequences [220]. In adult organisms, cell movement is instead essential in wound healing, where epidermal tissues are repaired by the progressive extension of tongues of epidermal cells, which completely close the lesions, as well as in the cases of inflammations. On the other hand, cell migration is involved in chronic inflammatory diseases and artheriosclerosis, and is responsible for the primary dissemination of tumors and its following metastatization [337]. It is also important in biomedical engineering applications, where bioactive scaffolds act as physical support structures and insoluble regulators for cell activity (i.e., adhesion, growth, and differentiation) in *in vivo* or *in vitro* regenerations of various tissues, such as cartilage, skin, or peripheral nerves [63, 178, 373, 418].

An accurate description of cell migratory capacity, as well as a deep analysis of the complex multilevel underpinning processes, are therefore pivotal issues in developmental biology, giving rise to a number of experimental models. In particular, given the outstanding amount and heterogeneity of approaches, a first and useful differentiation can be made following their scale of interest. At the cellular and subcellular level, a large set of experimental works have taken into account the dynamics of the movement of single individuals, providing a good understanding of the driving mechanical and molecular mechanisms, as the activation of integrin receptors [382], the activity of focal adhesion structures [188], or the cytoskeletal polarization [267, 311, 328]. At the multicellular level, the migration of populations has been analyzed, drawing the attention on both the migratory capacities of entire cultures and on the cooperative behaviors and interactions of single individuals.

In this regard, one of the most common experimental models is the *wound healing assay*, which measures the overall motility of a cell population by quantifying its invasion of a planar substrate. In particular, it consists of the incubation of a cell population toward confluence and the subsequent creation of an artificial scratch with a sharp object (for example, a pipette tip). The removal of cells from the wounded area acts as a stimulus for the remaining mass to invade and fill the open space, with a characteristic traveling-wave-like behavior. The rate of advancement of the wound edge, i.e., the quantification of the area recolonized, gives a measure of the migratory capacity of the population of interest. In particular, this technique is widely used to compare the motility properties of a cell line either in "resting conditions" (i.e., in the absence of external stimuli) or in response to specific chemical stimulations or modifications of expressions of molecules putatively involved in the migratory processes.

The wound healing experiment has been reproduced by several mathematical models, most of which are based on the Fisher equation. For instance, these methods have described the cell population as a density, while its motility has been prescribed by a diffusive flux, where the diffusivity has been considered either a constant [248, 249, 344, 387] or a function of an external chemical factor [98]. These authors, numerically calculating the traveling-wave solution of the Fisher equation, have therefore provided a simple relation between the speed of the moving front of the mass and both the diffusion constant and the mitotic time of cells. However, Fisher-like models have not included the effects of mutual interactions between cells and between cells and the extracellular environment and they cannot characterize and differentiate the dynamics of single individuals (i.e., cells at the front of the mass behave obviously differently from those at the center of the mass), assuming that all cells within the culture have the same migratory capacity.

A discrete approach has the potential to overcome these issues by retaining the identity of each cell. Indeed, we here simulate the healing process of a culture of ARO cells in response to the hepatocyte growth factor (HGF). The HGF is a potent growth factor that elicits multiple cellular responses, including scattering, motility, and morphogenesis [45, 92, 389, 394]. Such a combination of events, also known as *invasive growth*, is fundamental during the embryonic development of most epithelial tissues. When inappropriately activated, this genetic program confers an invasive ability on normal and neoplastic epithelial cells [101, 380, 407]. The high affinity receptor for HGF is the tyrosine kinase Met [233] . As explained in Section 2.1, Met activation causes both the disruption of intercellular adhesion complexes (cadherin–cadherin interactions) and the enhancement in cell motility. In particular, the latter is driven by a number of intracellular signaling pathways, which could not be reproduced in the standard model presented in Chapter 2, but which can now be described in full detail. Among others, we focus on phosphatidylinositol 3-kinases (PI3K) and mitogen-activated protein kinase (MAPK) cascades that have been intensively studied, and well characterized [91, 423]. Indeed, the activity of the multi-docking sites of Met triggers the biosynthesis in the cell sub-plasmamembrane regions of PI3K (via the production of Gab1) and of MAPK (via the activity of adaptor proteins Grb2 and Ras). PI3K and

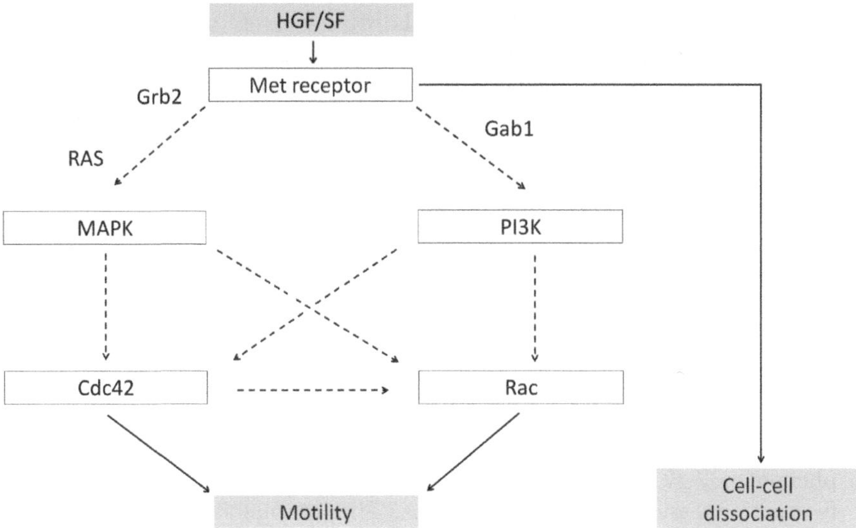

FIGURE 5.1: Simplified schematic representation of HGF/SF signaling cascades in the control of ARO cell motogenesis. Met receptors activate a series of intracellular events that induce the recruitment of phosphatidylinositol 3-kinase (PI3K) and mitogen-activated protein kinase (MAPK). Both intracellular messengers are able to activate Rac and Cdc42 molecules. Increases in Rac and Cdc42 activity trigger cell motility. The dashed arrows stand for indirect pathways not completely included in the model.

MAPK molecules then diffuse within the cell cytosol, where they induce the production of Cdc42 and Rac small GTP-ases [101, 423], which in turn stimulate the migratory capacity of the AROs. In particular, Cdc42, which can lead to the activation of Rac itself [18], is considered to be a central regulator of cell protrusive activity [125, 329]. Rac is instead required for lamellipodia and membrane ruffles [42]. Finally, both Cdc42 and Rac mediate actin polymerization by activating actin-related proteins, as Arp2/3 complexes [184, 328, 329]. Figure 5.1 diagrammatically represents the key biochemical processes just described.

The same phenomenon has already been modeled in Chapter 2. This choice allows a useful investigation of how the proposed CPM developments are able to affect and improve the representation of a given biological process. Indeed, each ARO individual can now be characterized by its own biophysical properties, such as its motility, that are realistically regulated by the external chemical stimuli via well-defined intracellular dynamics. The resulting model is highly flexible and able to characterize the healing of the cell population in different conditions consistently with the experimental counterparts, making clear the relevance of various mechanisms involved in determining the motile phenotype. Indeed, it describes the migratory phenotype of the whole cell culture as well as of single individuals by separately analyzing and quantitatively evaluating statistical parameters, such as cell velocity, directionality of their movement and distribution of their final displacement. In this way, the invasive capability of the population can be assessed both as a whole and as a sum of individual behaviors. Indeed, single cells can be specifically correlated according to their type of movement and grouped in well-defined subpopulations.

5.2 Mathematical Model

At the mesoscopic cellular level, a compartmentalized CPM represents the phenomenology of the ARO population, focusing on cell shape, membrane fluctuations, and adhesive interactions. The internal state of each individual is then explicitly approached with a continuous model that reproduces the biochemical signaling pathways activated, via Met receptors, by the HGF/SF, whose extracellular evolution is in turn described by a standard continuous equation. Finally, all the levels are inextricably linked, so that the behavior of the AROs is realistically driven by their microscopic, molecular dynamics.

5.2.1 Cell-Level Model

Since we wish to compare our simulations with experimental wound healing assays, we use a bidimensional domain $\Omega \subset \mathbb{R}^2$. The AROs are compartmentalized individuals η of type $\theta(\eta) = E$. In particular, each cell is subdivided in three units presented at the end of Section 4.2 and in Figure 4.3: the nucleus, a central cluster of type $\tau = N$, the surrounding, initially circular, cytosolic region, of type $\tau = C$, and the plasmamembrane (PM), of type $\tau = M$. For any cell η, we define the state vector of each compartment:

- If Σ_σ is such that $\tau(\Sigma_\sigma) = M$, $\mathbf{s}_{\eta,\Sigma_\sigma}(\mathbf{x}, t) = (m(\mathbf{x}, t)) \in \mathbb{R}_+$, where $m(\mathbf{x}, t)$ is the local concentration of activated surface receptors Met.
- If Σ_σ is such that $\tau(\Sigma_\sigma) = C$, $\mathbf{s}_{\eta,\Sigma_\sigma}(\mathbf{x}, t) = (p(\mathbf{x}, t), k(\mathbf{x}, t), c(\mathbf{x}, t), r(\mathbf{x}, t)) \in \mathbb{R}^4_+$, where $p(\mathbf{x}, t)$ corresponds to the local concentration of PI3K, $k(\mathbf{x}, t)$ of MAPK, $c(\mathbf{x}, t)$ of Cdc42, and $r(\mathbf{x}, t)$ of Rac.
- If Σ_σ is such that $\tau(\Sigma_\sigma) = N$, $\mathbf{s}_{\eta,\Sigma_\sigma}(\mathbf{x}, t) = (c(\mathbf{x}, t), r(\mathbf{x}, t)) \in \mathbb{R}^2_+$.

All these quantities are expressed in units of μM. The extracellular environment, i.e., the experimental Matrigel, is modeled as a special generalized cell Σ_0 of type $\tau = Q$. As usual, it is assumed to be static, passive, and homogenously distributed throughout the simulation domain, and therefore without volume or surface attributes. In particular, we neglect cell–matrix interactions or matrix remodeling mechanisms, such as the secretion of protein or degrading enzymes.

The system *Hamiltonian* is given by

$$H(t) = H_{constraint}(t) + H_{adhesion}(t). \tag{5.1}$$

$H_{constraint}$ includes the constraints regulating cell volume and surface which, in two dimensions, correspond to the cell area and perimeter. They have the same form as in Equation (4.6). The cell target dimensions, given in Appendix C, correspond to their initial measures, which reflect the most common morphologies observed in classical cultures of ARO cells [101] measured in [349]. Cells are not seen to grow during an experimental wound healing assay [101] and, as already seen in Section 2.1, HGF/SF stimulation does not cause a dramatic cytoskeletal reorganization: therefore, it is not restrictive to assume for any cell η and any unit Σ_σ high constant values of

$$\begin{cases} \lambda_{\eta,N}^{surface} = \lambda_{\eta,C}^{surface} = \lambda_{\eta,M}^{surface}; \\ \\ \lambda_{\eta,N}^{perimeter} = \lambda_{\eta,C}^{perimeter} = \lambda_{\eta,M}^{perimeter}. \end{cases}$$

$H_{adhesion}$ is straightforwardly differentiated into internal and external contact energy contributions, cf. Equations (4.1), (4.2), and (4.3). In particular, $J_{E,E}^{ext}$ is related to the capability of cells η and η' of creating local intercellular cadherin–cadherin complexes that are downregulated by the activation of Met receptors. Hence, adopting the same notation used in Section 4.3, for any ARO η and for $\tau(\Sigma_{\sigma(\mathbf{x})}) = M$, we obtain that

$$\mathbf{s}_{\eta,\Sigma_\sigma}^{J^{ext}}(\mathbf{x},t) = \mathbf{s}_{\eta,\Sigma_\sigma}^{J^{ext},I}(\mathbf{x},t) = (m(\mathbf{x},t)).$$

Therefore, for any couple of neighboring cells η and η' we have the following local relation (at the interface $(\partial\mathbf{x} \in \partial\eta) \cap (\partial\mathbf{x}' \in \partial\eta')$):

$$J_{E,E}^{ext}((\partial\mathbf{x} \in \partial\eta) \cap (\partial\mathbf{x}' \in \partial\eta'), t) = J_{E,E}^{ext}(\mathbf{s}_{\eta,\Sigma_\sigma}^{J^{ext},I}(\mathbf{x},t), \mathbf{s}_{\eta',\Sigma_{\sigma'}}^{J^{ext},I}(\mathbf{x}',t)) =$$

$$= g_{J^{ext}}(\mathbf{s}_{\eta,\Sigma_\sigma}^{J^{ext},I}(\mathbf{x},t), \mathbf{s}_{\eta',\Sigma_{\sigma'}}^{J^{ext},I}(\mathbf{x}',t)) = J_0 \exp(km(\mathbf{x},t)m(\mathbf{x}',t)).$$

The coefficient J_0 represents the adhesive force between resting AROs: it is a low value, qualitatively reproducing the high contact interactions between unstimulated AROs that, in the absence of external stimulations, tend to remain closely packed in circular islands, as it is possible to appreciate in the representative experimental images in [101, 349].

The above relations reproduce the fact that Met activation induces the local disruption of cell–cell adhesion junctions by the dispersal of E-cadherin and β-catenin from the intercellular complexes [101]. It is noteworthy that the basic CPM, without the compartmentalization technique, could have not defined a realistic relation for the local cell adhesive strength and for the local quantity of activated Met.

Since, obviously, cells should not break into small fragments, we fix very low bond energies $J^{int}_{C,N}$ and $J^{int}_{C,M}$. The contacts between the nucleus and the PM within the same cell and between the nuclei and the cytosolic regions of different cells do not occur because of the choice of interfacial energy and so the relative terms are not taken into account. Finally, the adhesion energy between the cell PM and the ECM is not considered since is not influenced by the activity of the growth factor (see the considerations described in Chapter 2).

Given the *Hamiltonian*, the transition probability of a spin flip has the form of Equation (4.24). In particular, we test the model using $p(T_{\eta,\Sigma_\sigma}(t)) = \tanh(T_{\eta,\Sigma_\sigma}(t))$. For any cell η and $\tau(\Sigma_\sigma) = N$, $T_{\eta,\Sigma_\sigma} = T_{\eta,N}$ is a constant low value that mimics the passive motion of the nucleus (the reader can refer to Chapter 6 for a more detailed explanation). If $\tau(\Sigma_\sigma) \in \{M, C\}$, T_{η,Σ_σ} represents instead, as seen, a measure of the real motility of η that is regulated by the intracellular quantity of the activated Rac and Cdc42 molecules (see the review article [329] and references therein). Adopting again the notation of Equations (4.9) and (4.10), for any ARO η and for $\tau(\Sigma_\sigma) = C$, we now have that

$$\mathbf{s}^{T}_{\eta,\Sigma_\sigma}(\mathbf{x}, t) = \mathbf{s}^{T,A}_{\eta,\Sigma_\sigma}(\mathbf{x}, t) = (r(\mathbf{x}, t), c(\mathbf{x}, t));$$

$$T_{\eta,\Sigma_\sigma}(t) = T_{\eta,\Sigma_\sigma}(\mathbf{s}^{T,A}_{\eta,\Sigma_\sigma}(\mathbf{x}, t)) = f_T(\mathbf{s}^{T,A}_{\eta,\Sigma_\sigma}(\mathbf{x}, t)) = T_0\left[\frac{1 + c_\eta(t)r_\eta(t)}{1 + hc_\eta(t)r_\eta(t)}\right],$$

where $r_\eta(t) = \sum_{\mathbf{x}\in\eta} r(\mathbf{x}, t)$ and $c_\eta(t) = \sum_{\mathbf{x}\in\eta} c(\mathbf{x}, t)$ are the total intracellular amounts of Rac and Cdc42, respectively, and T_0 is a measure of the basal ARO motility. T_0 is a low value, which has been qualitatively estimated by observing the negligible migratory capacity of unstimulated ARO cultures, as provided by wound healing assays and time-lapse video recording techniques in [101]. The Michaelis–Menten law has been chosen to model a dose-dependent enhancement of cell motility driven by the total intracellular level of the G-proteins Rac and Cdc42.

Finally, for any ARO η, the actual motility of the plasmamembrane is set equal to that of the cytosolic region of the same individual, i.e., $T_{\eta,M}(t) = T_{\eta,C}(t)$.

It is useful to emphasize that the characterization of a intrinsic motility specific for each single individual was not present in the basic model of the wound healing assay proposed in Chapter 2, but is now allowed by the nested approach introduced in the previous chapter.

5.2.2 Molecular-Level Model

The intracellular HGF/Met-driven pathways on which the model is based are depicted in the scheme in Figure 5.1. Although simplified, they are consistent with the experimental literature presented in [92, 423].

As seen in Chapter 2, Section 2.1, HGF/SF is uniformly added to the culture, diffuses in the ECM, where it decays at a constant rate and is taken up by ARO cells through Met receptors sited in the membrane compartment. Mathematically, the HGF/SF is a continuous object, whose extracellular spatial profile thus satisfies the following diffusion equation:

$$\frac{\partial h}{\partial t} = \underbrace{D_h \nabla^2 h}_{diffusion} - \underbrace{\lambda_h h}_{decay} - \underbrace{B(\mathbf{x}, t, h)}_{uptake} + \underbrace{S}_{addition}, \tag{5.2}$$

where $h(\mathbf{x}, t)$ denotes the local concentration of the peptide at medium site $\mathbf{x}$ (i.e., $\tau(\Sigma_{\sigma(\mathbf{x})}) = Q$). The coefficient of diffusivity, D_h, and of degradation, λ_h, are assumed to be homogeneous throughout the simulated Matrigel and have been derived from experimental measurements performed for a recent paper of us [349]. $S(\mathbf{x}, t)$ describes the addition of the growth factor at a constant rate ϕ_h outside the cells. The HGF/SF binding and uptake by AROs is defined as $B(\mathbf{x}, t, h(\mathbf{x}, t))$, and is limited to a maximum rate $\beta_h(t) > 0$ over the external surface of the cells:

$$B(\mathbf{x}, t, h) = \min\{\beta_h(t), z h(\mathbf{x}, t)\}, \tag{5.3}$$

where $\tau(\Sigma_{\sigma(\mathbf{x})}) = Q$ and $\exists \mathbf{x}' \in \Omega'_{\mathbf{x}} : \tau(\Sigma_{\sigma(\mathbf{x}')}) = M$. Relation (5.3) is realistic since the capacity of an ARO cell to locally uptake the growth factor through its boundary will saturate at a rate limit, β_h, which is the maximal amount of HGF/SF molecules that can be locally bound and internalized per unit of time. To compute $\beta_h(t)$, we have followed the discussion provided in [25] for the uptake of VEGF molecules by an endothelial cell. In particular, we have here considered a spatially homogeneous average number of Met receptors per cell membrane site: this number has been estimated by dividing the total number of HGF receptors in a generic ARO cell by the actual extension (i.e., at time t) of the cell membrane. Moreover, we have taken into account an instantaneous HGF/SF-receptor complex internalization rate of $4.3 \cdot 10^{-4}$ per second [152], and 95 kDa as the molecular weight for an HGF/SF molecule [91]. The local number of activated surface Met receptors (i.e., at site $\mathbf{x} : \tau(\Sigma_{\sigma(\mathbf{x})}) = M$) for any individual η is therefore defined as

$$m(\mathbf{x}, t) = \sum_{\substack{\mathbf{x}' \in \Omega'_{\mathbf{x}}: \\ \tau(\Sigma_{\sigma(\mathbf{x}')}) = Q}} \frac{1}{z} B(\mathbf{x}', t, h(\mathbf{x}', t)), \tag{5.4}$$

where z is the same as in Equation (5.3). In Equation (5.4), we have assumed a one-to-one interaction between the HGF/SF molecules and their surface receptors [91].

For each ARO cell η, the intracellular events initiating by the activation of Met receptors regulate the cytosolic biochemical kinetics of PI3K (defined by p) and MAPK (k) molecules:

$$\frac{\partial p}{\partial t} = \underbrace{D_p \nabla^2 p}_{diffusion} - \underbrace{\lambda_p p}_{decay} + \underbrace{\frac{\mu_p \Upsilon}{\mu_{p,0} + \Upsilon}}_{Met-induced\ production}, \tag{5.5}$$

$$\frac{\partial k}{\partial t} = \underbrace{D_k \nabla^2 k}_{diffusion} - \underbrace{\lambda_k k}_{decay} + \underbrace{\frac{\mu_k \Upsilon}{\mu_{k,0} + \Upsilon}}_{Met-induced\ production}. \tag{5.6}$$

The diffusion coefficients, D_p and D_k, and the degradation rates, λ_p and λ_k, are assumed to be constant and homogeneous within the cell cytosolic compartment, and are taken from theoretical works presented in [210, 243, 255]. The third terms in Equations (5.5) and (5.6) describe the production of the chemicals in the sub-plasmamembrane region, triggered by Met receptors, as for $\tau(\Sigma_{\sigma(\mathbf{x})}) = C$ and $\tau(\Sigma_{\sigma(\mathbf{x}')}) = M$

$$\Upsilon(\mathbf{x}, t) = \sum_{\mathbf{x}' \in \Omega'_{\mathbf{x}}} m(\mathbf{x}', t). \tag{5.7}$$

In the absence of specific determinations for AROs, the Michaels–Menten coefficients μ_k and $\mu_{k,0}$ are chosen to reproduce the curve of the HGF/SF-dependent MAPK phosphorylation provided for retinal endothelial cells (RECs) in [60]. μ_p and $\mu_{p,0}$ are instead estimated to fit the maximal amount of phosphorylated PI3K at a given concentration of HGF/SF, which has been measured in the same article.

Finally, the intracellular quantities of activated Cdc42 (c) and Rac (r) are regulated, respectively, by the equations

$$\frac{\partial c}{\partial t} = \underbrace{D_c \nabla^2 c}_{diffusion} - \underbrace{\lambda_c c}_{decay} + \underbrace{\mu_{p-c} p}_{PI3K-induced\ activation} + \underbrace{\mu_{k-c} k}_{MAPK-induced\ activation}; \tag{5.8}$$

$$\frac{\partial r}{\partial t} = \underbrace{D_r \nabla^2 r}_{diffusion} - \underbrace{\lambda_r r}_{decay} + \underbrace{\mu_{p-r} p}_{PI3K-ind.\ act.} + \underbrace{\mu_{k-r} k}_{MAPK-ind.\ act.} + \underbrace{\mu_{c-r} c}_{Cdc42-ind.\ act.}, \tag{5.9}$$

where the diffusion and decay rates are constant within the entire cell (i.e., $\tau(\Sigma_{\sigma(\mathbf{x})}) \in \{C, N\}$). The third and forth terms in Equation (5.8) describe the cytosolic (i.e., p and k do not diffuse in the nuclear region) activation of Cdc42 mediated by PI3K and MAPK, at constant rates μ_{p-c} and μ_{k-c}. In the literature, there are no available values for such parameters. However we set $\mu_{p-c} < \mu_{k-c}$, to describe a stronger dependency of Cdc42 on MAPK and a weaker dependency on PI3K. This is biologically consistent as PI3K activates Cdc42 only in a direct way, while MAPK interacts with the G-protein through distinct pathways [386]. The analogous terms in Equation (5.9) reproduce the triple regulation, by PI3K, MAPK, and Cdc42, on the activation of Rac at

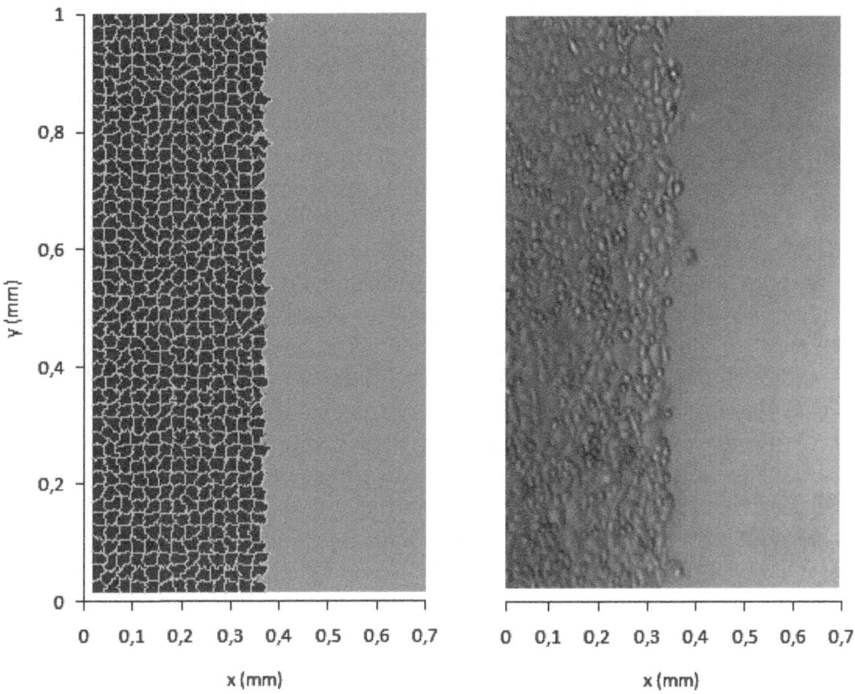

FIGURE 5.2: The initial condition (i.e., at $t = 0$) of the ARO cell mass for all sets of simulations. The cell monolayer is scraped at the right border by a wound of ≈ 0.35 mm (175 px). The dimensions of the cells, as well as all the other model parameters, are given in Appendix C. On the right panel is the initial frame of the experimental assay [349].

constant rates given by μ_{p-r}, μ_{k-r}, and μ_{c-r}, respectively. Given the value of μ_{c-r} provided in the case of a moving keratocyte in [254], in the absence of other specific published estimates, we set $\mu_{p-r} = \mu_{k-r} = \mu_{c-r}$.

5.3 Simulations

The simulation domain Ω is a bidimensional lattice of 350×500 sites. The characteristic length of each site is 2 μm: Ω therefore represents a section of a 24-well of size 0.7 mm $\times$ 1 mm. One MCS is set to correspond to 2 s. The initial conditions of all simulations consist of a mass of ARO cells, whose dimensions are given in Tables C.1 and C.5, scraped at one side border by a wound of $\approx$ 0.35 mm (175 px); see Figure 5.2. The dimension of the scratch is lower than in most experimental cases to avoid cell overlapping and to compensate for the absence of mitosis. In fact, to model a pure migration assay, the overall observation period is set equal to 12 h. This choice ensures a sufficient distance from critical events, such as culture splitting and cell cycle synchronization. For the same reason, we also neglect cell apoptosis.

The boundary conditions of domain Ω are zero flux at the lateral sides (i.e., at $x = 0$ and $x = 350$) and periodic at the others (i.e., at $y = 0$ and $y = 500$). The no-flux assumption at $x = 0$ is appropriate as the cell population has grown to confluence, and therefore there is no space for cells to move in that direction. The periodic boundary conditions at the top and the bottom of the lattice are also reasonable since the model is intended to examine a section of a much larger well. The no-flux condition at $x = 350$ is caused by the fact that this border reproduces the wall of the experimental dish.

Initially, the extracellular level of HGF is zero, as well as the intracellular amounts of messenger proteins. The basal properties of the cells, listed in Table C.5, have been evaluated following biological considerations listed both in the previous section and in Chapter 2. Indeed, they have been determined through preliminary simulations that showed the model consistency in a wide range of parameters. As usually done, the position of each ARO cell η is established by calculating the position of its center of mass. Similarly, its path is defined as the path of its center of mass and its speed, $\mathbf{v}$, as the speed of its center of mass.

The directional component of cell motion is measured by the linearity L, which is defined as the ratio between the x-component of the final displacement of a cell (i.e., of its center of mass, in the direction of the wound) and the total length of its path within the observation period [251]. Its value ranges from 0 to 1, being close to 0 when the cell movement is almost isotropic, with no directional trend, and getting larger for motions clustered toward the center of the wound.

In the circular charts, the final displacement of a cell is represented by the polar coordinates of the final position of its center of mass, as in [23]. As is typically done in experimental works, the healing capacity of the overall population is calculated by measuring the mean percentage of the recolonized scratch over time:

$$D(t) = \frac{1}{N} \sum_{i=1}^{N} \left(1 - \frac{d_i(t)}{d_0} \right) 100, \tag{5.10}$$

where d_0 is the initial width of the scratch (175 px $\approx$ 0.35 mm) and $d_i(t)$ is the distance across the wound between a cell at the front and the right border of the domain Ω. The number 100 is used to avoid the biases toward accounting for outlier individuals. Obviously, $D(t_{final})$ quantifies the invasive distance at the end of the observation period. Results of the healing capacity of the culture are shown as means with standard deviations (SDs) over 20 independent simulations.

The statistical analysis of the migratory properties of the three subpop-

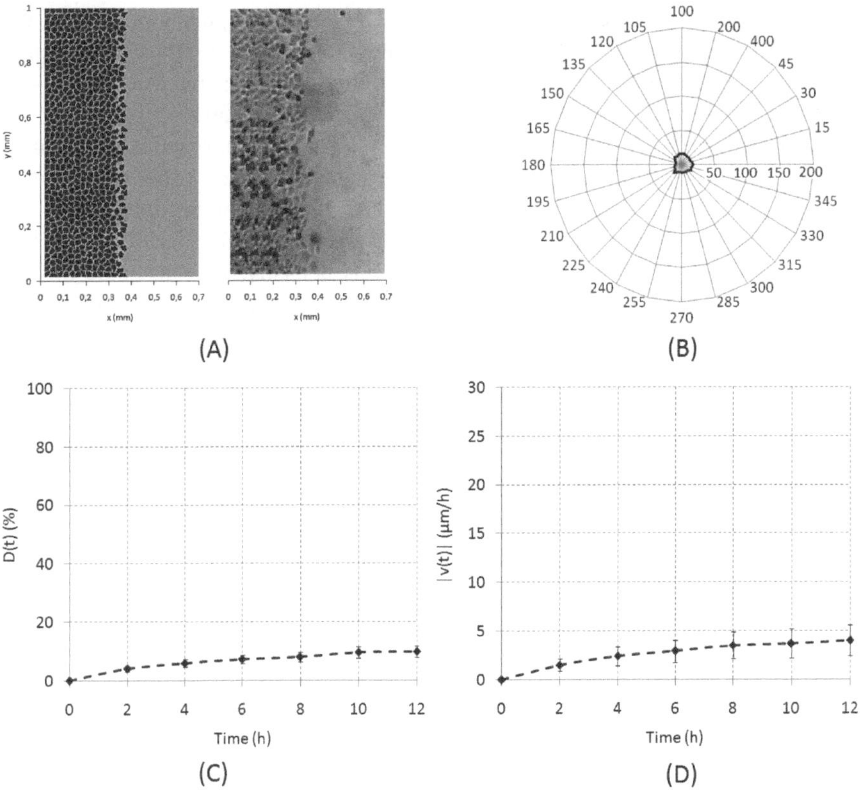

FIGURE 5.3: Healing process of the unstimulated ARO population. The cellular mass shows a negligible invasive capability within the observation period. (A) Final pattern configuration (i.e., at $t = 12$ h) for both simulation (left panel) and experiment (right panel). (B) Time evolution of D. (C) Polar plot of the net displacements of 20 representative cell randomly selected within the population. (D) Time evolution of the modulus of their velocity.

ulations in which we will differentiate the cell culture are instead performed in a single realization. This choice was made after observing small SDs over multiple simulations. In particular, for each subgroup, the instantaneous velocity (calculated for each cell as defined in Equation (1.11), where obviously the center of mass is established as

$$\mathbf{x}_{\eta}^{CM}(t) = \frac{1}{a_{\eta}^{volume}(t)} \sum_{\mathbf{x} \in \eta} \mathbf{x},$$

since we apply the new compartmentalization approach) and the linearity are averaged with SD over 30 individuals, while in the polar charts, we represent the final displacements of 20 representative cells.

The first set of simulations analyzes the healing capacity of the cell population in the absence of HGF stimulation (i.e., imposing $S(t) = 0$ in Equation (5.2)). The cultures displays barely no detectable invasion within the observation period ($D(t = 12$ h$) \approx 10\%$), which is indicative of the low basal motility of cells, given by T_0; see Figure 5.3(A-C). In particular, unstimulated individuals typically maintain close contacts with each other and feature a negligible random-oriented movement. Such a behavior is summarized in the polar plot in Figure 5.3(B), which represents the spatial distribution of their net displacement: the isotropy of directions is easily recognized by observing the circular shape of the delimited area of the chart, while its small size highlights the inability of cells to strongly move away from their starting point and invade surrounding spaces. Coherently, as represented in the (D) panel of the same figure, the instantaneous speed of the unstimulated cells tends to remain stable around a low initial value during the whole simulation period.

The evolution of the unstimulated ARO culture coincides qualitatively well with experimental evidence, as shown in the same figure and provided in [101]. This gives us confidence in the choice made for the parameters describing the basal properties of the simulated AROs, in particular, for T_0. Indeed, in the light of these considerations, simulations of wound healing assays in the case of unstimulated cell lines, along with comparisons with the relative experimental counterparts, will be used in the following chapters as basic migratory tests, which will allow one to infer the value of the intrinsic motility of the populations of interest.

After the addition of nanomolar concentration of chemical, the cell population is observed to have an evident enhancement of motility with the overall movement biased toward the gap, which is significantly invaded, as $D(t = 12$ h$) \approx 40\%$; see Figure 5.4.

In order to focus on the differences in the behavior of single individuals within the culture, we divide the cell mass into three regions from its front, namely, external, middle, and internal, as illustrated in Figure 5.4(A). We then consider the cells allocated to each color-coded region and analyze their movement in time, comparing the resulting migratory parameters. We find that the AROs placed at the edge of the scratch are characterized by a high short-range motility, which overcomes their adhesive interactions and allows them to shed; see Figure 5.5(A). In particular, as represented in the (B) panel of the same figure, such external individuals typically move according to angles that moderately cluster around the expected direction, i.e., the middle of the wound, as shown by a good value of linearity ($L \approx 0.6$).

The cells located in an intermediate position have instead decreased the average net displacement and linearity, while those far away from the front feature an almost negligible isotropic movement. As a further confirmation of the different dynamics of the three subpopulations, the evolution of the instantaneous speeds are reported in Figure 5.5(C): external cells move faster than middle cells, and the velocity is significantly further reduced for internal individuals. The migratory capabilities of cells therefore decrease with the

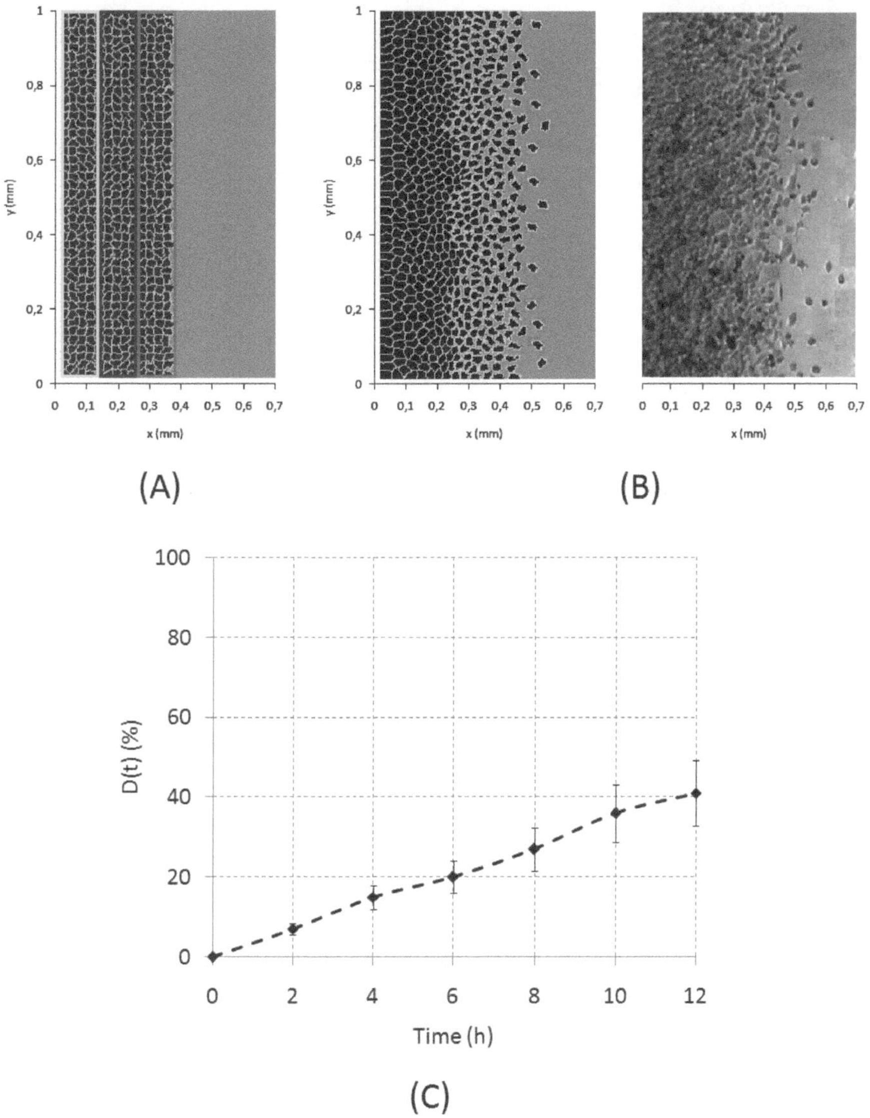

FIGURE 5.4: Healing process of the stimulated cell culture. (A) The three bordered regions in which the cell population is divided form its front, i.e., external, middle, and internal. (B) Final pattern configuration (i.e., at $t = 12$ h) for both the simulated and the experimental culture. (C) Time evolution of the invasive distance D.

distance from the edge of the wound, i.e., toward the zones of higher local cell density. This phenomenon, called *contact-inhibition of cell locomotion*, has

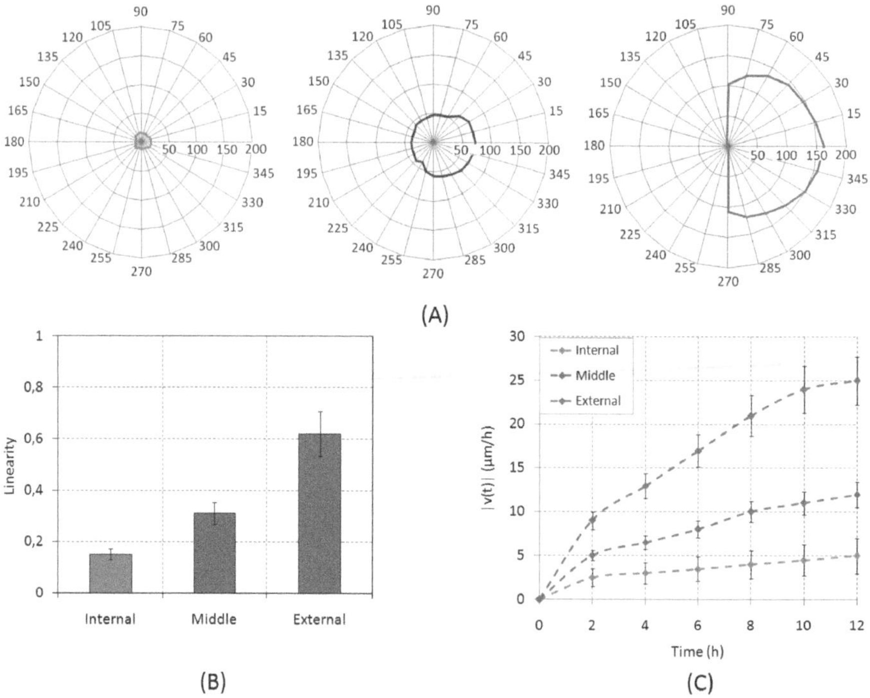

FIGURE 5.5: (See color insert.) Comparison of the dynamics of the three subpopulations. (A) Polar plots of the net displacements of 20 representative cells randomly selected within each subpopulation. (B) Mean linearity and (C) time evolution of the modulus of velocity for each subpopulation.

been described in detail for epithelial cell lines [420] and fibroblast cultures [393] and was first proposed by Abercrombie to describe the influence on cell crawling of the contact interactions between individuals [1].

It is finally useful to emphasize that the proposed model is highly flexible, as, in principle, it can be applied to any cell population stimulated by any motility agent only by changing either the parameters describing the cell basal properties or the microscopic models of chemical dynamics. For instance, as evident biomedical relevant examples, applying our approach to an epithelial cell culture allows one to test the physiological conditions in which the recovery of skin lesions is more efficient. On the contrary, using it for a malignant cell line, allows one to test potential therapies able to interfere with the ability of tumors to expand and invade.

Chapter 6

Effect of Calcium–Related Pathways on Single Cell Motility

6.1 Biological Introduction

Finely tuned spatiotemporal calcium events are a highly conserved and ubiquitous mode for the control of biochemistry and physiology in almost all eukaryotic cells [35, 40, 86]. Endothelial cells (ECs) are no exception, and intracellular calcium serves as a positive or negative regulatory signal for a wide range of cell functions, including survival, proliferation, motility, apoptosis, and differentiation [34, 36, 276]. A broad number of calcium-dependent enzymes are also associated with the progression through the cell cycle (the exit from quiescence in early G1 phase, the G1/S transition and other checkpoints during S and M phases [22, 204, 280]), and mediate the activation of several nuclear factors involved in the DNA division machinery, for example cdk and cyclins [108, 130, 340].

Under resting conditions, different control mechanisms maintain the concentration of free cytosolic calcium very low, nearly 10^{-7} M, with respect to the extracellular environment and to intracellular compartments (primarily the endoplasmic reticulum, ER [7, 330, 372]). In particular, the ion is extruded from the cytosol by calcium pumps, which are present in both plasma and ER membranes (respectively, PMCAs and SERCAs) and directly consume ATP energy, and by Na-Ca exchangers, which are located only in the plasmamembrane and use the energy of Na^+ electrochemical gradients [166, 191, 399]. On the other hand, calcium enters into the cytosol through permeable channels sited within the different membranes and activated by intracellular messengers [33, 292].

The correct Ca^{2+} concentration is thus strictly regulated by these active and passive fluxes. Alterations of this fine balance can be triggered only by specific extracellular agonists, and are typical chemical signals transducing information during specific phases of vascular progression in both physiological (such as vascularization of ovary and uterus during the menstrual cycle, of mammary glands during lactation and of granulation tissue after wound healing) and pathological (for example, chronic inflammatory diseases, vasculopathies, degenerative disorders, tissue injury occurring in ischemia, or tumor progression) conditions [64, 306].

In particular, vascular endothelial growth factors, VEGFs, which act as a positive chemotactic guidance for ECs (as widely demonstrated in literature [57, 66, 405], concomitantly mediate several calcium-dependent pathways. As characterized in [32, 133, 135, 211, 285], VEGF molecules, binding to their surface tyrosine kinase receptors, initiate a series of intracellular cascades, which results in the indirect production of arachidonic acid (AA) and nitric oxide (NO), as characterized in [133, 135]. Both these second messengers bind to plasmamembrane sites, opening cation channels and allowing the influx of extracellular calcium into the cytosol; see Figure 6.1. The process, also called noncapacitative (or non-store-operated) calcium entry (NCCE or NSOCE), causes localized and peripheral restricted accumulations of the ion [390], which regulates important biophysical properties of ECs, such as their intrinsic motility, elasticity, and chemotactic strength, ultimately enhancing their migratory capacity, which is fundamental in the initial phases of the vascular progression (see Figure 6.1 and refer to [32, 133, 135]), that will be further analyzed in the following chapter.

Indeed, we here aim at systematically characterizing the relationships between Ca^{2+} dynamics and selected cell migratory parameters (such as velocity and mean displacement) under different physiopathological conditions, as well as to reproduce varying morphologies of the motile endothelial cell. To do this, we simulate a simple and reliable motility assay with an EC (whose 3D morphology is based on experimental images and characterized by a realistic differentiation between the nucleus and the cytosolic region) placed on one side of a three-dimensional chamber and stimulated by an exogenous VEGF source. For instance, we focus on the model counterparts of biomedical approaches which, interfering with calcium machinery, are able to slow down cell migration; such strategies, in fact, may represent a useful starting point for pharmacological treatments that, inhibiting cell motion, have the potential to disrupt malignant neovascularization and, eventually, invasion. It is finally useful to underline that the proposed model is referred to a specific type of endothelial cell: the tumor-derived individual (TEC). TECs are isolated and cultured from human carcinomas on the basis of membrane markers and, whereas they display analogous calcium-dependent cascades with respect to their "normal" counterpart, exhibit altered genotype, phenotype, and function. They are often aneuploid, display chromosomal instability and express peculiar genes [56, 303, 359]. In addition, tumor-derived ECs avoid senescence *in vitro* and show enhanced proliferation, motility and overexpression of membrane receptors [21, 54, 55, 56, 170].

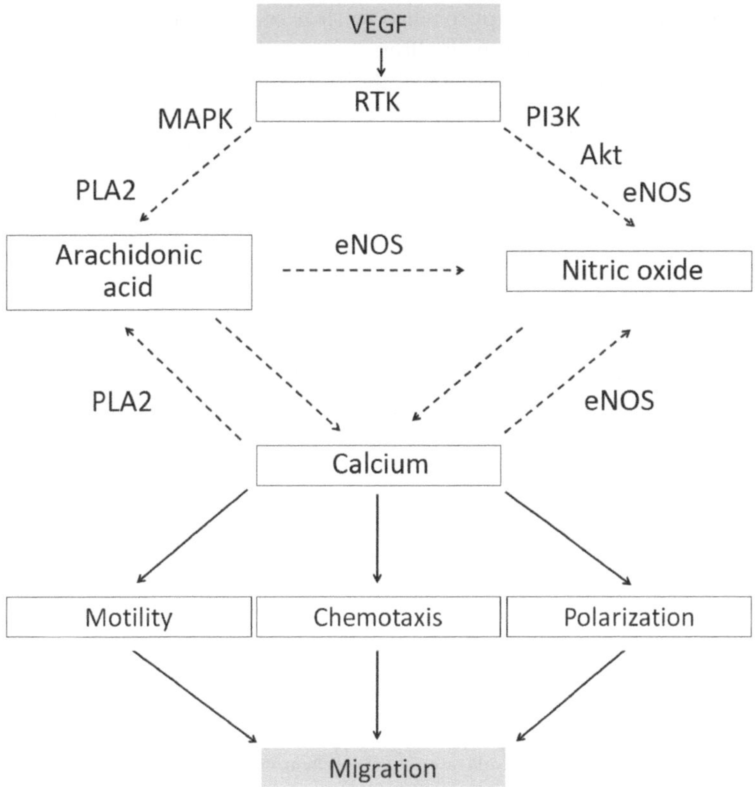

FIGURE 6.1: Simplified schematic representation of VEGF-induced calcium-dependent events during chemotactic migration of a vascular endothelial cell. VEGF molecules, binding to tyrosine kinase receptors, activate a series of intracellular signalling inducing the recruitment of enzymes phospholipases A2 (PLA2) and nitric oxide synthase (eNOS), and the subsequent production of arachidonic acid (AA) and nitric oxide (NO), respectively. Both these second messengers mediate calcium influxes from the extracellular environment through plasma membrane channels. Increases in the cytosolic calcium level trigger cell intrinsic motility, cytoskeletal reorganization and chemical response, crucial mechanisms for the regulation of cell migratory capacity. The dashed arrows stand for indirect reactions, which are bundled in the model.

6.2 Mathematical Model

In the hybrid-nested modeling environment, a compartmentalized CPM represents at the mesoscopic, cellular level the phenomenology of the TEC individual, while the microscopic biochemical VEGF-induced calcium-dependent dynamics are explicitly approached with a continuous model, which follows the approaches presented in the literature, based on experimental data mainly obtained from electrophysiological and fluorometric measurements, provided on different cell lines [132, 194, 217, 283, 371]. All the levels of the model are as usual explicitly integrated and fed back over the whole simulations.

6.2.1 Cell-Level Model

The TEC is a three-dimensional compartmentalized individual η of type $\theta(\eta) = E$, as depicted in Figure 6.2. In particular, it is differentiated in the nucleus, of type $\tau = N$ and in the surrounding cytosol, of type $\tau = C$. In this case, the plasma envelope is not explicitly identified. The extracellular medium is classically represented as a generalized cell $\tau = M$, which is assumed to be static, passive, and homogenously distributed throughout the simulation domain, forming no large-scale structures and thus without geometric constraints.

The internal state vectors of the sub-cellular compartments are as follows:

- for Σ_σ such that $\tau(\Sigma_\sigma) \in \{C, N\}$, $\mathbf{s}_{\eta,\Sigma_\sigma}(\mathbf{x}, t) = (a(\mathbf{x}, t), n(\mathbf{x}, t), c(\mathbf{x}, t)) \in \mathbb{R}^3_+$, where $a(\mathbf{x}, t)$ corresponds to the local concentration of AA, $n(\mathbf{x}, t)$ of NO, and $c(\mathbf{x}, t)$ of Ca^{2+}.

The system *Hamiltonian* is given by:

$$H(t) = H_{shape}(t) + H_{adhesion}(t) + H_{chemotaxis}(t) + H_{persitence}(t). \quad (6.1)$$

$H_{shape} = H_{volume} + H_{surface}$ takes into account of cell shape growth and deformations with elastic-like terms in the form of (4.6), where the target dimensions of the sub-units are their initial measures. Cell volume fluctuations are kept negligible, within a few percent, by high constant values for $\lambda^{volume}_{\eta,\Sigma_\sigma}$ for Σ_σ such that $\tau(\Sigma_\sigma) \in \{C, N\}$, as done in [259]. Moreover, because cell nuclei do not strongly deform, we set a high value also for

$$\lambda^{surface}_{\eta,\Sigma_\sigma} = \lambda^{surface}_{\eta,N}$$

when $\tau(\Sigma_\sigma) = N$. Instead, when $\tau(\Sigma_\sigma) = C$, $\lambda^{surface}_{\eta,\Sigma_\sigma}$ is a measure of the ease with which the TEC changes its shape due to cytoskeletal remodeling. As shown in [32], this is mediated by the actin–myosin interactions and reorganizations that is in turn facilitated by the presence of free calcium ions. Therefore, for $\tau(\Sigma_\sigma) = C$, we set:

$$\mathbf{s}^{\lambda^{sur}}_{\eta,\Sigma_\sigma}(\mathbf{x}, t) = \mathbf{s}^{\lambda^{sur},A}_{\eta,\Sigma_\sigma}(\mathbf{x}, t) = (c(\mathbf{x}, t));$$

$$\lambda^{surface}_{\eta,\Sigma_\sigma}(t) = \lambda^{surface}_{\eta,\Sigma_\sigma}(s^{\lambda^{sur},A}_{\eta,\Sigma_\sigma}(\mathbf{x}, t)) = f_{\lambda^{sur}}(s^{\lambda^{sur},A}_{\eta,\Sigma_\sigma}(\mathbf{x}, t)) = \lambda^{surf}_0 \; e^{-k\tilde{c}_\eta(t)},$$

where $\tilde{c}_\eta(t) = [c_\eta(t)/C_0] - 1$ is a positive value, since $c_\eta(t) = \sum_{\mathbf{x}\in\eta} c(\mathbf{x}, t)$ corresponds to the total concentration of the ion in cell η at time t, and

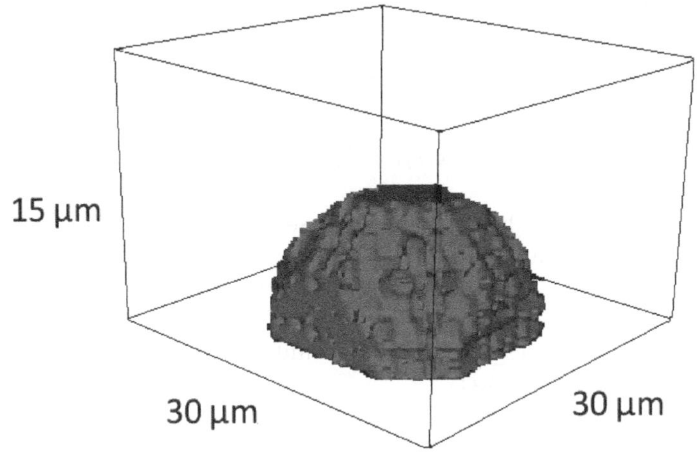

FIGURE 6.2: Representation of the initial 3D morphology of the simulated tumor-derived EC, which is 30 μm in length and width and about 15 μm in height.

$C_0 = \sum_{\mathbf{x}\in\eta} c_0$ is its total basal level, i.e., the level below which the cell dies. λ_0^{surf} is instead the intrinsic cell resistance to compression at the basal calcium amount C_0: observing that resting TEC maintain their initial geometrical configuration, with negligible changes of shape or cytoskeletal active reorganization, we have chosen a high value for λ_0^{surf}. In particular, for saturating levels of Ca^{2+} and for Σ_σ such that $\tau(\Sigma_\sigma) = C$, $\lambda_{\eta,\Sigma_\sigma}^{surface} \to 0$, and the cell can undergo dramatic changes in its morphology in response to the external stimulus.

$H_{adhesion}$ takes into account only the generalized adhesion between the nucleus and the cytosolic compartments, which, as seen, is a general extension of the Steimberg's Differential Adhesion Hypothesis (DAH), refer to Equation (4.2). To prevent the cell to split into disconnected patches, we assign a large negative energy penalty $J_{C,N}^{int}$.

Since vascular ECs have been demonstrated to migrate along gradients of VEGF concentration [157, 360], we add a classical linear-type chemotaxis term of the form (1.9):

$$\Delta H_{chemotaxis} = \mu^{ch}_{\eta,\Sigma_{\sigma(\mathbf{x}_{source})}}(\mathbf{x}_{source}, t)\,[q(\mathbf{x}_{target}, t) - q(\mathbf{x}_{source}, t)]\,, \quad (6.2)$$

where $\mathbf{x}_{source}$ and $\mathbf{x}_{target}$ are, as usual, the source and the final lattice sites randomly selected during a trial update in a MCS and $\tau(\Sigma_{\sigma(\mathbf{x}_{source})}) = C$. Obviously, $\mathbf{x}_{source}$ is a site belonging to the border of the cytosolic region, while $\mathbf{x}_{target}$ is a medium site. The parameter $\mu^{ch}_{\eta,\Sigma_{\sigma(\mathbf{x})}}$ represents the local chemical sensitivity of the cell and evolves according to a Michaelis–Menten function, that is, when

$$\mathbf{s}^{\mu^{ch}}_{\eta,\Sigma_{\sigma(\mathbf{x})}}(\mathbf{x},t) = \mathbf{s}^{\mu^{ch},A}_{\eta,\Sigma_{\sigma(\mathbf{x})}}(\mathbf{x},t) = (c(\mathbf{x},t)),$$

$$\mu^{ch}_{\eta,\Sigma_{\sigma(\mathbf{x})}}(t) = \mu^{ch}_{\eta,C}(s^{\mu^{ch},A}_{\eta,\Sigma_{\sigma(\mathbf{x})}}(\mathbf{x},t)) = f_{\mu^{ch}}(s^{\mu^{ch},A}_{\eta,\Sigma_{\sigma(\mathbf{x})}}(\mathbf{x},t)) = \mu^{ch}_0\left[\frac{c_0 + c(\mathbf{x},t)}{c_0 + jc(\mathbf{x},t)}\right],$$

where μ^{ch}_0 is a basal chemotactic response. With this relation, we model the experimental evidence that the local redistribution of VEGF receptors from the Golgi apparatus to the plasmamembrane, and their following activation, is caused by a local elevation of cytosolic calcium ions [264].

In (6.2), q measures the local extracellular VEGF concentration sensed by the moving membrane site:

$$q(\mathbf{x}, V) = \sum_{\mathbf{x}' \in \Omega'_{\mathbf{x}}} v(\mathbf{x}', t), \tag{6.3}$$

where $\mathbf{x} \in \{\mathbf{x}_{source}, \mathbf{x}_{target}\}$, and the $\mathbf{x}'$ are all the matrix first-nearest neighbors of $\mathbf{x}$. Such a spatial approximation is in agreement with experimental results given in [157], where a direct chemotactic guidance by VEGF has been provided to require the existence of precisely shaped extracellular gradients in the close proximity of cell surfaces.

The local characterization of cell chemical responses is a key advantage of the proposed CPM extensions with respect to similar published works [259, 260, 261]. Therein, the cells experience in fact a homogeneous chemotactic strength over the entire membrane, hiding relevant microscopic inhomogeneities such as the clusterization of VEGF receptors or their local agonist-induced activation which, as we will see in next section, is fundamental for polarization mechanisms. Moreover, in classical descriptions, the chemical strengths are constant over time, without adapting during the simulated process. This is an unplausible situation, since real cells constantly change their biophysical and biomechanical properties as a consequence of continuous internal and external stimuli: this can be obtained from the modeling point of view connecting the value of the Potts parameters to the internal concentration of chemical factors and activation of related pathways, as commented in Chapter 4. It is also useful to underline that the chemotactic response of TEC individuals is similar, from a mechanical view point, to that of "normal" ECs, except from an enhanced magnitude, due to the overexpression to VEGF receptors [55, 170]. This feature is taken into account in the above relation by setting an asymptotic chemical sensitivity which is twice its basal level.

Besides the directional chemotactic migration, real tumor vascular cells, as well as their normal counterparts, are seen to have a persistent motion dictated by their longer axes. It is a consequence of their polarization, i.e., their differentiation in a leading and a trailing edge. This inertial, shape-dependent motion can be modeled with an energy term of type (2.4), which coherently takes into account of the running mean over the cell past movements [20, 349].

Also in this case $\Delta t = 60$ MCS, whereas μ^{pers}, which controls the persistence time, is no longer a constant, but is given by:

$$\mu_{\eta}^{pers}(t) = \mu_{0}^{pers}\left[\frac{L(t)}{L_0} - 1\right]. \tag{6.4}$$

L is the current length of the longer axis of the cell, which is approximated with an ellipsoid, and L_0 its initial value (the initial cell diameter). Obviously $L \geq L_0$, since we have assumed that the cell deforms but does not grow during motion. Equation (6.4) is based on the experimental evidence provided in [192], where, after analogous chemical stimulations, elongated vascular cells have seen to have a longer persistence time than more rounded individuals. This is explained with the observation that cell polarization is a cytoskeleton-driven process: the more a cell is polarized (i.e., the longer its main axis is), the more its actin filaments are oriented in such a direction, requiring therefore a longer time to reorient toward a new direction, and change the direction of motion of the whole cell. As seen, if $\mu_{\eta}^{pers} = 0$ the TEC undergoes uncorrelated Brownian motion, while if μ_{η}^{pers} is very large its motion is almost ballistic.

Given the *Hamiltonian*, the transition probability of a spin flip has the form of Equation (4.24). In particular, we use $p(T_{\eta,\Sigma_\sigma}(t)) = \tanh(T_{\eta,\Sigma_\sigma}(t))$. For $\tau(\Sigma_\sigma) = N$, $T_{\eta,N}$ is a low constant positive value and mimics the passive motion of cell nucleus, which is dragged by the surrounding cytosolic region (in the next section we will give a more complete discussion). For $\tau(\Sigma_\sigma) = C$, $T_{\eta,C}$ represents instead the measure of the frequency of the biased extensions and retractions of cell plasmamembrane, which result from the nonequilibrium state between outward and inward forces. Such an intrinsic motility, that is the primary driving force of cell nondirectional movements, is mediated by the overall intracellular calcium level in a dose dependent manner, as provided in [135], where wound healing assays have been performed on several types of vascular cells stimulated by different agonists (such as AA and NO themselves). Indeed, for $\tau(\Sigma_\sigma) = C$, we use a Michaelis–Menten law, that is

$$\mathbf{s}_{\eta,\Sigma_\sigma}^{T}(\mathbf{x},t) = \mathbf{s}_{\eta,\Sigma_\sigma}^{T,A}(\mathbf{x},t) = (c(\mathbf{x},t));$$

and

$$T_{\eta,\Sigma_\sigma}(t) = T_{\eta,\Sigma_\sigma}(\mathbf{s}_{\eta,\Sigma_\sigma}^{T,A}(\mathbf{x},t)) = f_T(\mathbf{s}_{\eta,\Sigma_\sigma}^{T,A}(\mathbf{x},t)) = T_0\left[\frac{1+\widetilde{c}_\eta(t)}{1+h\widetilde{c}_\eta(t)}\right],$$

where T_0 corresponds to the basal motility of the cell (i.e., in resting conditions at the basal calcium level C_0), and T_0/h to the asymptotic value (i.e., for saturating concentrations of the ion). The specific form of the above relation results in significant increments of cell motility in response to high intracellular accumulations of calcium (i.e., the asymptotic motility of individuals is twice its basal value). This choice reflects the enhanced migratory capacities of TECs with respect to their "normal" counterparts upon agonist stimulations.

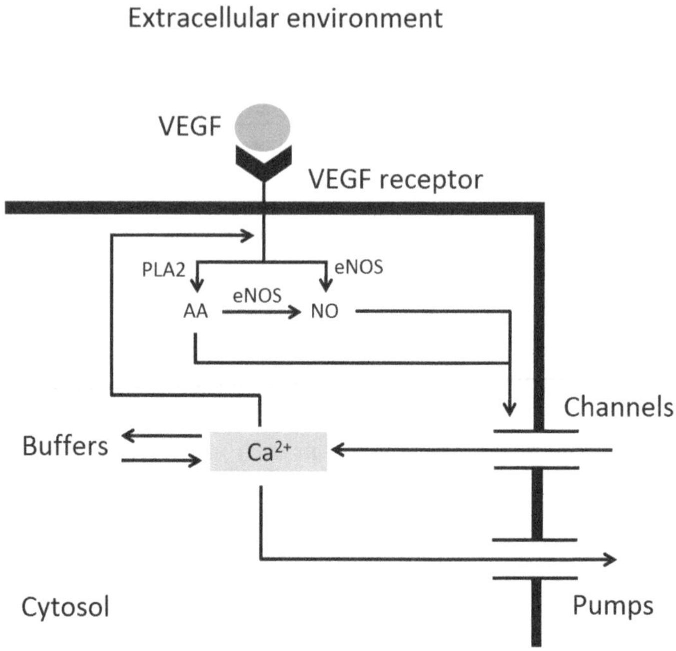

FIGURE 6.3: Diagram of the key VEGF-induced intracellular Ca^{2+} processes included in the mathematical model. Cell surface receptors activated by VEGF molecules stimulate the biosynthesis of AA and NO which, in turn, mediate calcium entry from the extracellular environment into the cytosol. Cytosolic Ca^{2+} is reversibly buffered by proteins and extruded back from the cell.

6.2.2 Molecular-Level Model

The key biochemical processes incorporated in the model of the VEGF-induced calcium dynamics are as follows (see Figure 6.3 for a diagrammatic representation):

- The exogenous VEGF diffuses throughout the extracellular medium, where it decays with a characteristic half-life.
- Single molecules of the morphogen reversibly combine with their tyrosine kinase receptors on the cell surface and form receptor–ligand complexes. These complexes decompose into a free receptor (which is recycled back) and some products, that initiate a sequence of reactions (i.e., the activation of enzymes PLA2 and eNOS) culminating in the production in the sub-plasmamembrane regions of second messengers AA and NO [137, 211, 276, 390].
- AA and NO open the relative and independent calcium channels in the

plasma membrane, leading to extracellular calcium entry into the cytosol. The rates of calcium influx are increasing and saturating functions of AA and NO concentration, respectively [133, 134, 274, 279, 281, 390].

- NO release is further mediated by AA itself within cell cytosol [274, 276, 402].
- Calcium enhances the rates of cytosolic AA and NO biosynthesis, with a positive feedback mechanism [236, 274, 276].
- calcium ions, which are reversibly buffered to proteins, such as calmodium, or to mitochondria [24, 33, 212, 200], are extruded from the cell by plasmamembrane ATPase and Ca^{2+}-Na^{+} exchangers [166, 191, 399].

The evolution of the exogenous VEGF satisfies the following reaction-diffusion equation:

$$\frac{\partial v}{\partial t} = \underbrace{D_v \nabla^2 v}_{diffusion} - \underbrace{\lambda_v v}_{decay} - \underbrace{B(\mathbf{x}, t, v)}_{uptake} + \underbrace{S}_{addition}, \tag{6.5}$$

where v denotes the actual concentration of the VEGF at medium site $\mathbf{x}$ (i.e., $\tau(\Sigma_{\sigma(\mathbf{x})}) = M$). The coefficient of diffusivity, $D_v > 0$, and of degradation, λ_v, are assumed to be constant and derived from previous experimental determinations [360]. $S = S(\mathbf{x}, t)$ describes the input of VEGF at a constant rate ϕ_v per unit of time by a discrete source, whose location and extension will be discussed in the different sets of simulations. B is the amount of VEGF molecules per unit of time that are locally sequestered by the cell. As done in Chapter 5 and following [25, 246], we assume that receptor binding occurs very rapidly compared to the time-scale of cell migration, and that the local quantity of VEGF molecules instantly bound by the TEC at time t is equal to the smallest between the actual available molecular concentration v and the actual maximal amount of VEGF that can be internalized by the cell surface receptors, which is defined with $\beta_v(t)$. Thus, we set:

$$B(\mathbf{x}, t, v) = \min\{\beta_v(t), zv\}, \tag{6.6}$$

where $\mathbf{x}$ belongs to the external surface of the cell PM (i.e., $\tau(\Sigma_{\sigma(\mathbf{x})}) = M$ and $\exists \mathbf{x}' \in \Omega'_{\mathbf{x}} : \tau(\Sigma_{\sigma(\mathbf{x}')}) = C$). $\beta_v(t)$, the maximal amount of VEGF molecules that can be locally (i.e., per site) bound and internalized by the TEC for unit of time has been estimated following [25, 246]. In particular, for the computation of $\beta_v(t)$, we have considered a spatially homogeneous average number of VEGF receptors per cell membrane site, which has been estimated by dividing the total number of VEGF receptor in a generic endothelial cell, considered constant and equal to 311.200 (as measured in [405] for human colonic ECs), for the actual extension of the TEC membrane, which is initially equal to 2700 μm^2 [390]. The local number of VEGF receptor thus varies in time: however, to avoid further overcomplications, we have not taken into account of other time-dependent phenomena, such as receptor clusterization. Moreover, we have taken into account an instantaneous VEGF-receptor complex internalization rate of $4.3 \cdot 10^{-4}$ per second [152], and 45 kDa as the molecular weight for a VEGF molecule [64].

For the construction of a mathematical model of VEGF-mediated intracellular events, we use a PDE-type approach based on the model by L. Munaron for the proangiogenic calcium signals in a static EC [283]. However, with respect to that model, we here explicitly describe the VEGF-induced production of second messengers AA and NO, as well as the calcium feedback mechanism for AA biosynthesis. Moreover, we use slightly different flux distributions for both calcium entry and extrusion (see below for details).

Summing up, the current levels of AA and NO at cell location $\mathbf{x}$ (i.e., $\tau(\Sigma_{\sigma(\mathbf{x})}) \in \{N, C\}$) are controlled by standard reaction-diffusion equations consisting in terms for their production, homogeneous diffusion and natural decay:

$$\frac{\partial a}{\partial t} = \underbrace{D_a \nabla^2 a}_{diffusion} - \underbrace{\lambda_a a}_{decay} + \underbrace{\frac{k_a R}{K_a + R}}_{VEGF-induced\ production} + \underbrace{\frac{v_c c}{c_a + c}}_{Ca-induced\ production}, \tag{6.7}$$

$$\frac{\partial n}{\partial t} = \underbrace{D_n \nabla^2 n}_{diffusion} - \underbrace{\lambda_n n}_{decay} + \underbrace{\frac{k_n R}{K_n + R}}_{VEGF-induced\ production} + \underbrace{v_{ca} \frac{c}{c_n + c} \frac{a}{a_a + a}}_{AA-\ and\ Ca-induced\ prod.}. \tag{6.8}$$

The diffusion parameters have been estimated with laboratory-based techniques [226, 276]. The degradation of both the intracellular messengers (and thus the following production, respectively, of eicosanoids [209] and peroxynitrites [300]) is considered constant, at rates λ_a, $\lambda_n > 0$ and measured in [283], where it has been estimated fitting experimental measurements made by the same group on bovine aortic endothelial cells [274]. The third terms in Equations (6.7) and (6.8) describe the VEGF-induced production rate of AA and NO at the cell submembrane region, as

$$R(\mathbf{x}, t) = \sum_{\mathbf{x}' \in \Omega'_{\mathbf{x}}} B(\mathbf{x}', t), \tag{6.9}$$

where $\tau(\Sigma_{\sigma(\mathbf{x})}) = C$, $\tau(\Sigma_{\sigma(\mathbf{x}')}) = M$ and $B(\mathbf{x}', t)$ is defined in (6.6). We apply Michaelis–Menten laws supposing that the intermediate receptor-ligand complexes are in a pseudo-steady state. In particular, the Michaelis–Menten coefficients of the VEGF receptor-mediated production of AA have been chosen for the best fit with experimental measures made for the same theoretical paper [283]. Note that such single simplified reactions, we bundle multiple signal transduction cascades that are known intermediates in AA and NO generation, such as the activity of PLA2 and eNOS enzymes. The last term in Eq. (6.7) implements the calcium-dependent feedback mechanism in AA biosynthesis, while the analogous term in Equation (6.8) accounts for the double regulation of NO production (both AA- and Ca^{2+}-mediated). They are described by others saturating functions of Michaelis–Menten type, work within the entire cell cytosol and agree with electrophysiological measurements [283], which have proved that they happen at comparable rates and time scales. In particular, the maximal rate of AA- and Ca-dependent NO release has been estimated by fitting the experimental measures made using fluorescent probes and selective electrodes in [37, 274].

The local intracellular level of calcium $c(\mathbf{x}, t)$, where $\tau(\Sigma_{\sigma(\mathbf{x})}) \in \{N, C\}$, is determined by a balance between the influxes through either AA- and NO-activated PM channels, respectively denoted by F_{AA} and F_{NO}, extrusion of the ion from the cytosol, F_{out}, and its buffering. It is thus controlled by the following diffusion equation:

$$\begin{cases} \dfrac{\partial c}{\partial t} = K_{buff} \underbrace{D_c \nabla^2 c}_{diffusion} & \text{in } \eta; \\ \\ n_{\partial \mathbf{x}} \cdot \nabla c(\partial \mathbf{x}, t) = \underbrace{F_{AA}(\partial \mathbf{x}, t) + F_{NO}(\partial \mathbf{x}, t)}_{influxes} - \underbrace{F_{out}(\partial \mathbf{x}, t)}_{efflux} & \text{at } \partial\eta, \end{cases} \tag{6.10}$$

where $n_{\partial \mathbf{x}}$ is the unit outward normal to $\partial \mathbf{x} \in \partial\eta$. The coefficient of diffusion, D_c, is assumed to be homogeneous throughout the cell (the diffusion of calcium across the nuclear envelope is taken to occur through non selective pores, whose permeability is proportional to the diffusion constant of the ion in the cytosol [89, 283]). The scaling factor

$$K_{buff} = \frac{K_{off}}{K_{off} + K_{on} b_T}$$

models the activity of intracellular endogenous buffers (proteins including calmodulin and others, cytoskeleton, mitochondria), which have a significant impact on the overall calcium dynamics, influencing its effective diffusion and contributing to decrease its level [24, 33, 212]. b_T is the total concentration of buffer sites (considered constant and experimentally estimated in different cell types [132, 283]), K_{on} is the rate of calcium uptake, K_{off} the rate of its release. This approximation works under the assumption that the buffering reactions take place on a faster timescale than the flux dynamics described above, and thus they can be considered in a quasi-steady state. This is the case of immobile buffers, characterized by low affinity and fast kinetics [371, 372].

F_{out} represents the overall rate of calcium efflux from the cell at time t and incorporates the extrusion of the ion both via PM ATP-ase and Ca^{2+}-Na^{2+} exchangers, which is given in several models as a sum of Hill functions [283, 399]. However, in order to avoid over-complications while remaining close to the curves set in literature in the considered agonist concentration range, we here approximate F_{out} as a single Michaelis–Menten form:

$$F_{out}(\partial \mathbf{x}, t) = k_{Ca} \left[\frac{c(\mathbf{x}, t)}{Ca_{out} + c(\mathbf{x}, t)} \right], \tag{6.11}$$

where $\partial \mathbf{x} \in \partial \eta$, k_{Ca} is the maximal rate of calcium extrusion and Ca_{out} the calcium concentration at which the rate of efflux is half maximal.

The Ca^{2+} influx distributions from the extracellular environment through independent and either AA- and NO-activated channels, respectively given by F_{AA} and F_{NO}, are assumed to be saturably dependent on the relative second messenger concentration:

$$F_{AA}(\partial \mathbf{x}, t) = F_{AA,max} \left[\frac{a(\mathbf{x}, t)}{q_a + a(\mathbf{x}, t)} \right]^m , \tag{6.12}$$

$$F_{NO}(\partial \mathbf{x}, t) = F_{NO,max} \left[\frac{n(\mathbf{x}, t)}{q_n + n(\mathbf{x}, t)} \right]^m , \tag{6.13}$$

where, as usual $\partial \mathbf{x} \in \partial \eta$, and $a(\mathbf{x}, t)$ and $n(\mathbf{x}, t)$ are derived by Eqs. (6.7) and (6.8). We select a quadratic $m = 2$ sigmoidal activation reflecting the fact that two AA or NO molecules are supposed to be bound to the respective Ca^{2+} channel in order to open it. In particular, the flux parameters have been chosen to fit the currents obtained by path clamp experiments in [274, 283].

The level of calcium in the extracellular environment (i.e., at site $\mathbf{x}$: $\tau(\Sigma_{\sigma(\mathbf{x})}) = M$) evolves according to:

$$\begin{cases} \dfrac{\partial c}{\partial t} = \underbrace{D_c \nabla^2 c}_{diffusion} & \text{outside } \eta; \\[2ex] n_{\partial \mathbf{x}} \cdot \nabla c(\partial \mathbf{x}, t) = \underbrace{-F_{AA}(\partial \mathbf{x}, t) - F_{NO}(\partial \mathbf{x}, t)}_{fluxes\ to\ cells} + \underbrace{F_{out}(\partial \mathbf{x}, t)}_{flux\ from\ cells} & \text{at } \partial \eta; \\[2ex] n_{\partial \mathbf{x}} \cdot \nabla c(\partial \mathbf{x}, t) = 0 & \text{at } \partial \Omega, \end{cases} \tag{6.14}$$

where, $\partial \Omega$ is the boundary of the domain and, as usually, $n_{\partial \mathbf{x}}$ is the unit outward normal to $\partial \mathbf{x}$, and the fluxes F_{AA}, F_{NO}, and F_{out} are specified in Equations (6.11), (6.12), and (6.13).

The model of VEGF-induced calcium signals, although very simple, in the absence of more specific evidence [134, 282], is acceptable for low agonist concentrations and it is experimentally validated [351, 372]. Indeed, it is based on the hypothesis of two independent calcium channels (AA- or NO-activated), but we cannot exclude either the existence of channels co-modulated by AA and NO or potential cross regulations between different channel types: for example, several member of the TRP family (in particular TRPC and TRPV sub-families) can be also involved in the proangiogenic signals [221, 282, 292].

There would be also two additional fluxes that influence the level of cytosolic calcium: the rate of its release from the ER and the rate of its re-sequestration back in the endoplasmic reticulum [8, 330]. We neglect both contributions, since the former is not stimulated by low concentrations of AA and NO [133, 274, 390], and the latter is typically involved only in the recovery of calcium response after agonist removal [265]. Finally, it is useful to

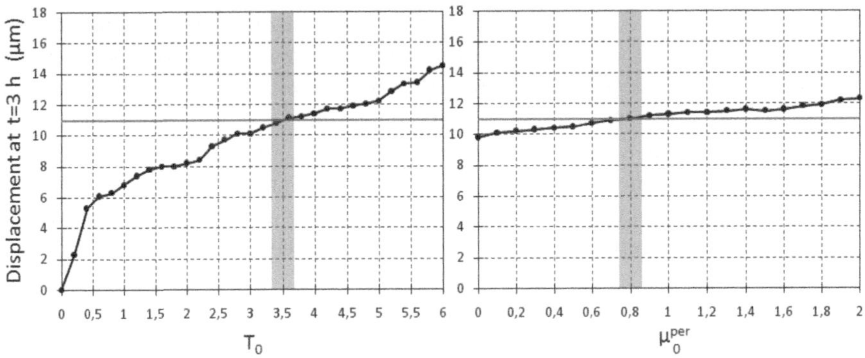

FIGURE 6.4: Effective displacement ($d = \mathbf{x}(t = 3\text{ h}) - \mathbf{x}(t = 0h)$, where $\mathbf{x}$ is the position of the cell center of mass) of unstimulated vascular endothelial cell in a 3 h time lapse for experimental (horizontal line, see [309]) and simulated case. Mean values over 10 simulations. Left plot has been used to estimate the intrinsic motility of the cell T_0, right plot to estimate its basal persistence time μ_0^{pers}.

underline that PM channels and pumps have been assumed to be uniformly distributed over the entire cell surface, and that several membrane-localized signal transduction proteins, that are known intermediates in AA and NO generation, such as PLA2 and eNOS, have not been included.

6.3 Simulation Details and Parameter Estimates

The simulated domain consists in a $600 \times 200 \times 80$ square lattice. One lattice voxel is equivalent to 0.125 μm^3: thus the lattice represents a 300 μm $\times$ 100 μm $\times$ 40 μm *motility chamber* with a volume of $1.2 \cdot 10^6$ μm^3. One Monte Carlo Step corresponds to 10 s.

The TEC initially is an hemisphere with a diameter of 30 μm, while the nucleus is a full sphere with a diameter of 10 μm, as estimated from experimental images obtained by the Department of Life Sciences and Systems Biology of the Università degli Studi di Torino in classical cultures of TECs resting conditions, i.e., in the presence of adequate nutrients and in the absence of external forces [133, 390]. The biological variables are initiated at their typical levels: in particular, there are no AA and NO within the cell. The local resting cytosolic Ca^{2+} concentration, $c_0 = 0.05$ μM, and the external calcium concentration, $c_{ext,0} = 2000$ μM, have been assumed uniform and taken from experimental [33] and theoretical [283] works. The model parameters related to the basal motility properties of the cell, T_0, and the basal persistence parameter, μ_0^{pers}, have been estimated with a quantitative comparison with experimental analysis. In particular, we have computed the displacement d covered by the cell in a 3 h time-lapse by the cell in *control conditions* (i.e., in the unstimulated case) $d = \mathbf{x}_\eta^{CM}(t = 3 \text{ h}) - \mathbf{x}_\eta^{CM}(t = 0)$, where $\mathbf{x}_\eta^{CM}$ is, as seen, the position of its center of mass, and have fitted this value with analogous measurements for EC autonomous motility in embryonic mouse allantoides, made in [309]. The values in the parameter space that have lead to the optimal empirical fitting to *in vitro* data were $T_0 = 3.5$ and $\mu_0^{pers} = 0.8$; see Figure 6.4.

6.4 Simulations in Standard Conditions

The hemispheric vascular TEC is initially placed in the right side of the motility chamber. We run a first set of simulation in *control conditions*, i.e., without the application of external chemical stimuli in Figure 6.5.

As expected, no calcium events take place ($c_\eta(t = 6 \text{ h}) = C_0$), according to experimental observation made by Munaron and coworkers for their model in [283]. In addition, we do not observe any spontaneous change in the shape of the TEC, that remains more or less hemispherical and starts a slow isotropic motion; see Figure 6.5. Such a cell behavior is qualitatively consistent with *in vitro* experiments, where a block in VEGF signaling causes normal endothelial cells to keep a round shape [116] and to have a substantially negligible autonomous motility [309].

A point source of VEGF (S in Equation (6.5)) is then applied in $\mathbf{x} = (600, 100, 0)$. Figure 6.6 shows time lapse images of both the phenomenological migration of the cell and the relative calcium events. As represented in Figure 6.6(A), the motile tumor-derived EC undergoes a gradual transition from the initial symmetric stationary state to a polarized migratory state, characterized by clearly distinguishable leading and trailing edges. In particular, a long and thin membrane-bound cytoplasmic pseudopodium emerges at the front of the cell, defining the direction of migration toward the chemotactic source, as the elongated cell moves by constantly protruding at the leading edge, while retracting at the rear.

Such a phenomenology is coherent with *in vitro* realizations performed in [116], where VEGF-stimulated normal vascular ECs are found to assume an extended, bipolar morphology. In the presented model, cell elongation results from the interplay between the chemotactic-induced membrane extension at the leading front of the cell and the mechanical properties given to its compartments. In fact, the nucleus is characterized by a constant stiffness and the cytoplasm by variable calcium-dependent elasticity and motility.

In particular, the exogenous stimulus causes the cell plasmamembrane to locally protrude in the direction of increasing VEGF gradients, with a speed of protrusion proportional to the modulus of the local chemical strength $\mu_{\eta,C}^{ch}$

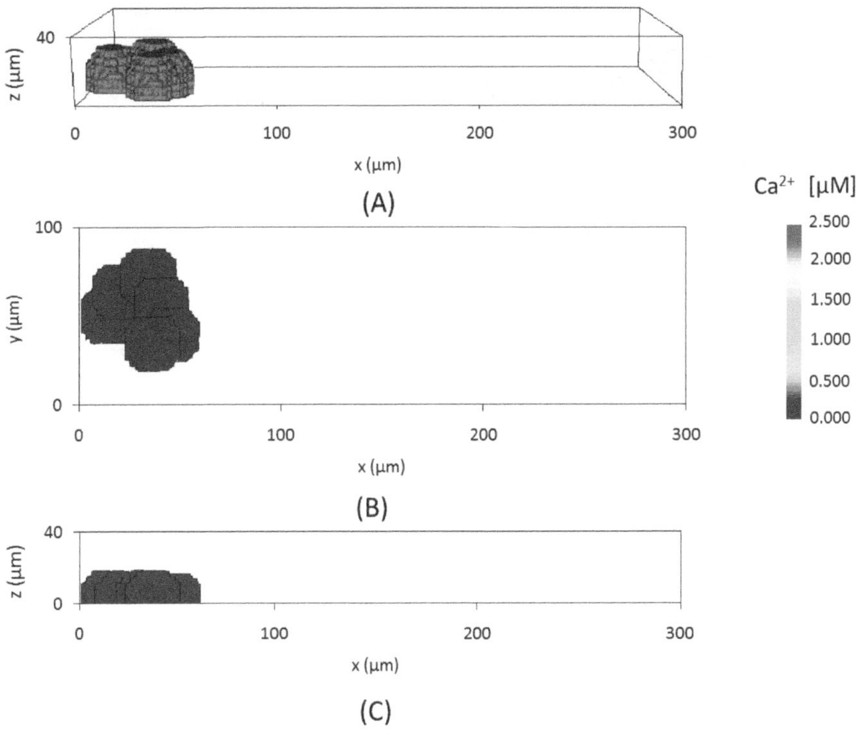

FIGURE 6.5: (See color insert.) Endothelial cell phenomenology and relative calcium events in *control conditions* (without external stimuli) at 30 min intervals until $t = 3$ h. (A) Three-dimensional view of cell migration. Contour plot showing cell position and intracellular calcium concentration in (B) $z = 1$ μm plane and (C) $y = 50$ μm plane. Calcium concentration has been pseudocolor-scaled so that the maximum value is red. The simulated cell almost maintains its initial hemispheric morphology and undergoes an essentially random motion. No calcium signals are observed.

itself. We recall that in any CPM model the simulated objects experience an implicit drag force from the lattice, and thus they have Aristotelian dynamics rather than a Newtonian dynamics (refer to [20, 254] for more detailed comments). Pulled by the leading front, the overall cytosolic region, whose elasticity and mobility increase due to the concomitantly calcium accumulation, then deforms and moves forward, while pulling onto the nucleus with the same force. The nucleus (which, as a CPM object, also follows Aristotelian dynamics) is instead basically stuck to the underlying lattice, as a consequence of its rigidity: therefore it moves with a lower velocity than the surrounding cytosol and lags behind, creating the characteristic polarized morphology of the TEC.

These mechanisms are a simplified picture of the biological processes un-

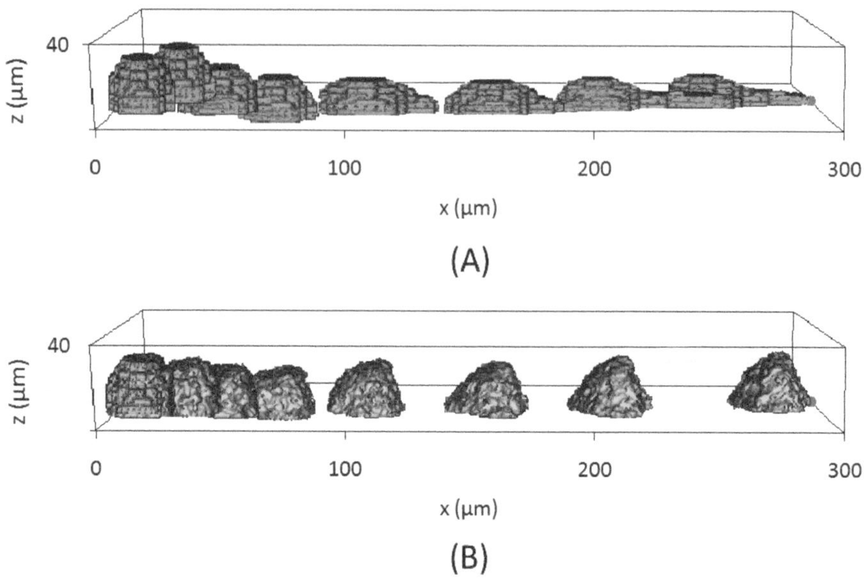

FIGURE 6.6: VEGF-induced chemotactic migration and cell polarization. Representative three-dimensional views taken at 45 min intervals until $t = 6$ h. Red dot represents the VEGF source. (A) The cell, after a short latency, rapidly polarizes establishing a leading edge, a long and thin pseudopodium, which gives the direction of motion. (B) For comparison purposes, model results in the case of a monocompartmental endothelial cell (i.e., formed by a unique unit representing an undifferentiated cytoplasm). All the other parameters are the same as in the previous case. It is worth to notice that this approach is unable to reproduce cell polarization, as the cell is only a deformed mass that moves toward the chemical source.

derlying cell movement. The relative literature in this field is vast and we refer the reader to the comprehensive biochemical reviews [311, 323] and the classical books [7, 190]. In reality, the external chemical stimulus, via surface receptors, triggers in fact the polymerization of the cell cytoskeleton, which results in the constant abutting of the cell PM in the direction of motion and in the coordinated development and release of focal adhesions (FAs). During the motion of the overall cell, the nucleus is unable to have an autonomous directional movement, as it only negligibility fluctuates in the cytoplasmic fluid. However, the nucleus is anchored to intermediate actin filaments and microtubules, which are in turn linked to the extracellular matrix through the focal adhesion clusters: therefore, it is passively pulled by active forces transmitted by the substrate via the cytoskeletal components (see the review [404] and the references therein). Such indirect mechanical interactions between the nucleus and the matrix environment are implicitly reproduced by the model

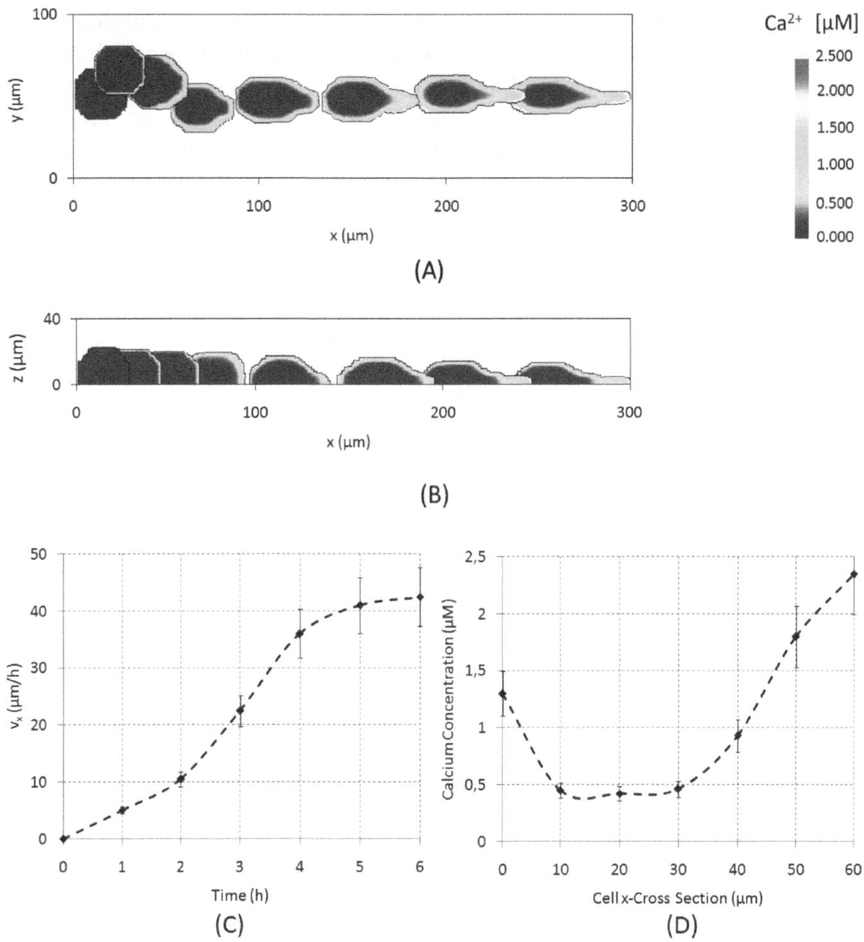

FIGURE 6.7: (See color insert.) VEGF-induced calcium events. Contour plots showing intracellular calcium concentration during cell motion in (A) $z = 1\ \mu m$ plane and (B) $y = 50\ \mu m$ plane. Representative images taken at 45 min intervals until $t = 6$ h as in Figure 6.6(A). (C) Cell directional velocity v_x (i.e., toward the chemotactic source). (D) Ca^{2+} concentration profile at the end of migration, $t = 6$ h, along cell x-cross-section of $z = 1\ \mu m$ plane from the trailing edge to the leading edge of the polarized cell. Higher Ca^{2+} accumulation ($> 2\ \mu M$) is observed in the thin pseudopodium. Negligible calcium events are seen in the central-nuclear area ($< 0.5\ \mu M$). Mean over 10 simulations; the error bars represent standard deviation.

artifact of staking the nuclear compartment to the underlying lattice with a finite rigidity, which is sufficiently high to results in a slow movement of the nucleus but low enough to avoid the total block of its locomotion.

A calcium-dependent elasticity of the cell cytosol, in conjunction with the presence of a rigid nucleus and of a local chemical stimulus, constitutes therefore a sufficient set of minimal and simplified requirements for a cell transition from a symmetric stationary morphology to a polarized mobile state. Obviously, a more realistic model should explicitly include the dynamics of the cell cytoskeleton and its signal transduction (this topic, often approached in the literature with multiphase models, see for example the book [77], could represent a fundamental improvement of CPM applications, since it has received little though increasing attention). In fact, the proposed approach does not focus on the dynamics of actin filaments, but rather considers the cell cytosol as a single elastic body undergoing local mechanical stresses, due to thermodynamical forces and chemical stimuli, on its membrane. A more detailed multiscale model would combine both approaches, using the stress distributions at the PM as a signaling input for the subsequent polimerization process of the actin cytoskeleton. The introduction of the cell cytoskeleton dynamics would result also in a more accurate description of the movement of cell nucleus, which is in fact mediated by the interactions with the matrix substrate, via intermediate filaments and microtubules. These could be more realistically described by an explicit model of the actin component of the cell cytoskeleton.

However, the assumptions made do not strongly influence the final outcomes of the model. The dynamics of the nucleus are in fact kept constant in all the proposed sets of simulations by fixing the value of its stiffness, $\lambda_{\eta,N}^{volume}$ and $\lambda_{\eta,N}^{surface}$, and of its mobility, $T_{\eta,N}$. In this way, the migratory properties of the TEC are mainly determined by the other mechanisms involved (such as molecular processes), which instead vary in each case.

The importance of the compartmentalized approach is also underlined by studying the phenomenology of the cell in the case of a monocompartmental representation (i.e., the TEC is formed by a single, undifferentiated cytoplasm, while all the other model assumptions are not changed). As reproduced in Figure 6.6(C), the polarization process does not emerge and the cell is a deformed mass which moves in the direction of the chemical source (notice that also the top of the cell unrealistically protrudes).

For a quantitative characterization of cell motility, we define cell average *directional displacement* and *velocity* (i.e., along the x-axes, in the direction of the chemical source), which facilitate the comparison between the analyzed experimental conditions. In particular, the cell directional displacement is simply the x-coordinate of the cell center of mass at a given time, while the directional velocity is the relative component of its average velocity, $v_x = (\mathbf{v}_{\eta}^{CM})_x$ defined as in Equation (1.12), where $\Delta t = 60$ MCS, and thus ≈ 10 min.

After an initial stage, when the cell is still round and does not strongly sense the VEGF stimulus, the directional velocity rapidly increases until it stabilizes at 42 μm/h (when the EC is completely polarized); see Figure 6.7(C, left panel). This agrees with the range of speeds measured in [309] for real ECs in embryonic mouse allantoides (28–40 μm/h). In particular, the cell velocity has been therein evaluated relative to the motility of the surrounding environment in which the cells reside. The observed lower values can be partly explained by the fact that the experiments are performed in a more realistic, noisy environment, that reduces the persistence length of cell motion. Moreover, the cell takes ≈ 6 h to reach the opposite side of the chamber. In this case, obviously, $x_{\eta}^{CM}(t = 6 \text{ h}) = 300$ μm: this value is used as a reference value.

Figure 6.7 also illustrates also the time course of the intracellular calcium signals in two perpendicular sections of the cell. As reference sections we choose the $z = 2$ vx (≈ 1 μm) plane, which includes the development of the pseudopodium, and the $y = 100$ vx (≈ 50 μm) plane, the main transversal section of the cell. After a short latency (≈ 30 min), surface receptor activation by VEGF stimulates the full activation of downstream cascade leading to the subsequent production of second messengers AA and NO, which, in turn, start to induce calcium influx from the extracellular environment through the relative channels. The resulting intracellular calcium events, initiating at the sub-plasma membrane regions, rapidly propagate in every direction, to inhomogeneously fill the whole cell volume. The final pattern of Ca^{2+} accumulation is strongly localized: higher concentrations (> 2 μM) occur in the leading front of the polarized cell (i.e., at the tip of the pseudopodium), and decay within a distance of 15–20 μm (< 1 μM). They are undetectable in the peri-nuclear regions (< 0.5 μM in the range (10 μm, 30 μm) of the cell x-cross section). Moreover, the rate of the final intracellular accumulation of the ion is $c_{\eta}(t = 6 \text{ h}) \approx 3.6$.

These findings are in agreement with both experimental [390] and modeling [283] data on the spatial propagation of calcium events in endothelial cells stimulated with angiogenic factors. Indeed, as described in more detail in the next chapter, the heterogeneity of accumulation of the ion is highly relevant during the overall angiogenic progression, as several reports have suggested that the transcriptional pattern of specific genes during vascular formation critically depends on the spatiotemporal calcium dynamics [108, 130]. Finally, if the external VEGF stimulus is subsequently removed, imposing $S = 0$, the production of AA and NO ceases and they rapidly decay. The cytosolic calcium, extruded by pumps and exchangers, then quickly returns to the baseline level C_0 (not shown).

6.5 Interfering with Calcium Machinery

To establish a direct qualitative and quantitative relation between intracellular calcium events and cell motion, in Figure 6.8, we study the consequences of a gradual inhibition of the Ca^{2+} influx distributions.

In particular, we decrease in a graded fashion the overall VEGF-induced

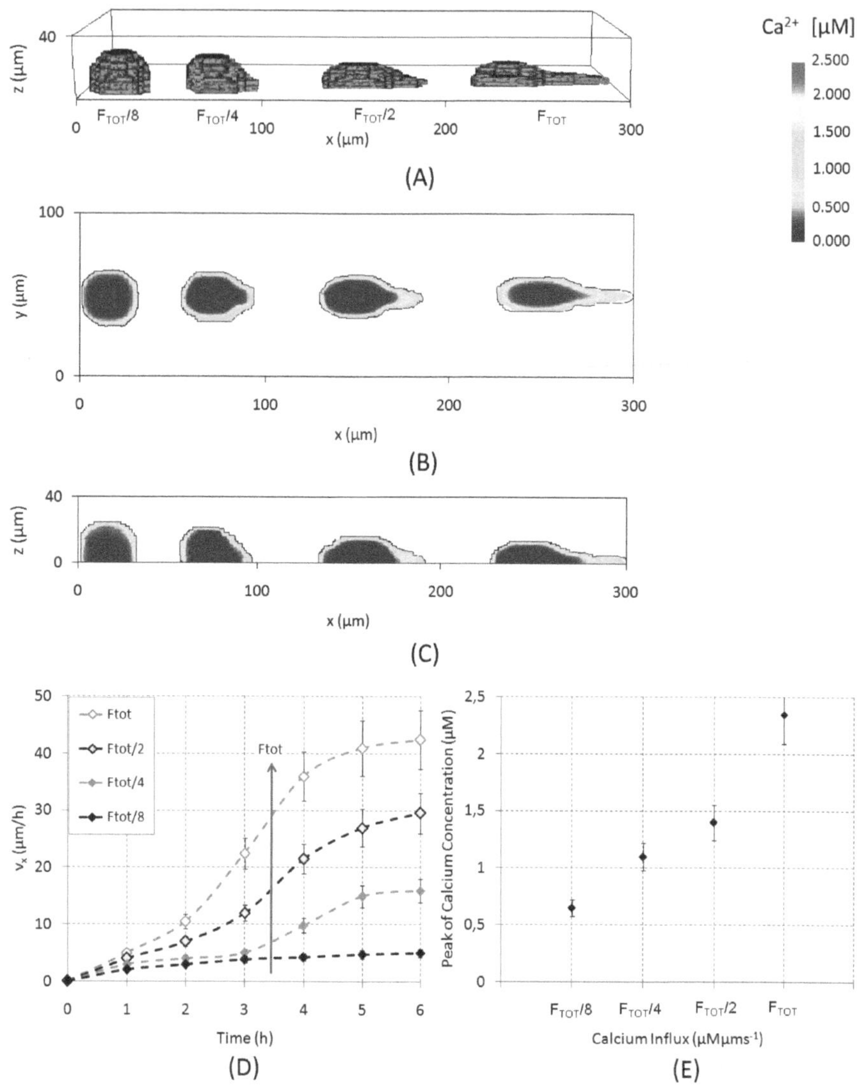

FIGURE 6.8: (See color insert.) Partial inhibition of overall VEGF-induced calcium influx $F_{TOT} = F_{AA} + F_{NO}$. (A) Three-dimensional view of cell position at $t = 6$ h for each F_{TOT}-decrement. Red dot represents the VEGF source. Contour plots showing the final intracellular calcium profile in (B) $z = 1$ μm plane and (C) $y = 50$ μm plane. Calcium concentration has been pseudocolor-scaled so that the maximum value is red. (D) Evolution in time of cell directional velocity v_x. The complete inhibition of calcium responses ($F_{TOT} = 0$) dramatically stops cell directional motility ($x_\eta^{CM}(t = 6 \text{ h}) = 20$ μm and $v_x \approx 3.3$ μm/h). (E) Maximal Ca^{2+} concentration at $t = 6$ h. The error bars show standard deviation over 10 simulations.

calcium influx, defined as

$$F_{TOT} = F_{AA} + F_{NO}.$$

Each decrement is the model counterpart of a treatment with an increasing concentration of carboxyamidotriazole (CAI) drug. CAI is in fact an anti-invasive and anti-angiogenic agent, which alters calcium-mediated signal transduction in ECs by blocking agonist-activated calcium entry in a dose-dependent manner [26, 213, 277]. In particular, it is currently under investigation as an orally administered tumoristatic agent in Phase II and III clinical trials for different tumors, as explained in [214, 284].

Decrements in F_{TOT} result in marked decrements in the cell directional velocity v_x, which are followed by the relative decrements in the final cell displacement ($x_\eta^{CM}(t = 6 \text{ h}) = 150\ \mu\text{m}$ for $F_{TOT}/2$ and $x_\eta^{CM}(t = 6 \text{ h}) = 70\ \mu\text{m}$ for $F_{TOT}/4$), as shown in Figure 6.8. Furthermore, an approximately complete inhibition of the mitogen-induced calcium responses, $F_{TOT} \longrightarrow 0$, causes the cell to remain in its unpolarized stationary state, dramatically stopping its directional locomotion, as for $F_{TOT}/8$, v_x is almost a constant $\approx 4\ \mu\text{m/h}$ and $x_\eta^{CM}(t = 6 \text{ h}) < 20\ \mu\text{m}$.

The cell behavior is explained by the fact that partial inhibitions of F_{TOT} cause decrements in the intensity of calcium responses: in fact both the maximal peaks and the overall accumulation of the ion decrease (i.e., $c_\eta(t = 6 \text{ h}) \approx 1.4$ for $F_{TOT}/2$, $c_\eta(t = 6 \text{ h}) \approx 0.6$ for $F_{TOT}/4$, and $c_\eta(t = 6 \text{ h}) \approx 0.2$ for $F_{TOT}/8$); see Figure 6.8(E). The consequence is a downregulation of the biophysical properties of the EC, such as the intrinsic motility, elasticity and chemotactic strength, involved in its migratory capacity. However, the interferences in the influx distributions do not affect the spatial propagation of the ion, as represented in Figure 6.8(B-C).

Disruptions in the second messenger machinery are of high biological interest as well. In particular, we exclude the production of arachidonic acid (similarly nitric oxide) by imposing $k_a = v_c = 0$ in Equation (6.7) (similarly $k_n = v_{ca} = 0$ in Equation (6.8)). It is the model counterpart of the activity of PLA2 (eNOS) inhibitor (such as AACOCF3 or, respectively, L-NAME drugs, which are well studied anti-angiogenic compounds both *in vitro* and *in vivo*, see again the excellent medical review [284]). Both exclusions result in an incomplete transition of the TEC toward the motile phenotype, and in the consequent decrement of its directional velocity and final displacement ($x_\eta^{CM}(t = 6 \text{ h}) = 110\ \mu\text{m}$ in the case of AA inhibition, and $x_\eta^{CM}(t = 6 \text{ h}) = 180\ \mu\text{m}$ in the case of NO inhibition), as shown in Figure 6.9(A-D). The explanation is that the disruption in the biosynthesis of each of the two molecules extinguishes its intracellular presence and, consequently, abolishes the relative calcium influx, as also provided by the experimental system in [274]. Indeed, the final outcome of both proposed treatments is a downregulation of the VEGF-induced calcium signals similar to those obtained from the direct block of the influxes of the ion (reproduced in Figure 6.8) which, as seen, have caused partial inhibitions of the cell migratory capacity. In particular, also with the exclusion of either AA or NO production, the spatial propagation of calcium responses appears relatively unchanged w.r.t. the standard parameter setting, while the peaks have a lower amplitude (i.e., ≈ 1.1 μM in the case of AA inhibition and ≈ 1.6 μM in the case of NO inhibition) and the total intracellular amount a significant decrement as well (i.e., $c_\eta(t = 6 \text{ h}) \approx 1.2$ in the case of AA inhibition and $c_\eta(t = 6 \text{ h}) \approx 1.6$ in the case of NO inhibition), as shown in Figure 6.9(B-C-E). Taken together the results in Figures 6.8 and 6.9 confirm the efficacy of two families of anti-angiogenic compounds, that target different calcium-dependent mechanisms involved in cell migration (i.e., either directly the fluxes through PM channels or the dynamics of the second messengers).

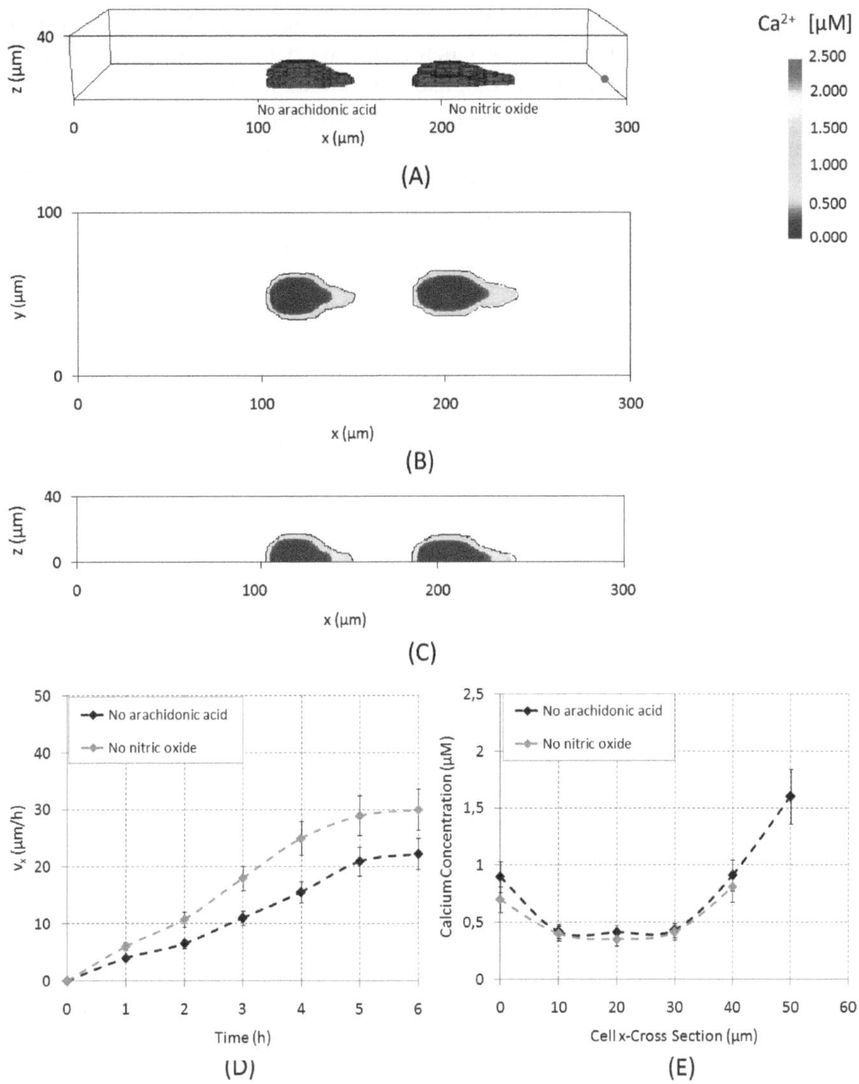

FIGURE 6.9: (See color insert.) Modeling interference in the production of either arachidonic acid ($k_a = v_c = 0$) and nitric oxide ($k_n = v_{ca} = 0$). (A) 3D view of cell position at $t = 6$ h for the two cases. Red dot represents the VEGF source. Contour plots showing the final intracellular calcium profile in (B) $z = 1$ μm plane and (C) $y = 50$ μm plane. Calcium concentration has been pseudocolor-scaled so that the maximum value is red. (D) Evolution in time of cell directional velocity v_x. (E) Ca^{2+} concentration profiles at the end of migration, $t = 6$ h, along cell x-cross-section of the $z = 1$ μm plane from the trailing edge to the leading edge of the polarized cell. Higher Ca^{2+} accumulations are again observed in the thin pseudopodium, even if with decreased maximal amplitudes. Negligible calcium events are seen in the central-nuclear area (< 0.5 μM). Mean over 10 simulations; the error bars represent standard deviation.

The model has been able to characterize the chemotactic migration of a vascular cell and the relative calcium events, to reproduce the activity of some anti-angiogenic pharmacological compounds, and to suggest the use of already existing drugs to inhibit cell locomotion. We now can turn to perform simulations that provide additional insights of the calcium-dependent factors that control the migratory properties of the tumor-derived vascular cell. In particular, we focus on some conditions that would be difficult or impossible to actually establish *in vitro*, but that can suggest nonintuitive, but potentially very effective ways to interfere with the motility capacity of the TEC.

The diffusion is clearly a crucial element in the intracellular propagation of calcium, and therefore affects a number of localized biochemical processes, as the feedback mechanism included in the kinetic scheme of AA and NO and the cell chemical sensitivity. A quantification of its importance is tested by setting $D_c = 0$ in Equation (6.10). Without diffusion, the calcium ions entered from the medium are completely sequestered in the inner surface of the PM, and therefore calcium signals do no longer occur throughout the entire cell volume; see Figure 6.10(B-C-E). In particular, the cell quickly reaches the peaks of calcium accumulation, which have the same intensity and localization (i.e., at the tip of the pseudopodium) as in the standard case. However, the TEC undergoes only a partial elongation, which results in a lower directional velocity and a consequent decrement in the final displacement, as $x_{\eta}^{CM}(t = 6$ h$) = 165$ μm; see Figure 6.10(A-D). These results have a clear biological relevance, since they represent a definitive demonstration that the cell migratory properties are not established by the maximal amplitude of calcium responses (which, as seen, remains unchanged), but by the overall intracellular concentration of the ion, which is obviously strongly reduced by the exclusion of its diffusive behavior, as $c_{\eta}(t = 6$ h$) \approx 0.9$.

Modeling allows us to assess the perturbing effect of endogenous buffers on calcium signaling and on the overall migratory capacity of the cell. In the absence of buffers (i.e., by imposing $K_{buff} = 1$ in Equation (6.10)), the spatial dynamics of Ca^{2+} events are not affected, but, as expected, the local concentrations are greater than control, as shown in Figures 6.11(B-C) and 6.12(H). In particular, the intracellular region in which calcium signals are not detectable decreases in size, as it is now restricted in the close proximity of the nucleus. In this regards, it is useful to compare the spatial profiles of the ion in Figures 6.7 and 6.11. The consequent increment in the total intracellular calcium level (i.e., $c_{\eta}(t = 5$ h$) \approx 3.9$) enhances the migratory capacities of the EC, as the effective velocity grows up to ≈ 46 μ/h, and the cell reaches the opposite side of the chamber in nearly 5 h, as shown in Figure 6.11(A) and 6.12(G).

We finally model a spatially inhomogeneous buffer distribution: in particular, an increasing clusterization of buffers is set toward the central regions of the cell, where the perinuclear mitochondria are more concentrated (as widely shown in [7] and confirmed in [283] by staining with mitotracker). Mathemat-

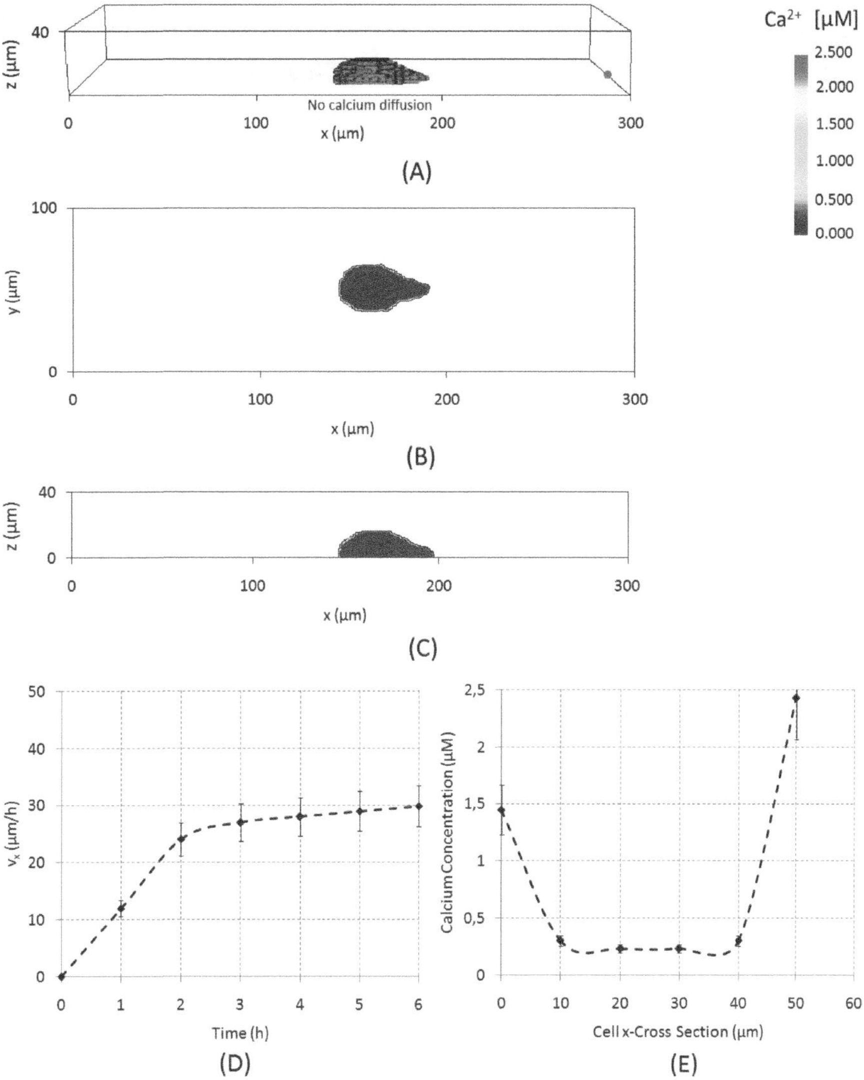

FIGURE 6.10: (See color insert.) Cell migration with the exclusion of calcium diffusion ($D_c = 0$). (A) Three-dimensional view of cell position at $t =$ 6 h, red dot represents the VEGF source. Contour plots showing the final intracellular calcium concentrations in (B) $z = 1$ μm plane and (C) $y = 50$ μm plane. Calcium concentration has been pseudocolor-scaled so that the maximum value is red. (D) Evolution in time of cell directional velocity v_x. (E) Ca^{2+} concentration profiles at the end of migration, $t = 6$ h, along cell x-cross-section of the $z = 1$ μm plane from the trailing edge to the leading edge of the polarized cell. The cell reaches more quickly the maximal directional velocity. which has a lower maximal peak. Mean over 10 simulations; the error bars represent standard deviation.

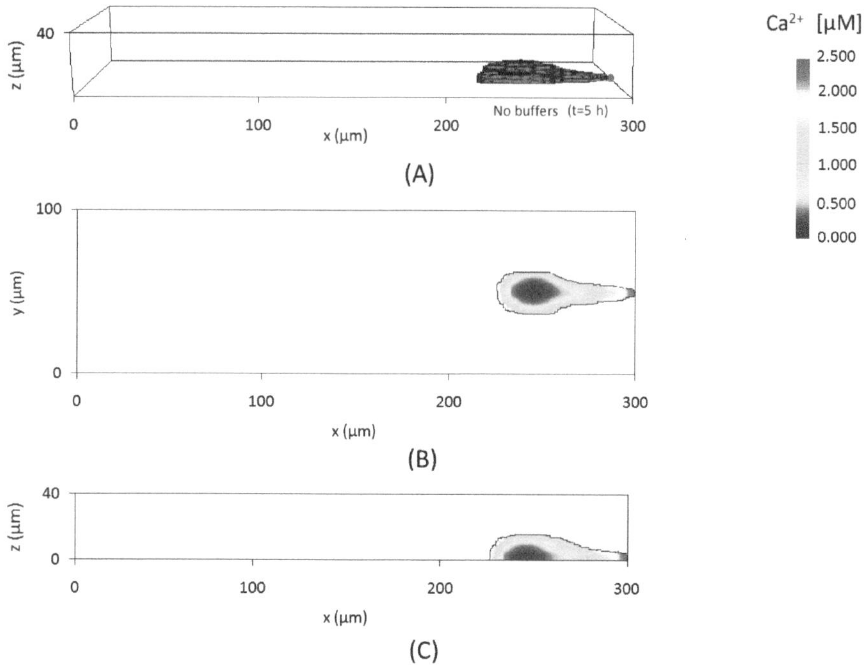

FIGURE 6.11: (See color insert.) Cell migration with interference in the buffering process. (A) Three-dimensional view of cell position at $t = 5$ h in the case of no buffers with the relative intracellular calcium profiles in (B) $z = 1$ μm plane and (C) $y = 50$ μm plane. In (A) red dot represents the VEGF source.

ically, we locally vary the factor K_{buff}:

$$K_{buff}(\mathbf{x}) = \frac{K_{off}}{K_{off} + K_{on}b_T} \frac{d(\mathbf{x})}{d_0}, \tag{6.15}$$

where $d(\mathbf{x})$ and d_0 are the measures of the net Euclidean distance from the PM of, respectively, site $\mathbf{x}$ and of the cell center of mass $\mathbf{x}_\eta^{CM}$. With respect to the standard case, also the proposed buffer clusterization does not strongly change the final spatial profile of calcium, but causes an increment of the maximal peak of responses at the tip of the motility structures (whose value is obviously the same as in the absence of buffers) and a decrement in the concentration of the ion in the more internal regions (i.e., the elevate presence of buffers almost blocks the diffusion of calcium, so that the zone with a basal level of the ion is bigger); see Figure 6.12. Indeed, we have a steeper gradient of the intracellular Ca^{2+} level along the polarized axes of the cell, as illustrated in detail in Figure 6.12(H). However, the total final calcium concentration remains elevate ($c_\eta(t = 6 \text{ h}) \approx 3.2$), as well as the motile capability of the cell, as we do not observe a significant reduction in the maximal value of v_x and, consequently, a delay in the cell migration time; see Figure 6.12(D).

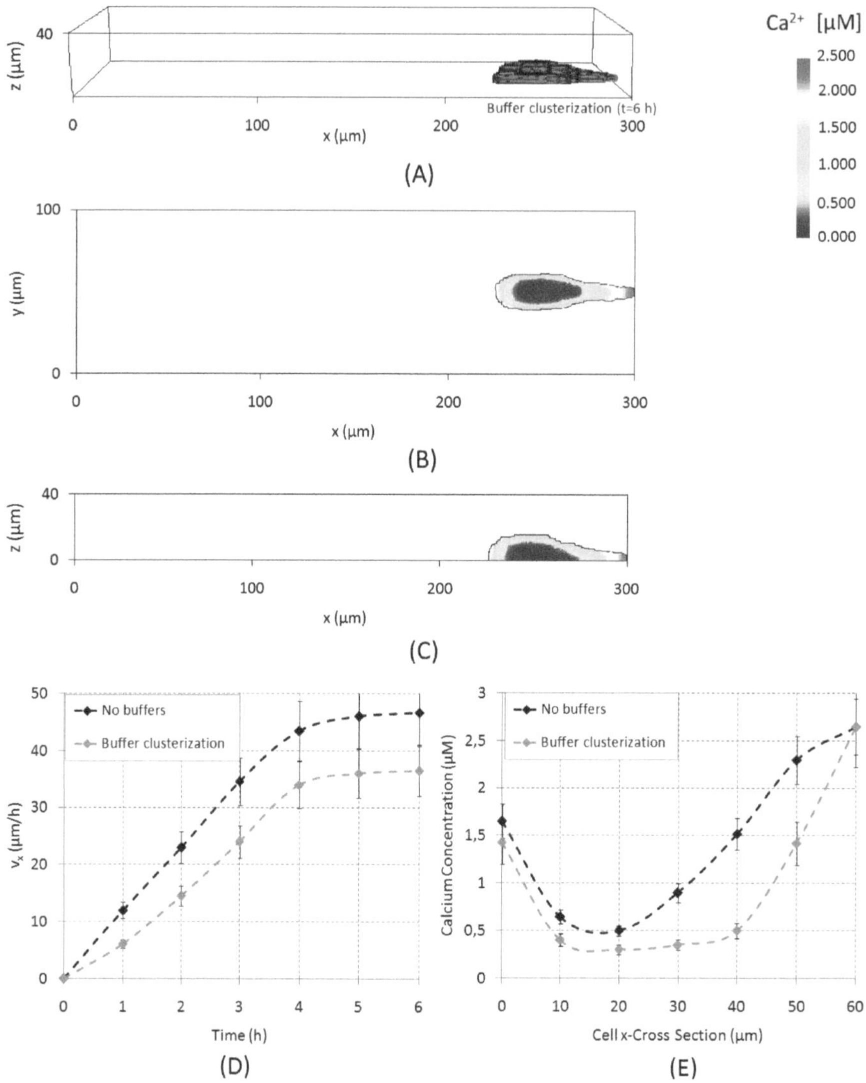

FIGURE 6.12: (See color insert.) Cell migration with interference in the buffering process. (A) Three-dimensional view of cell position at $t = 6$ h in the case of buffer clusterization, and relative contour plots of the final intracellular calcium concentration in (B) $z = 1$ μm plane and (C) $y = 50$ μm plane. In (A) red dot represents the VEGF source. (D) Evolution in time of cell directional velocity v_x in both cases. (E) Ca^{2+} concentration profiles at the end of migration, $t = 5$ h in the absence of buffers and $t = 6$ h in the case of clusterized buffers, along cell x-cross-section of the $z = 1$ μm plane from the trailing edge to the leading edge of the polarized cell. Mean over 10 simulations; the error bars represent standard deviation.

6.6 Altering Cell Morphology

So far, we have provided the fact that differently modulated calcium signals influence the final morphology of the TEC and its migratory properties. We now turn to the opposite by analyzing how a constrained cell shape affects cell migration, and the relative calcium events. We thus perform simulations forcing the vascular cell to keep its initial hemispheric morphology, with a high constant value of $\lambda_{\eta,C}^{surface} = 50$. This is the model counterpart of the activity of phalloidin-like compounds, which block the calcium-dependent reorganization of actin cytoskeleton by inhibiting actin–myosin interactions. As shown in Figure 6.13, the simulated tumor-derived endothelial cell, loosing its capacity to differentiate and polarize, is characterized by a shortly persistent motion, with a predominant random component, which results in a low directional velocity v_x and a short final displacement, as $x_\eta^{CM}(t = 6\text{ h}) = 85\ \mu\text{m}$. These observations agree well with the experimental results provided in [192], where more rounded and symmetric ECs are seen to have an isotropic migration, and suggest a possible pharmacological way to reduce cell motility. The constrained cell morphology affects also the pattern of calcium accumulation; see Figure 6.13(E): the peaks remain localized in the sub-plasma membrane region of leading front of the cell (and almost negligible, $\ll 1\ \mu$M, within a distance of 5–10 μm from the cell surface), but they occur slightly later, and have a significantly lower amplitude ($\approx 1.6\ \mu$M), as resulted also experimentally in [390] and theoretically in [187, 283]. Consistently, the overall accumulation of calcium ions is significantly decreased, as $c_\eta(t = 6\text{ h}) \approx 1.1$. These observations could provide an additional explanation of the great heterogeneity of calcium responses in the same population; not only calcium machinery expression can vary from cell to cell, but also in cells expressing the same amount of channels the response could significantly change as a function of the their geometry. Moreover, with another view point, the same cell can vary its response to the same external chemical stimuli only varying its shape, which is of particular relevance during vascular development, when TECs constantly remodel.

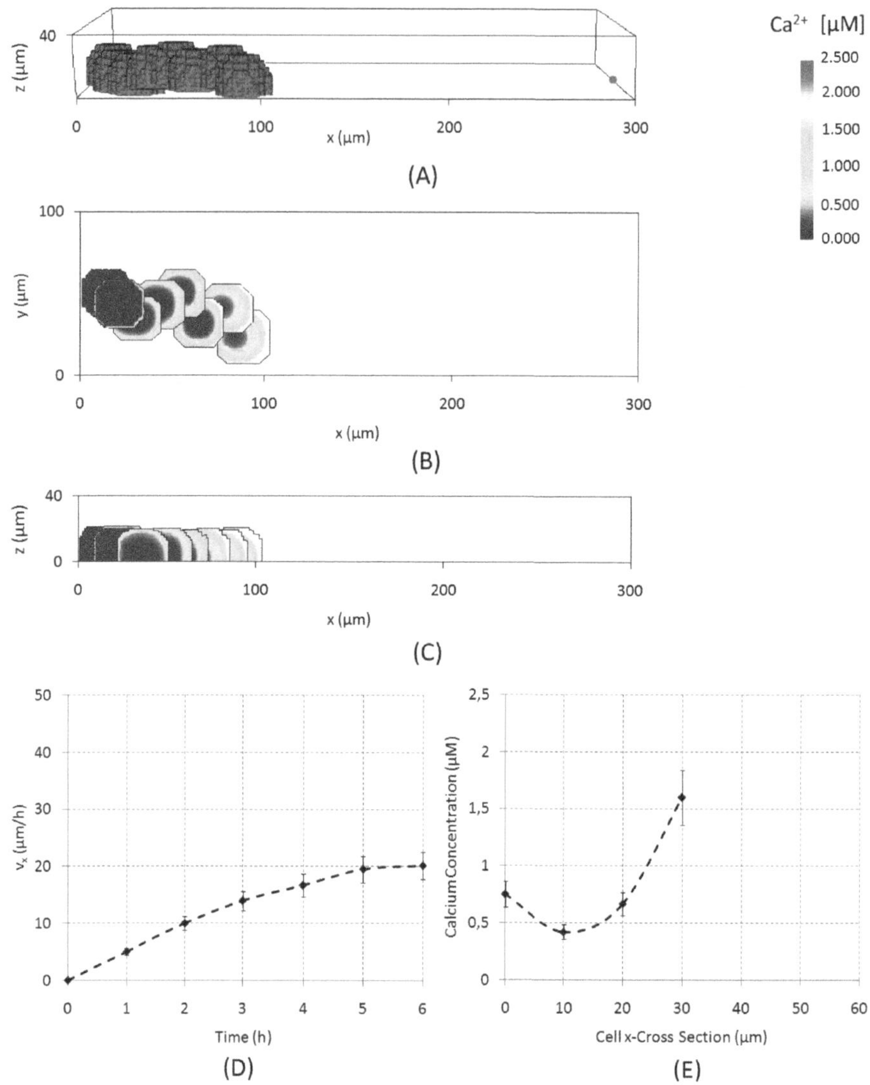

FIGURE 6.13: (See color insert.) Inhibition of active cytoskeletal reorganization, $\lambda_{\eta,C}^{surface} = 50$. (A) Three-dimensional view of cell position (red dot represents the VEGF source), (B) and (C) contour plots showing intracellular calcium concentration during cell motion in, respectively, $z = 1$ μm plane and $y = 50$ μm plane. For (A), (B), and (C) representative images taken, as usually, at 45 min intervals until $t = 6$. (D) Evolution in time of cell directional velocity v_x. (E) Ca^{2+} concentration profiles at the end of migration, $t = 6$ h, along cell x-cross-section of the $z = 1$ μm plane from the trailing edge to the leading edge. Mean over 10 simulations; the error bars represent standard deviation.

6.7 Varying the Chemical Source

When the external chemical signal is extended to the whole right lower edge of the migration chamber, with the same intensity ϕ_v, we observe a different cell shape reorganization; see Figure 6.14(A). After an initial stage ($\approx$ 1 h), in which its phenomenology resembles the standard case in Figure 6.6, the motile TEC adapts to the new VEGF profile, featuring a flat and thin ($\approx$ 2 μm) lamella in the direction of motion. An interesting consideration that emerges here is that such a different morphological transition is completely self-generating, and only due to the new spatial profile of the external stimulus as, beside the extension of the VEGF source, we have not changed any other model assumptions. The mechanical explanation is that, as reproduced in Figure 6.15(A), the planar chemical source results in a planar front of maximal concentration, which chemotactially stimulates a larger part of the cell membrane to protrude, thus forming the flat lamella. On the opposite, the standard VEGF point source features a curved profile: consequently, the maximal chemical force, given by the maximal chemical gradient, is concentrated on a restricted part of the cell leading surface, from which the thin pseudopodium emerges; see Figure 6.15(B).

A further confirmation of this mechanism is given by the fact that, at the beginning of both simulations, when the curvature effects are not so strong, the morphology of the cell is very similar. As far as we know, there are no *in vitro* evidences replicating such a model outcome, probably because it is very difficult to experimentally vary the local spatial extension of a chemical source. However, this result is supported by plausible biomechanical implications. External chemical signals are in fact demonstrated to be locally transmitted from membrane-bound receptors to the nearest "central nodes" of cell cytoskeleton, including small G-proteins Cdc42, Rac and Rho [180, 149, 388]. The activity of the rho-family molecules, in turn, directly drives actin dynamics, regulating and fine-tuning polimerization and nucleation processes of cytoskeletal filaments, which determine the remodeling of the cytosol and the protrusion of the plasma-membrane [125, 258, 293]. An extended (i.e., not punctual) exogenous signal may therefore cause the growth of the actin filament network in a larger part of the cell leading front, eventually resulting in the formation of a flat motility structure, as the lamella emerged in our simulations. Interestingly, as shown in Figure 6.14, the evolution of the cell migratory properties (the overall displacement, as $x_\eta^{CM}(t = 6 \text{ h}) = 300$ μm, and the directional velocity) remains almost unaltered with respect to the standard case as well as the VEGF-induced calcium dynamics. In particular, the peak of maximal calcium response, that remains localized in the thinner part of the cell (in the lamellipodium), measures ≈ 2.2 μM, while $c_\eta(t = 6 \text{ h}) \approx 3.4$. This consideration provides the fact that proangiogenic Ca^{2+} signals are typically localized in the motility structures (either pseudopodia or lamellipodia) of vascular cells:

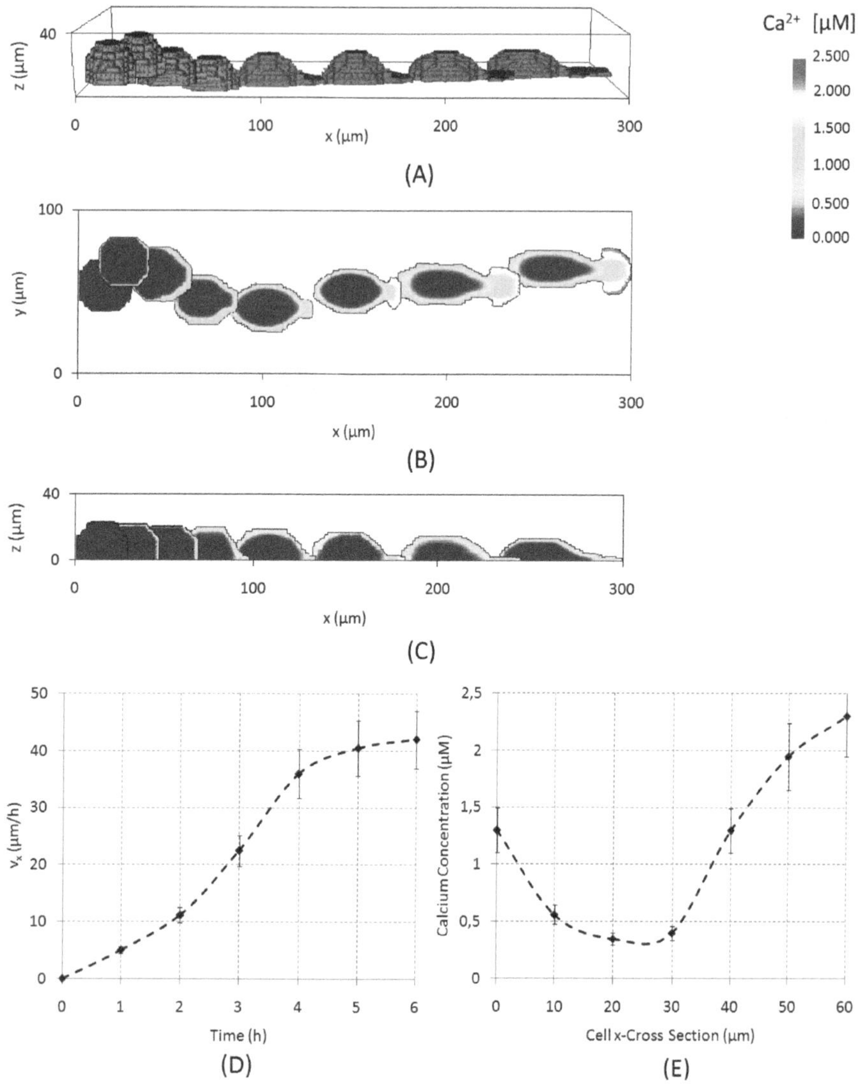

FIGURE 6.14: (See color insert.) VEGF source extended to the whole right lower edge of the migration chamber. (A) Three-dimensional view of cell position (red line represents the VEGF source), (B) and (C) contour plots showing intracellular calcium concentration during cell motion in, respectively, $z = 1$ μm plane and $y = 50$ μm plane. For (A), (B), and (C) representative images taken, as usually, at 45 min intervals until $t = 6$. (D) Evolution in time of cell directional velocity v_x. (E) Ca^{2+} concentration profiles at the end of migration, $t = 6$ h, along cell x-cross-section of the $z = 1$ μm plane from the trailing edge to the leading edge. Mean over 10 simulations; the error bars represent standard deviation.

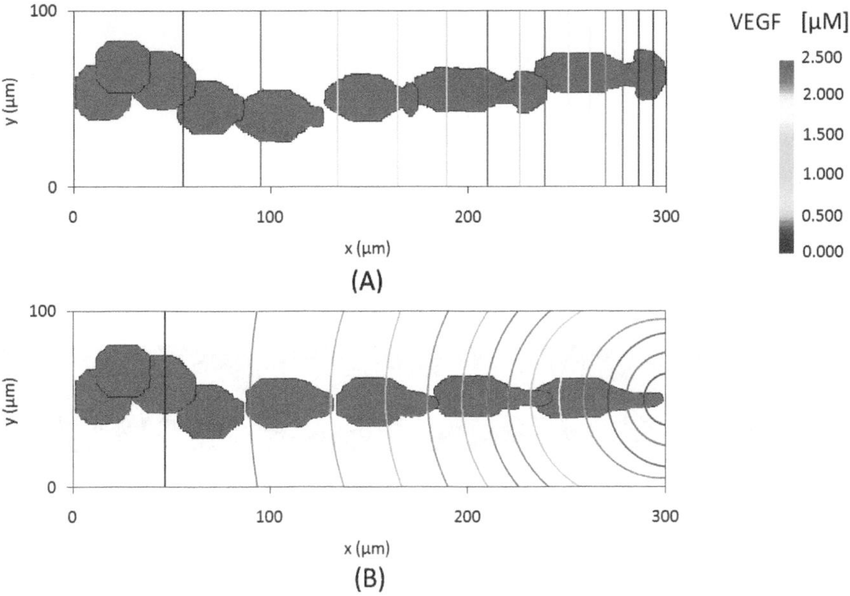

FIGURE 6.15: Cell morphological transition in the case of (A) planar (B) punctual VEGF source. Representation of the $z = 1$ μm plane. Isolines indicate chemical levels.

such a peculiar pattern could in principle carry information of the specific role of the ion in other cellular functions, such as the adhesive interactions with neighboring cells during the formation of vascular chords.

In Figure 6.16, we review all the proposed strategies to influence the phenomenology of the motile cell. In particular, we summarize the values of critical parameters quantifying both the VEGF-induced calcium signals and the resulting migratory properties of the TEC.

Molecular mechanism	Model parameter change	Peak of Ca^{2+} accumulation [μM]	Rate of total Ca^{2+} accumulation	Maximal directional velocity [μm/h]	Final cell displacement [μm]
Ca^{2+} influx	$F_{TOT}/2$	1.45	1.4	29.1	150
	$F_{TOT}/4$	1.02	0.6	17.8	70
	$F_{TOT}/8$	0.62	0.2	4.9	18
AA production	$k_a = v_c = 0$	1.06	1.2	22.7	110
NO production	$k_n = v_{ca} = 0$	1.56	1.6	30.5	180
Cytoskeletal remodeling	$\lambda^{surf}_{\eta,C} = 50$	1.57	1.1	20.3	85
Extended source		2.23	3.4	41.3	300
Ca^{2+} diffusion	$D_c = 0$	2.39	0.9	29.7	165
NO buffers	$K_{buff} = 1$	2.67	3.9 (in 5 h)	46	300 (in 5 h)
Buffer clusterization	$K_{buff} = K_{buff}(\mathbf{x})$	2.67	3.2	36.8	300
Control		2.28	3.6	42	300

FIGURE 6.16: Review of the proposed strategies interfering with cell motility. Second column gives the relative modified model parameters. The last four columns quantify each treatment by summarizing the critical parameters describing both the resulting calcium signals and the overall migratory ability of the EC. All values are mean over 10 simulations. Control solution represents the no treatment case, with no changes in the standard parameter setting, and corresponds to the simulations in Figures 6.6 and 6.7.

Chapter 7

Tumor-Derived Vasculogenesis

7.1 Biological Introduction

Blood vessel formation and development is a multiscale process, driven by the activation of endothelial cells (ECs, the main bricks of the capillary walls) induced by the action of suitable biochemical stimuli which are released both by environmental cells and by ECs themselves. Vascular progression involves two different mechanisms: *vasculogenesis* and *angiogenesis* (for a review, see [57, 358]). The former process mainly consists in the formation of a primitive vascular network, that emerges from a directed and autonomous migration, aggregation and organization of a dispersed population of endothelial cells. The latter consists instead of the formation of new vessels from an existing capillary or post-capillary venule. Angiogenic remodeling co-ordinates with the establishment of blood flow and can occur through sprouting, i.e., by the formation of new branches from the sides of existing capillaries, pruning, resizing of the capillary volume of the thickness of the capillary wall, or intussusception, i.e., by internal division of the vessel lumen.

Entering into more detail, in the embryo, the process of vasculogenesis starts with the assembly of mesoderm-derived precursors of ECs in polygons having well determined topological characteristics, dictated by the principal and paradigmatic function of vasculature: the oxygen transport to the tissues. After remodeling, these geometrical properties are more or less maintained in the adult body, where the capillary network embedded in the tissues and stemmed by the vascular tree has the same shape as the minimal unit participating in the formation of the embryo vascular net. Though angiogenesis already intervenes in the embryo to remodel the initial capillary network into a mature and functional vascular bed comprised of arteries, capillaries, and veins, its main role is played during the adult life, when it is involved in many physiological processes, for instance, the vascularization of the ovary and the uterus during the female cycle, of the mammary gland during lactation, and of granulation tissues after wound healing.

However, vascular progression is a complex and highly regulated phenomenon, controlled and balanced by coordinated molecular and cellular events operating at different levels. When their equilibrium is disrupted vascularization becomes pathological, as in the cases of chronic inflammatory

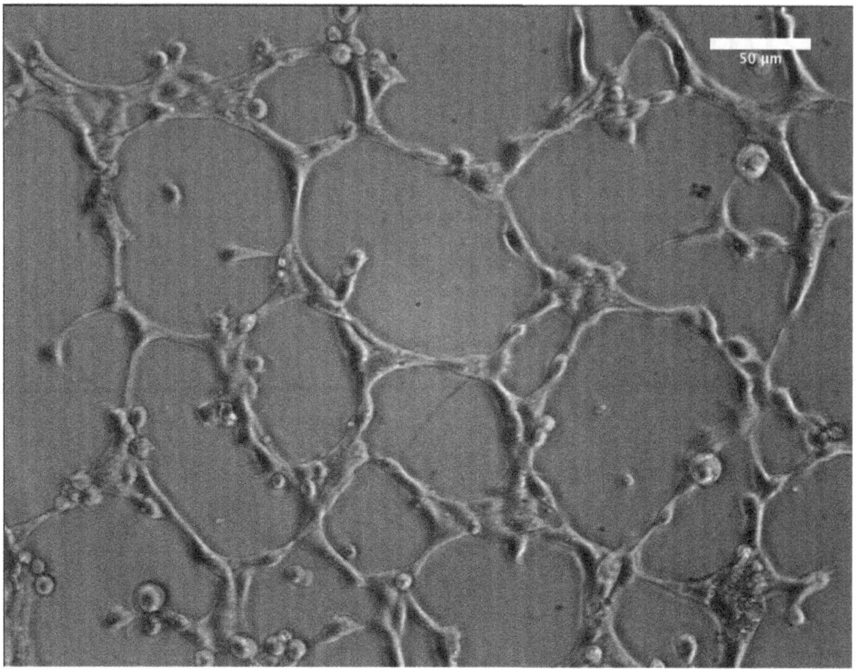

FIGURE 7.1: *In vitro tubulogenesis* of tumor-derived endothelial cells (TECs). Morphometry of the capillary-like network formed in 12 h after cell incubation.

diseases, vasculopathies, degenerative disorders, and tissue injury occurring in ischemia. Indeed, vascular formation is a pivotal transition also in cancer development. In fact, by providing the nutrition and oxygen, it allows malignant cells to grow and remain viable, and, eventually, to cause metastases and enter in the circulatory system [65]. Moreover, it is also active in determining the translation of dormant metastases to an aggressive status [64]. The switch to the angiogenic phenotype leads therefore to a fast progression and to a potentially fatal stage of the disease and represents an important target for therapeutic interventions in most types of malignancies.

However, despite the major progresses and promising successes achieved over the past few years in anti-angiogenic pharmacological therapies, several limitations still occur due to different factors, as commented in [64, 65, 129] and references therein.

A continuous effort in the development of biomedical therapies is advanced by several *in vitro* models, which are providing a deeper understanding of selected underpinning molecular and cellular events coordinated to control tumor-induced vessel formation, as reviewed in [9, 57, 282, 285, 358]. One of the best known is the *tubulogenic* experiment, the laboratory counterpart of *in vivo* vasculogenesis.

Tubulogenic assays can be obtained using different experimental set-ups, substrata (e.g., Matrigel, fibronectin, collagen, fibrin, semisolid methilcellulose), and endothelial cell-lines (e.g., human umbilical vein endothelial cells (HUVECs), human dermal microvascular endothelial cells (HDMECs), human capillary endothelial cells (HCECs), human marrow microvascular endothelial cells, bovine aorthic endothelial cells (BAECs), bovine capillary endothelial cells (BCECs), bovine retin endothelial cells (BRECs), rat capillary endothelial cells (RCECs), embryonic stem cells (ESCs), calf pulmonary aortic endothelial cells (CPAECs), adrenal capillary endothelial cells (ACECs)).

In particular, an increasing number of vasculogenic experiments has been recently performed with tumor-derived endothelial cells (TECs): laboratory investigations have in fact demonstrated that, according to a macroscopic morphological analysis, tumor blood vessels are irregular and dilated and the vessel hierarchy is not well defined, so that distinct venules, arterioles, and capillaries are undistinguishable [100, 148]. Moreover, they differ from their "normal" counterpart by altered blood flow and permeability, and by abnormalities in pericytes and in the basement membrane. Therefore, as also seen in the previous chapter, vascular endothelial cells deriving from tumors (TECs) represent a more adequate model for studying the mechanisms of malignant vascularization [21, 54, 170].

In spite of such a large variety of laboratory protocols mentioned above, it is possible to point out a unified illustration of the common features of the experimental process. The selected EC population is initially dispersed in a physiological solution and then poured on the top of a specific substrate, which typically favors cell motility and has biochemical characteristics similar to living tissues. The cells sediment by gravity onto the surface and then move on it, giving rise to the mechanisms of aggregation and pattern formation. In more detail, the overall process, which commonly lasts 9–15 h, consists of the following steps:

1. Cells initially undergo an isotropic motion around their initial position, maintaining a round shape. Then, it seems that they choose a direction, which is correlated with the location of areas characterized by higher cellular densities, and display an independent migration, with a small random component, until they collide with their closest neighbors (3–6 h). This motile phenotype is called in biology *cell persistence* and is related to the inertia of a cell in rearranging and repolarizing its cytoskeleton, maintaining its own direction of migration.

2. After collision, ECs attach to their neighbors eventually forming a continuous and structure multicellular network, which can be represented as a collection of nodes connected by capillary chords (see Figure 7.1).

3. The network slowly moves as a whole, undergoing a slow thinning process, which however leaves the structure mainly unaltered.

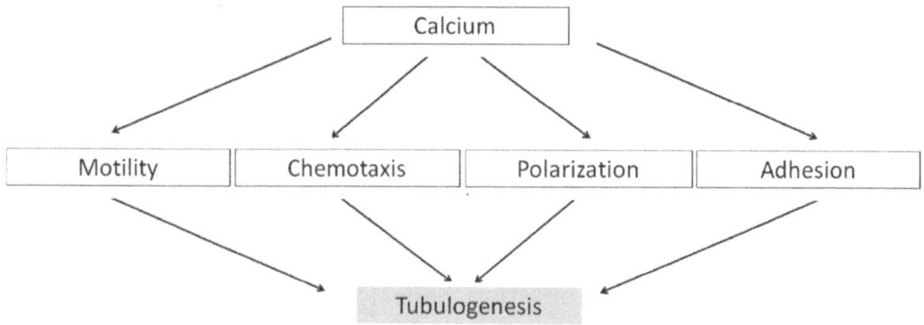

FIGURE 7.2: Effect of VEGF-induced calcium signals in the process of *in vitro* vasculogenesis. The accumulation of the ion triggers motility, chemical response, adhesion and cytoskeletal reorganization of TECs, crucial events for the capillary-like network formation. The specific pathways involved in cytosolic calcium entry are reproduced in Figure 6.1.

4. Finally, individual cells fold up to origin the lumen of the capillary, so that the resulting vascular network forms along the lines of the previous.

The early patterning is complete in a characteristic time sufficiently distant from critical events, such as cell mitosis or death, that allows to assume the cell density conserved. The final structure is no longer substantially modified and resembles a primitive *in vivo* capillary-like plexus [133]. The overall process is largely mediated by the redundant activity of multiple chemical morphogens (such as vascular endothelial growth factor isoforms, acidic and basic fibroblast growth factors, epidermal growth factor, transforming growth factor-α, transforming growth factor-β and several more, as reviewed in [128]) that, behaving as signal exchangers signals by their release and absorption, exert a chemotactic action and, concomitantly, activate both in normal ECs and in tumor-derived individuals a series of calcium-dependent cascades, regulating cells phenotypical behavior as already seen in Chapter 6 and described in [134, 216, 274, 276, 277, 282, 283, 366] and in the references therein (Figure 7.2).

The analysis performed on the tubulogenic system and on its driving molecular mechanisms has revealed the role of different factors operating in malignant vascular progression. However, innumerable other component processes, acting at different scales, are far from being completely elucidated. In particular, their complexity presents a number of components that could be in principle interfered in multiple ways to inhibit malignant vascular progression. This large combinatorial space of possible therapies is obviously unfeasible to search using only laboratory-based biological methods. A modeling approach able to replicate selected features of an *in vitro* TEC *tubulogenesis* is therefore useful in this respect.

Indeed, in this chapter, the CPM used in the previous chapter to simulate the motility assay of the individual tumor-derived endothelial cell is extended to the multicellular situation, including cross-talk among cells. The proposed model aims at reproducing the formation of the overall tumor capillary structure and at characterizing its topological properties, with a close comparison with published experimental observations. In particular, through different sets of numerical realizations, we show that the initial density of the overall culture, as well as selected biophysical properties of single individuals, such as their calcium-dependent chemical and adhesive strength and the persistent component of their migration, play a fundamental role in determining the dynamic and the final configuration of the overall network formation. If, on one hand, the systematical analysis of the TEC pattern ability confirms the efficiency of some currently available anti-angiogenic therapies, on the other hand it may suggest novel and experimentally testable strategies that have the potential to disrupt TEC capillary formation.

7.2 Mathematical Model

The model for the overall TEC tubulogenesis contains exactly the same ingredients as that for the single cell motion, with some slightly modifications regulating intercellular interactions that will be clarified below.

7.2.1 Cell-Level Model

First, since we wish to compare simulations with experimental cultures, where the vascular patterns are essentially monolayers, we turn to use a bidimensional domain $\Omega \subset \mathbb{R}^2_+$. Each simulated TEC, η, is therefore defined as a planar bi-compartmental unit, composed of the cell nucleus, in this case a round cluster of type $\tau = N$ and the surrounding cytosol, $\tau = C$ (see Figure 4.3). Indeed, given the bidimensional approach, λ^{volume} and $\lambda^{surface}$ now correspond to, respectively, $\lambda^{surface}$ and $\lambda^{perimeter}$. The TECs reside again in a generalized substrate, a further discrete object of type $\tau = M$, which represents the experimental Matrigel. The internal state vector of each subcellular compartment is the same as in the case of the single cell, while the *Hamiltonian* functional H reads again as in (6.1).

However, with respect to the model for the single TEC, $H_{adhesion}$ has to consider also the effective adhesion between membranes of different cells, given by (4.3). In particular, $J^{ext}_{E,E}$, where $\theta(\eta) = \theta(\eta') = E$, represents the local adhesive strength acting between neighboring cells, a measure from a statistical view point, of the probability of the formation of local intercellular cadherin-cadherin complexes, which depends on the quantity of active exposed molecules on either sides of the interface. Indeed, since VE-cadherin activity is enhanced by calcium ions, which generate clusters of activated molecules, for any TEC η and for $\tau(\Sigma_\sigma = C)$, we set

$$\mathbf{s}_{\eta,\Sigma_\sigma}^{J^{ext}}(\mathbf{x},t) = \mathbf{s}_{\eta,\Sigma_\sigma}^{J^{ext},A}(\mathbf{x},t) = (c(\mathbf{x},t)),$$

and therefore for each couple η and η',

$$J_{E,E}^{ext}((\partial\mathbf{x} \in \partial\eta) \cap (\partial\mathbf{x}' \in \partial\eta'), t) = J_{E,E}^{ext}(\mathbf{s}_{\eta,\Sigma_\sigma}^{J^{ext},A}(\mathbf{x},t), \mathbf{s}_{\eta',\Sigma_{\sigma'}}^{J^{ext},A}(\mathbf{x}',t)) =$$

$$= g_{J^{ext}}(\mathbf{s}_{\eta,\Sigma_\sigma}^{J^{ext},A}(\mathbf{x},t), \mathbf{s}_{\eta',\Sigma_{\sigma'}}^{J^{ext},A}(\mathbf{x}',t)) = J_0 \exp\left(-p\,\frac{c(\mathbf{x},t)c(\mathbf{x}',t)}{c_0(\mathbf{x})c_0(\mathbf{x}')}\right),$$

where $c(\mathbf{x},t)$ and $c(\mathbf{x}',t)$ are the local calcium levels, and $c_0(\mathbf{x}) = c_0(\mathbf{x}') = c_0$ the basal local concentration of the ion. J_0 represents the typical adhesive force of resting TECs, estimated by qualitative observations of experimental cultures (see Appendix C).

The contact energy between the cells and the simulated Matrigel is taken to be negligible (i.e., $J_{E,M}^{ext} = 0$). Even though it should obviously be included in a more detailed model, its exclusion does not strongly influenced the final outcomes of the approach. The contact interactions between vascular cells and homogenous and isotropic matrix substrates have been in fact demonstrated to play a major role in the stabilization of an *in vitro* vascular network, rather than in its early formation. In particular, the theoretical works presented in [10, 215], supported by the biological literature therein, have proved that the intercellular interactions and their chemotactic and persistent migration represent the minimal set of biological mechanisms sufficient to guarantee the emerging of the capillary structure.

On the other hand, cell–ECM adhesion only sustains the preservation of its morphology under shear stresses and perturbations. Consistently, in [391], the traction/adhesion between the cell population and the gel layer starts acting when a sufficient density is locally obtained, i.e., when cells are no longer isolated but already connected. Indeed, cell-substrate adhesive strengths could dominate the early stages of vascular patterning only in extreme cases. In particular, as observed again by the authors of [391], a too-strong cell–ECM adhesion inhibits the movement of cells, which remain isolated and dispersed in their initial position. An overly adhesive substrate causes in fact the formation of clusters of integrin-ligand bounds at cells surfaces, that do not detach, stopping further cell locomotion. On the contrary, a too weak cell–ECM adhesion results in the packing of cells in big islands, as they try to minimize the parts of their membrane in contact with the extracellular medium.

Therefore, the exclusion of cell–matrix adhesion from our model is not too restrictive, since we focused on the initial formation of the network, and not on its following stabilization, and we implicitly assume that cell–matrix contacts are characterized by an intermediate strength. Following the same line of thought, the presented approach could obviously be further developed with the inclusion of other interactions between the cell population and the substrate, that have been analyzed in other computational models. For instance, in [250], the movement of cells has been biased towards direction of principal strains of the underlying matrix, while in [289], the authors have added a cell sensitivity to matrix density gradients (i.e., the haptotactic mechanism) and long-range interactions in matrix stretching, due to its fibrous nature. In this regard, the mechanical and topological properties of the substrate could also be characterized. For example, dense protein matrices, trapping a number of growth factor molecules, slow the diffusion of the morphogen and, causing steeper gradients, alter cell chemical responses. Moreover, anisotropic substrates drive cell migration and interfere with the network formation, especially in a three-dimensional patterning.

The constitutive relations linking the Potts parameter to the internal state of cell compartments are the same as in the case of the motion of the single cell described in the previous chapter. In particular, it is useful to underline that, in a multicellular system, the relation assumed for the chemotactic coefficient represents a further strong improvement of this work with respect to classic CPMs (see, for example [259, 260, 261]). In fact, therein, all the cells of the same type feature the same chemical sensitivity, despite of their individuality and their internal state (here defined by the intracellular calcium level).

7.2.2 Molecular-Level Model

VEGF evolution is again described by Equation (6.5). However, the growth factor is no longer produced by an exogenous source, but is autocrinally secreted by TECs at a constant rate ϕ_v (see [360]). The VEGF production term therefore reads as

$$S = S(\mathbf{x}, t) = \phi_v, \tag{7.1}$$

where $\mathbf{x}$ belongs to the external surface of the cell PM (i.e., $\tau(\Sigma_{\sigma(\mathbf{x})}) = M$ and $\exists \mathbf{x}' \in \Omega_{\mathbf{x}}^{'} : \tau(\Sigma_{\sigma(\mathbf{x}')}) = C$). The model of the agonist-induced intracellular dynamics is also the same as in the case of the single cell. In particular, Equations (6.7), (6.8), and (6.10) work for each individual, and the extracellular calcium profile is again given by (6.14), with the flux exchanges to be considered obviously with all cells.

7.3 Simulations in Standard Conditions

The domain Ω consists in a 500×500 square lattice, where each site is equivalent to 4 μm^2. Therefore, Ω represents a section of a 24-well plate of size $L_\Omega = 1$ mm. The experimental time is set to 10 s per MCS: for this choice the simulated TECs move with nearly the experimental velocity of vascular cells (≈ 25 μm/h [261, 309]) and the overall patterning has a comparable time scale (≈ 12 h [133]).

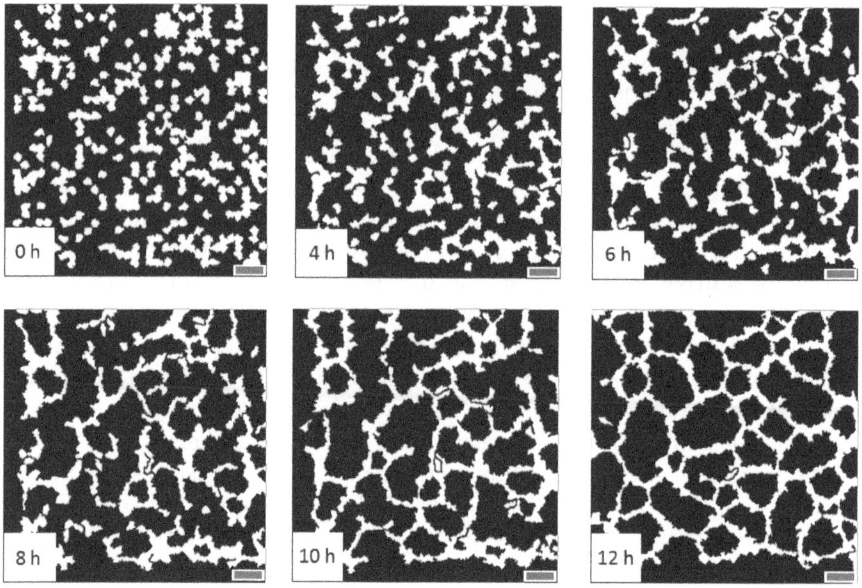

FIGURE 7.3: Tumor-derived capillary-like network formation. The simulation initiates with 150 quiescent TECs over Ω, a square lattice of size $L_\Omega = 1$ mm. Representative images show a time-sequence of the tubule organization. The scale bar is 100 μm long.

We randomly distribute $n = 150$ TECs in Ω, reproducing a typical experimental cell density [133]. Each individual has initially an unpolarized morphology with a diameter of 40 μm (20 lattice sites), while its nucleus is a central and round cluster with a radius of 6 μm (3 sites). These measures agree with the most usual geometries observed in *in vitro* TEC cultures [133, 135]. A single cell is therefore a moving subdomain, that extends over an area of nearly 320 sites ($\approx$ 1280 μm^2). Such a spatial resolution allows to have a good approximation in the resolution of the spatiotemporal intracellular dynamics without slowing down the computational speed too much.

Cells do not undergo mitosis or death during the entire simulations. In this way, consistently with the experimental observations given in the previous section and with other similar approaches [259, 260, 261], the density of the culture is conserved during the overall process.

The biological variables are initiated at their physiological levels. In particular, as in the simulations of Chapter 6, there are no activated second messengers within the cells, and the intracellular basal calcium concentration, c_0, is assumed to be spatially homogeneous. In the extracellular domain, the initial level of Ca^{2+} is again set equal to $c_{ext,0}$, while there are no VEGF molecules. The model parameters are also the same as in the case of the single cell, see Table C.6 in the Appendix. Moreover, $J_0 = 5$ represents the low

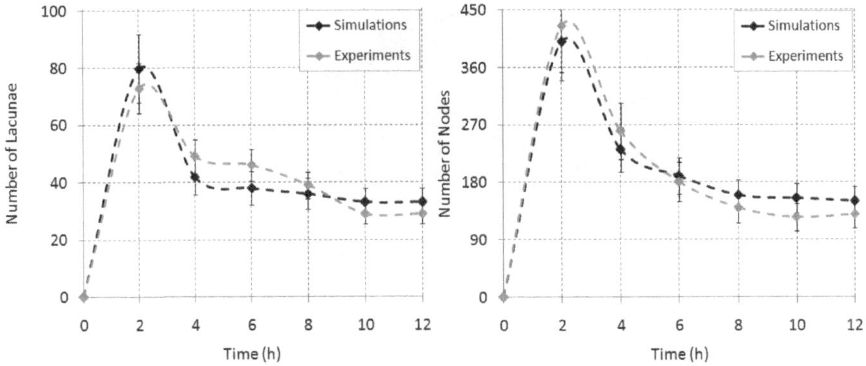

FIGURE 7.4: Dynamical evolution of TEC network formation. Time-evolution of the number of lacunae and the number of nodes during patterning in both simulations and experiments. Error bars for computational values show standard deviations over 20 simulations, while errors bars for experimental values show standard deviations over 8 realizations. *In vitro* data have been evaluated from experimental images periodically recorded (i.e., at 2 h-intervals) during tubulogenic processes of cultures of endothelial cells derived from human breast carcinomas (B-TECs), seeded onto Matrigel-coated wells and stimulated with angiogenic factors, as provided in [133].

basal adhesive strength of TECs, which have been demonstrated to adhere only in confluence (see again [135]).

As shown in Figure 7.3, the kinetics and the temporal dynamics of the *in silico* morphogenesis, as well as the final pattern configuration, clearly resemble experimental cultures (see Figure 7.1 and refer to [133]). In particular, the tumor-derived endothelial cells organize into a structured network similar to a capillary plexus, where cords of cells enclose lacunae. Vascular branches typically are one or two cells wide, while lacunae are almost uniform in size. It is useful to underline that the pattern emerges autonomously, without prescribed phenomenological rules. In fact, its formation is realistically driven by the complex and coordinated interplay of multilevel mechanisms, i.e., the calcium-dependent increase in cell motility, adhesion, and cytoskeletal remodeling and the VEGF-mediated chemotactic migration.

More information about the overall pattern formation are obtained by counting the numbers of nodes and lacunae at different times. As represented in Figure 7.4 both of them quickly drop and then slowly stabilize. This is due to the fact that the smaller lacunae shrink and collapse and the unstructured cellular branches adhere and align to form a single network.

As a further quantitative analysis of the formed structure, we measure its average metric and topological properties. The pattern features typical intercapillary distances (i.e., the mean diameters of lacunae) ranging from

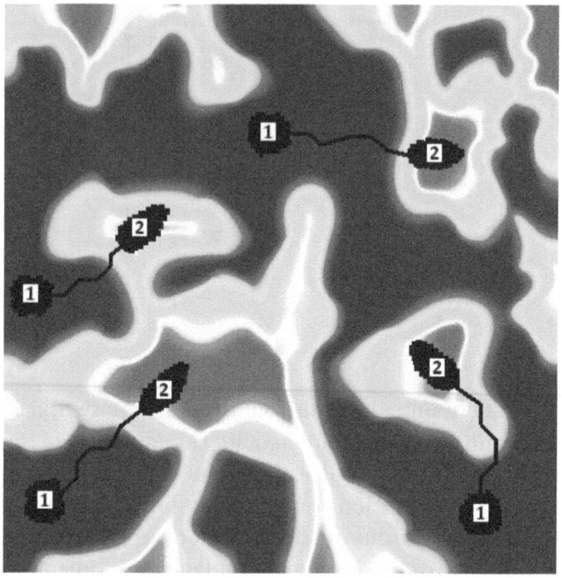

FIGURE 7.5: (See color insert.) Magnification of four representative single-TEC motion trajectories obtained from a time-lapse simulation. The initial (i.e., at 0 h) position of individuals is labeled with 1, while the final (i.e., at 12 h) with 2. Each segment tracks a displacement of 2 h. In the background the concentration field of VEGF is pseudocolor-scaled so that the maximum level is red and minimum level is blue. The TECs move in the directions of higher morphogen concentrations, displaying a persistent directional migration. The image also captures the polarization process typical of vascular cells: it is useful to underline that cell elongation is realistically reproduced in the model, as it emerges as the natural consequence of the interplay between the calcium-induced reorganization of the cell cytoskeleton and the VEGF chemotactic stimulus.

120 μm to 250 μm, and a mean chord length (measured as the segment from one node to another) of 180 $\pm$ 10 μm.

The geometrical description of the emerging structure is in good agreement with the experimental analysis provided in [133] on a culture of endothelial cells derived from human breast carcinomas (B-TECs), plated on a growth factor-reduced Matrigel and stimulated with angiogenic factors. Moreover, the network sizes captured by the model are surprisingly consistent also with those measured in the case of tubulogenic assays performed with "normal" endothelial cells, i.e., human umbilical vein ECs (HUVECs) [360]. In particular, such natural length scales have been demonstrated to be dictated by the effective range of cell-to-cell interactions, which is mediated by the activity of the soluble VEGF (in particular by its diffusion coefficient and decay rate,

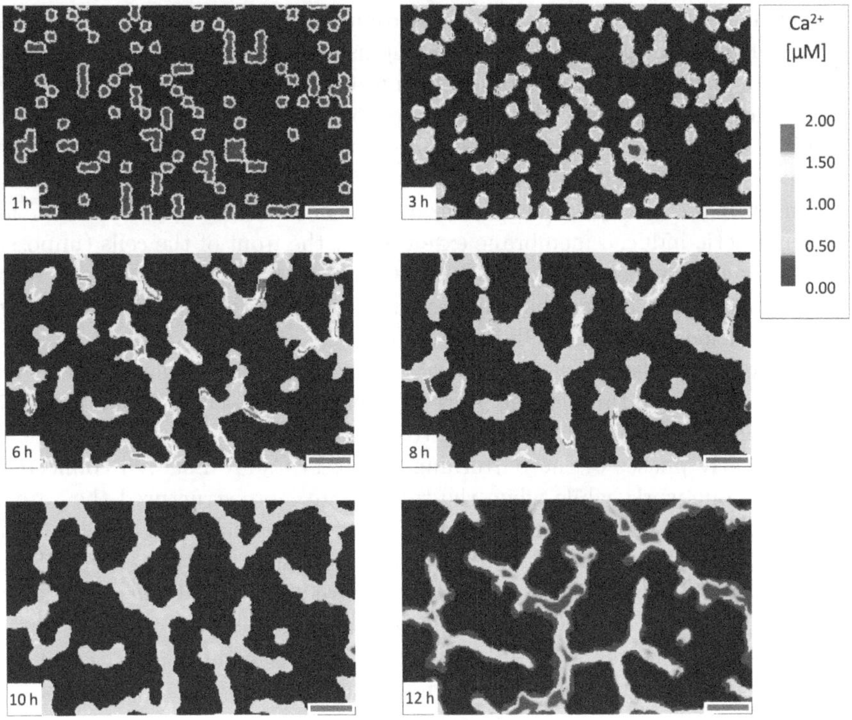

FIGURE 7.6: (See color insert.) Time-evolution of VEGF-induced intracellular calcium dynamics. Calcium signals are clearly detectable at the early stages of tumor capillary-like formation, while they are downregulated in the final phases of the process, when the network is more structured. Scale bar is 100 μm long.

see again [360]). Moreover, they are biologically functional and instrumental for an optimal metabolic exchange: a coarser capillary pattern would be in fact unable to differentiate to form the lumen, while an immature and finer structure would be obviously useless (refer to [82, 175]).

Similarly to the experimental pictures in [360] for HUVECs, we now focus on the trajectories of single cells. Figure 7.5 shows that their motion is directed towards zones of higher concentrations of morphogen, displaying a directional persistence. A random component is of course present, however it is not predominant. Such a characteristic migration, which is the combination of chemotaxis and persistence, has been experimentally demonstrated to be fundamental to produce capillary chords in physiological cases as well. On the opposite, the slower sideaway fluctuations of cells have been shown to be instead responsible of pattern coarsening, when single branches connect each other forming a structured network [261, 360].

Figure 7.5 also captures the elongation process of TECs, which undergo a gradual transition from the initial round stationary state to a migratory bipolar morphology, with clearly distinguishable leading and trailing surfaces dictating the direction of motion. It is useful to underline again that, also in this bidimensional case, the polarization of migrating individuals has not required any *a priori* assumption or prescribed rule on their length (as done in similar models, see [260]). As commented in the case of the single cell simulations, it in fact emerges as the natural result of the interplay between the local chemotactic-induced membrane extension at the front of the cells (impossible with the homogeneous chemical strength used in those approaches, as already commented) and the mechanical properties given to their compartments, with a stiff nucleus and a more fluid cytoplasm.

The model also allows an accurate analysis of intracellular calcium events during the overall tubulogenic process. In particular, as depicted in Figure 7.6, the simulations show that VEGF-evoked Ca^{2+} signals are clearly detectable in the early phases of the patterning, when TECs are not well connected in a mature network, while when the tubules are more structured they are significantly downregulated. The peculiar temporal evolution of calcium signals has been observed also in *in vitro* assays [133] and may play a specific role in transducing information during the different phases of the process [282]. In this regard, even though experimental models have not yet shed light on the molecular mechanisms underlying such a decay in calcium events, our approach could suggest a simple and plausible functional explanation. The initial dramatic increase in intracellular calcium levels is in fact necessary for the enhancement of cell migratory properties, which is in turn fundamental in the first stages of the patterning, as TECs have to efficiently move and interact. Subsequently, when the network is mature, the positions of cells are stabilized in the structure and, without the necessity of a significant further locomotion, their migratory capacities can be partially inactivated and, therefore, the cytosolic Ca^{2+} can decrease until almost its basal level. It is useful to notice that the use of a constant chemotactic strength (i.e., as in classical CPMs) would have unrealistically affected the stabilization of the structure, since the cells would have continued to move following the external chemical stimulus, likely detaching from each other. The regulation of Ca^{2+}-dependent mechanisms may thus represent an important target for therapeutic strategies, which aim at disrupting malignant vascularization and, eventually, at inhibiting tumor development as experimentally provided in [284, 306].

7.4 Varying Cell Density

Up to now we have shown the model ability to start from realistic initial data and foresee the dynamics of tumor tubulogenesis consistently with respect to published experimental works. We now focus on the pattern dependence

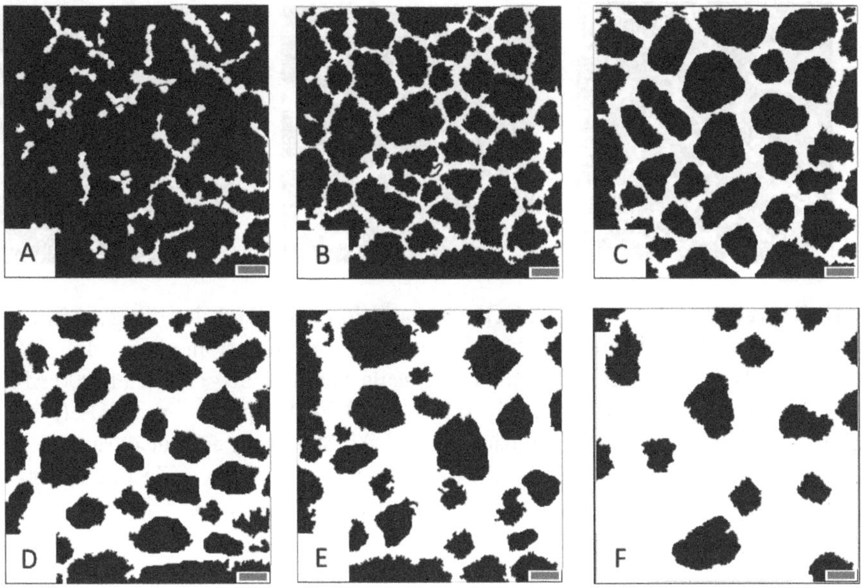

FIGURE 7.7: Percolative property of tumor-derived endothelial cells. Final patterns (i.e., after 12 h) formed by initially different numbers of cells. (A) $n = 50$, (B) $n = 150$, (C) $n = 200$, (D) $n = 300$, (E) $n = 400$, and (F) $n = 500$. Except from the initial cell density all simulations use the same standard parameter setting of the realization in Figure 7.3. The scale bar is 100 μm long.

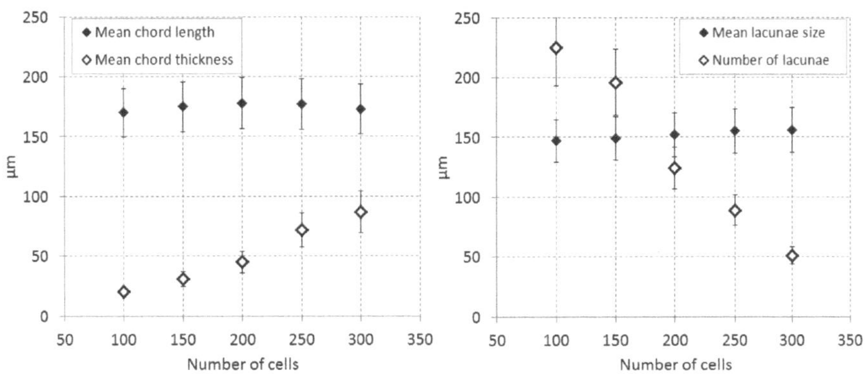

FIGURE 7.8: Percolative and "Swiss cheese" transition of tumor-derived endothelial cells. Topological properties of the pattern as a function of the initial cell density. The error bars show standard deviations over 20 simulations.

from some critical quantities. First, we perform simulations by varying the initial cell number; see Figures 7.7 and 7.8. Below a critical value, $n \leq n_{c1} =$

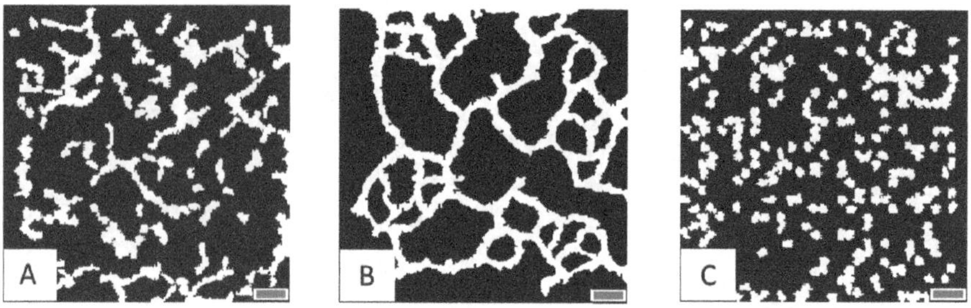

FIGURE 7.9: Increasing VEGF degradation rate (i.e., doubling λ_v in Equation (6.5)): (A) downregulation of network organization at $\approx$ 24 h (12000 MCS), (B) formation of a reduced-in-scale network at $\approx$ 36 h (18000 MCS). (C) disruption of patterning by inhibiting VEGF uptake by TECs (i.e., imposing $B(\mathbf{x}, t, v) = 0$ in Equation (6.5)). All the other model parameters are the same as for the basic simulation in Figure 7.3. The scale bar is 100 μm long.

100, we observe the formation of groups of disconnected structures, while, by increasing n, a single connected network forms. In particular, for a wide range of cell densities, $n \in (100\ ,\ 300)$, the typical length of chords and size of lacunae remain approximately fixed, while the chord thickness grows to accomodate the increasing number of cells, and, consequently, the number of lacunae slightly decreases. Finally, for $n > n_{c2} = 300$, we find a continuous carpet of cells intersparsed by holes. Interestingly, this phenomenon resembles the *percolative and Swiss cheese transitions* observed in the development of physiological vasculatures [154, 360]: therefore such a topological property is of biological relevance also for tumor-derived vascular cells, and may be directly linked to the efficiency of the nascent capillary plexus.

7.5 Testing Anti-Angiogenic Therapies

Having demonstrated the model consistency with experimental observations, we are now in the position to predict the effect of potential anti-angiogenic therapies, as already done in the case of single cell motion. As a quantitative measure characterizing the pattern at the end of the simulations (i.e., at 12000 MCS $\approx$ 24 h), we introduce the total rate of tubule length (in μm): it is defined as the sum of the lengths of all mature cords, i.e., the structures formed by a minimum of three detectable cells and connected through a node with a similar one. In the standard parameter setting it is labeled as L_T and, on average, is $2.423 \times 10^3 \pm 5 \times 10$ μm ($\approx 2.423 \times 10^3$ px, mean of 20 simulations $\pm$ standard deviation), agreeing well with the measurements reported in the

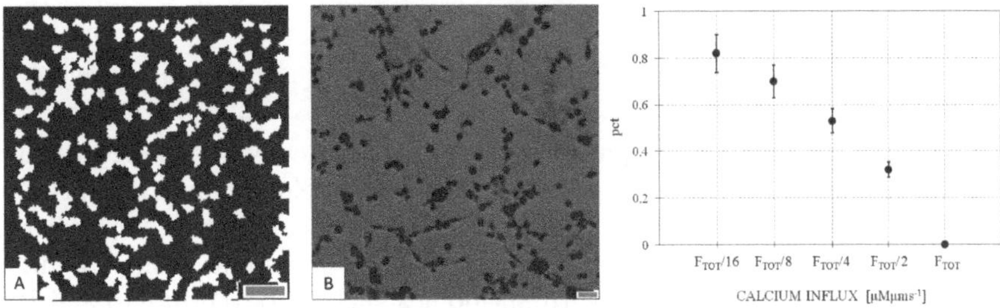

FIGURE 7.10: Complete inhibition of tubule formation in a TEC culture pretreated with saturating amounts of (A) virtual (i.e., $F_{AA} = F_{NO} = 0$ in Equation (6.10)) and (B) experimental (i.e., 1 μM) CAI. Figure (B) courtesy of the Department of Animal and Human Biology, Università degli Studi di Torino. Scale bars 100 μm long. Right panel: *pct* (i.e., percentage of reduction of the total tubule length) as a function of the dose of the virtual CAI, as $F_{TOT} = F_{AA} + F_{NO}$. All the other model parameters are the same as for the basic simulation presented in Figure 7.3. Error bars show standard deviations over 20 simulations.

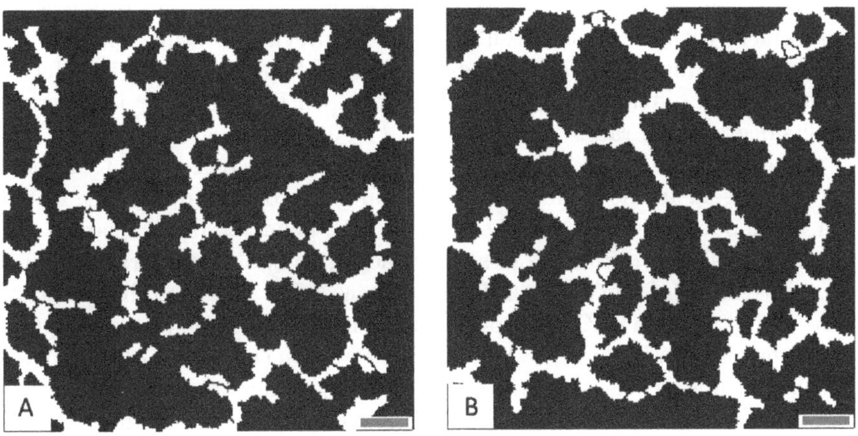

FIGURE 7.11: Partial disruption of TEC tubule formation by inhibiting (A) AA production (by imposing $k_a = v_c = 0$ in Equation (6.7), which is the model counterpart of the activity of AACOCF3 compound) and (B) NO biosynthesis (by imposing $k_n = v_{ca} = 0$ in Equation (6.8), which is the model counterpart of the activity of L-NAME or L-NMMA compounds). All the other model parameters are the same as for the basic simulation in Figure 7.3. The scale bars are 100 μm long.

literature for similar cell densities [133, 282, 360]. L_T is used as the reference value: The effect of the simulated therapies, whose relative total tubule length values are identified with l_T, is in fact quantified by the percentage of reduction of L_T, i.e.,

$$pct = \frac{L_T - l_T}{L_T}.$$

Obviously, all the model parameters can theoretically be changed, but we will only detail biologically reasonable (and therefore most influential and predictive) variations. We first test the actual clinically viable biomedical interventions. Most current drugs target the VEGF system, either by directly attacking VEGF molecules (with antibodies such as Bevacizumab and Ranimizumab or aptamers such as Pegaptanib), or by preventing the activation of its tyrosine kinase receptors (with small inhibitor molecules such as sorafenib, sunitinib, and vatalanib) [366]. We implement such interferences with VEGF activity by increasing its degradation rate (doubling λ_v) and, respectively, by reducing the effectiveness of its uptake (i.e., imposing $B(\mathbf{x}, t, v) = 0$ in Equation (6.5)). In the first case we observe a phenomenon of high interest: at the standard final time (i.e., $\approx$ 24 h) the tubulogenic process is dramatically downregulated, as shown in Figure 7.9(A), confirming the effectiveness of the treatment, as $l_T \approx 0.24 \cdot L_T$ ($pct = 0.76$). However, at much longer times ($\gg 30$ h) a pattern emerges, formed by a number of connected networks characterized by small lacunae and short chords; see Figure 7.9(B). This is somewhat in agreement with theoretical results from both previous continuous [9, 154, 360] and discrete models [261]: the reduced scale of these structures is in fact dictated by the diffusion of the morphogen, which is affected by the increment of its degradation rate.

However, the delay in the formation of such a reduced-in-length network is not present in those published works, and is probably due to the fact that in our model VEGF uptake has a threshold behavior (see Equation (6.5)). In fact, a slower diffusion of the morphogen causes a temporal shift in the full-activation of the calcium-dependent cascades. The overall inhibition of the activity of cell VEGF receptors results instead in a complete disruption of vascular progression, as $l_T \approx 0.08 \cdot L_T$ ($pct = 0.92$); see Figure 7.9(C). Without sequestrating VEGF molecules, in fact, TECs do not completely activate the downstream intracellular cascades and therefore almost remain in their resting conditions, characterized by a typical random movement.

Analogous efficient results are obtained by blocking the overall mitogen-induced calcium entry (with $F_{AA} = F_{NO} = 0$ in Equation (6.10)). As already seen, this is the model counterpart of the function of carboxyamidotriazole (CAI) compound [133, 279]. In particular, a comparison of experimental observations and numerical results is given in Figure 7.10(A-B): in both cases we observe a complete disruption of tubule formation, as the TECs remain almost scattered ($l_T \approx 0.12 \cdot L_T$ and, consequently, $pct = 0.88$). This is due to the fact that a dramatic interference in mitogen-induced calcium machinery results in a clear inhibition of cell adhesive and migratory properties. Moreover, a modulated reduction of calcium influxes confirms the dose-dependent efficiency of CAI, see right panel of the same figure.

By imposing $k_a = v_c = 0$ in Equation (6.7) (respectively, $k_n = v_{ca} = 0$

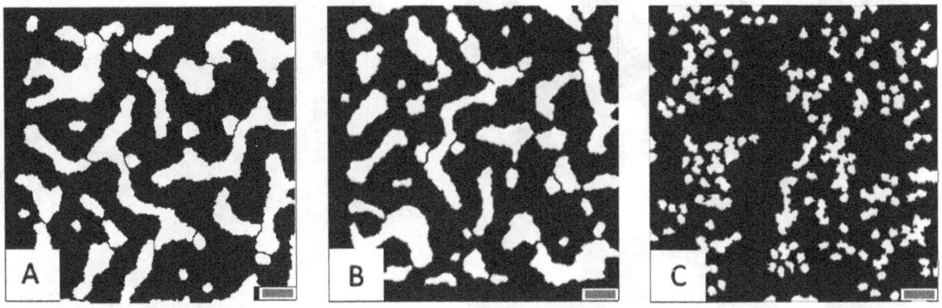

FIGURE 7.12: Partial disruption of TEC tubule formation by (A) inhibiting cytoskeletal reorganization (with a high constant $\lambda_{\eta,C}^{perimeter}$ for each η in Equation (6.4)), (B) interfering with the persistence component of cell motion (by setting $\mu_{\eta}^{pers} = 0$ for all cells in Equation (6.4)) and (C) disrupting intercellular adhesion (with a high spatially homogeneous value of $J_{E,E}^{ext}$). All the other model parameters are the same as for the basic simulation in Figure 7.3. The scale bars are 100 μm long.

in Equation (6.8)), we simulate the exclusion of arachidonic acid (respectively, nitric oxide) biosynthesis, resembling cells pre-treated with widely used PLA2 (respectively, eNOS) inhibitors (AACOCF3, respectively L-NAME or L-NMMA [274, 284]). In both cases, VEGF-mediated intracellular calcium events are not completely abolished and the relative microscopic mechanisms (i.e., enhancement of cell adhesion, motility, and chemical sensitivity) still occur, but with a significant delay and a lower intensity. Consequently, the ultimate pattern morphologies feature an immature network shown in Figure 7.11, where several branches have partially formed, but have not been able to organize in a single structure. In particular, the disruption of AA production leads to $l_T \approx 0.61 \cdot L_T$ (and thus *pct* = 0.39), while the disruption of NO biosynthesis results in $l_T \approx 0.73 \cdot L_T$ (*pct* = 0.27). This difference is caused by the fact that AA partially regulates the recruitment of NO itself, see Equation (6.8).

A potentially more efficient intervention strategy consists in blocking the calcium-dependent cytoskeletal reorganization of TECs: in the model with a high constant $\lambda_{\eta,C}^{perimeter}$ for each η and, experimentally, with phalloidin-like compounds. The resulting capillary morphology, illustrated in Figure 7.12(A), features in fact clumped, stunted and somewhat shorter and thicker sprouts, as l_T is $\approx 0.13 \cdot L_T$ and *pct* = 0.87. In particular, the vascular cords are 3–4 cells wide, with larger intervascular spaces. This phenomenology is consistent with typical vascular hyperplasia [25] and is caused by the fact that TECs are forced to keep their initial round morphology and, consequently, loose the capacity to differentiate and polarize. Consequently, the TECs do not have the persistent migration needed for the formation of a functional network, as they can only form small, disconnected, branches along the gradients of VEGF concentrations (see [192] for detailed comments).

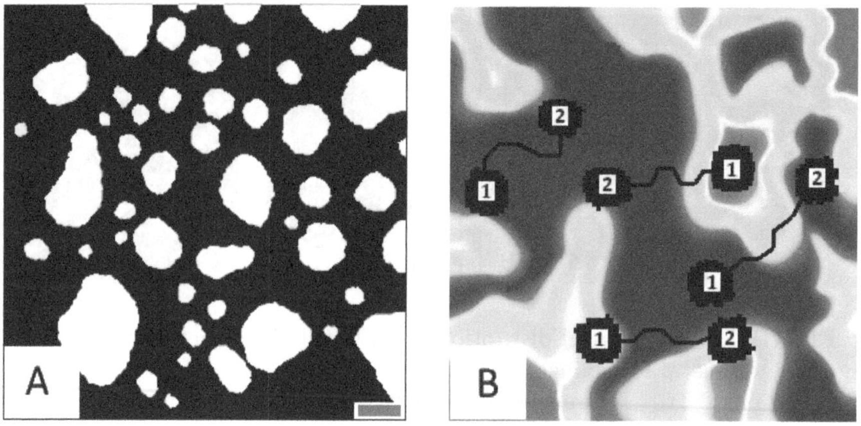

FIGURE 7.13: (See color insert.) Inhibition of network organization by disrupting cell chemotactic response (i.e., imposing $\mu^{ch}_{\eta,C} = 0$ for each η in Equation (6.2). In panel (A), we observe the formation of poorly differentiated vascular islands. As usually, the initial cell number is $n = 150$ while all the other model parameters are the same of the basic simulation in Figure 7.3. Scale bar is 100 μm long. Panel (B) reproduces the magnification of four representative single-TEC motion trajectories obtained from the time-lapse simulation. The initial (i.e., at 0 h) position of individuals is labeled with 1, while the final (i.e., at 12 h) with 2. Each segment tracks a displacement of 2 h. In the background the concentration field of VEGF is pseudocolor-scaled so that the maximum level is red and minimum level is blue, as in Figure 7.6. The motion of cells is uncorrelated from the directions of high VEGF concentration, as they do not experience the chemical force. Moreover, in the absence of the chemotactic stimulus, the individuals do not elongate, remaining in a stationary un-polarized morphology.

Analogous results are observed with the direct exclusion of the cell directional persistence ($\mu^{pers}_{\eta} = 0$ for all cells in Equation (6.4)), which is counterintuitive and currently hard to experimentally handle. The formed sprouts are again immature and swollen. They are typically thick (3–4 cells wide) and characterized by large intervascular spaces and by $l_T \approx 0.27 \cdot L_T$ and $pct = 0.73$, as depicted in Figure 7.12(B).

A further trivial, but possible, anti-angiogenic strategy deals with the interference in the cell adhesive strengths. In particular, a high spatially homogeneous value of $J^{ext}_{E,E}$ is the computational counterpart of the function of anti-VE-cadherin antibodies, which block VE-cadherin receptors preventing the formation of contact junctions between cells. This intervention produces aggregates of dispersed TECs, which are located in the zones of higher VEGF concentrations ($l_T \approx 0.06 \cdot L_T$, $pct = 0.94$), as reproduced in Figure 7.12(C).

Another strategy of high interest is the disruption of the chemotactic strength (by setting $\mu^{ch}_{\eta,C} = 0$ in Equation (6.2)). As represented in Figure 7.13(A), the resulting simulation reproduces a clear disruption of network assembly, with the formation of poorly structured vascular islands ($l_T \approx 0.14 \cdot L_T$ and $pct = 0.86$), similar to those experimentally obtained by extinguish VEGF gradients, for example, in the case of "normal" ECs, by adding saturating amounts of exogenous morphogen [260, 261, 360]. This result shows that a stimulation of cell adhesion and motility, without a chemotactic mechanism, does not suffice for the maturation of a capillary-like structure. In particular, an accurate analysis of cell tracks shows that, although maintaining a certain degree of directional persistence (due to the term $H_{persistence}$ in (6.1)), cell movement is completely uncorrelated from the directions of chemical gradients (see Figure 7.13(B)). Migrating individuals feature also an unpolarized morphology during the entire patterning, as reproduced in the same representative image: without experiencing the chemical stimulus, their membranes are in fact not able to protrude in a preferred direction, but isotropically fluctuate in the environment. This result shows that a stimulation of cell adhesion and motility, without a chemotactic mechanism, does not suffice for the maturation of a capillary-like structure, as also demonstrated experimentally by extinguish morphogen gradients in [360], and theoretically in other Potts models [260, 261].

In order to review and compare the different results, in Figures 7.14 7.15, the percentage of reduction of the total tubule length (i.e., *pct*) has been compared in the different cases using a one-sided equal-variance t-test with a Bonferroni correction as a scoring method [333]. The level of significance has been taken $P < 0.005$. Indeed, it is satisfying to see that the model has realistically reproduced the effects of currently available therapies (focusing on the interferences with VEGF activity or with calcium machinery) and has been also able to suggeste novel and interesting interventions. In particular, we have made some predictions about the possible success of therapies blocking the mechanisms of either the cytoskeletal remodeling or the increment in cellular adhesion. Development of interventions inhibiting cell chemotaxis and persistent movement might be also optimal strategies. All the proposed solutions have the potential to dramatically reduce the angiogenic phenotype of tumor-derived endothelial cells in unexpected multiple ways and have emerged by opportune and biologically reasonable variations in model parameters: even though it is possible to provide a *post hoc* explanation the effectiveness of each of them, it is unlikely that they would have been manually discovered. Obviously, more experiments, focused on the relevance of our findings, are then needed to shed light and validate our conclusions. Eventually, it would be also interesting the relevance of our results *in vivo*. However, in this case, it would be very difficult to have a complete control of all the involved biological mechanisms, and the relative investigations would imply obvious complications due to the recruitment of cell types other than endothelial cells and to potential unexpected side effects [57, 65].

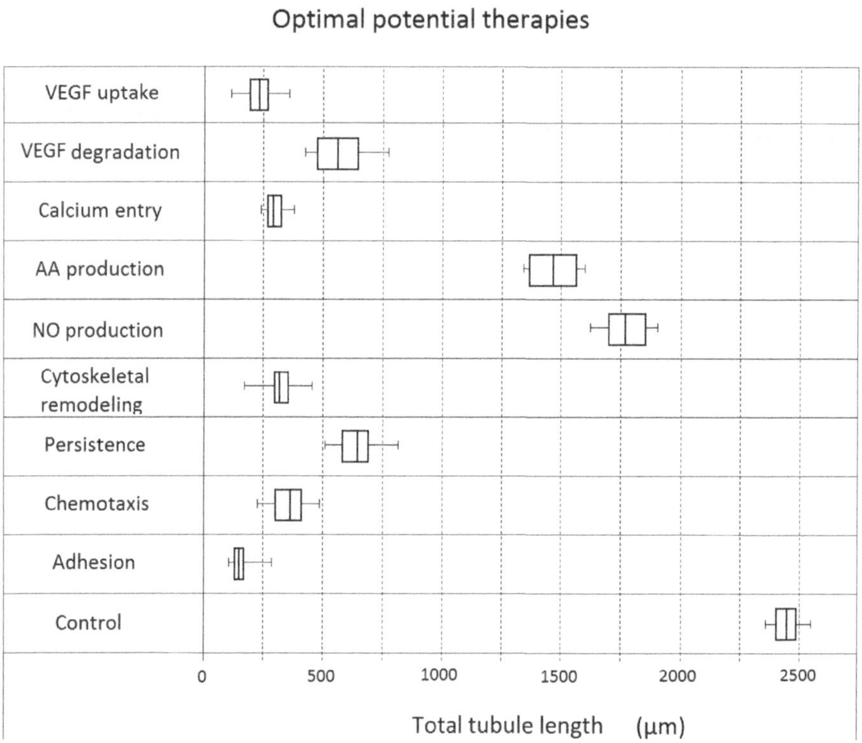

FIGURE 7.14: Optimal potential anti-angiogenic therapies found by model disruptions (control solution represents the no treatment case, with no changes in the parameter, and corresponds to simulation of Figure 7.3). Box plot of the total tubule length (in μm), with 20 simulations for each simulated strategy.

Optimal potential therapies

	Model parameter	Existing compound	P-value of pct	Efficiency
VEGF uptake	β_v	Sorafenib, Sunitinib, Vatalanib	0.00082	+++
VEGF degradation	λ_v		0.00234	++
Calcium entry	F_{AA}, F_{NO}	CAI	0.00085	+++
AA production	k_a, V_c	AACOCF3	0.00425	+
NO production	k_n, V_{ca}	L-NAME, L-NMMA	0.00473	+
Cytoskeletal remodeling	$\mu^{perimeter}{}_{\eta,C}$	Phalloidin	0.00092	+++
Persistence	$\mu^{pers}{}_{\eta}$		0.00139	++
Chemotaxis	$\mu^{ch}{}_{\eta,C}$		0.00097	+++
Adhesion	$J^{ext}{}_{E,E}$	Anti-VE-cadherin antibodies	0.00069	+++

FIGURE 7.15: Optimal potential anti-angiogenic therapies found by model disruptions. Table shows the proposed therapies with their relative modified parameters and, whenever currently available, the existing experimental drugs. Forth column of the table gives a statitistical test for each intervention: the percentage of reduction of the total tubule length (*pct*) has been evaluated with a one-sided equal-variance t-test with a Bonferroni correction, which tests the hypothesis $\overline{pct} > 0$ against the alternative $\overline{pct} = 0$, where $\overline{pct}$ is the mean over 20 simulations of pct, with a level of significance $P < 0.005$ [333]. In particular, for $P < 0.001$ we have an optimal solution (labeled with +++ in the table), for $0.001 < P < 0.003$ we have a suboptimal solution, ++, while for $0.003 < P < 0.005$ we have a bad solution, +.

Chapter 8

Different Morphologies of Tumor Invasion Fronts

8.1 Biological Introduction

Solid tumors are thought to arise from small nodes of cells (which are typically somatic stem cells, as recently suggest in [240]) that have undergone genetic mutations and/or epigenetic alterations. They are able to escape from DNA repair mechanisms and to cause abnormal growth regulatory mechanisms [7, 297], surviving and evolving even under extreme conditions, e.g., hypoxia and acidosis [155]. Such primary malignant colonies go through a relatively simple, avascular stage of growth, with nutrient and growth factor supply by diffusion from the local microenvironment [7, 275, 401]. However, a further search of available quantities of critical substrates results in a subsequent aggressive phase, with the invasion of the surrounding tissue [7, 47]. In particular, as reproduced in Figure 8.1, a part of the malignant mass remains densely packed, while a number of isolated cells detach and begin to invade the neighboring spaces. These individuals are less adhesive, highly mobile and metabolically active, due to the fact that they experience a high level of chemical factors, and are able to secrete an enhanced quantity of matrix degrading enzymes (MDEs) [193, 379]. In particular, the production of proteolytic enzymes, such as matrix metalloproteinases (MMPs), is essential during the invasive phase: the dissolution of the ECM provides in fact both a space into which aggressive cells can move and a gradient which can be used by the cells themselves to direct their movement (i.e., haptotaxis); see [74, 286, 379]. The scattered individuals, evading destruction by the immune system, may subsequently enter the host bloodstream or lymphatics, extravasate at a distant site, and establish secondary colonies with devastating consequences for the wellbeing of the patient, as the likelihood of success of therapeutic interventions strongly decreases [162, 324].

The invasiveness of a primary cancer mass can be therefore a quantifiable function of the altered biophysical characteristic of malignant cells, such as their enhanced motility and metabolism and the downregulation in their cell–cell adhesion, as well as of their aggressive interactions with the local microenvironment, such as the enhanced consumption of available nutrients and degradation of ECM proteins. A deep understanding of the regulatory

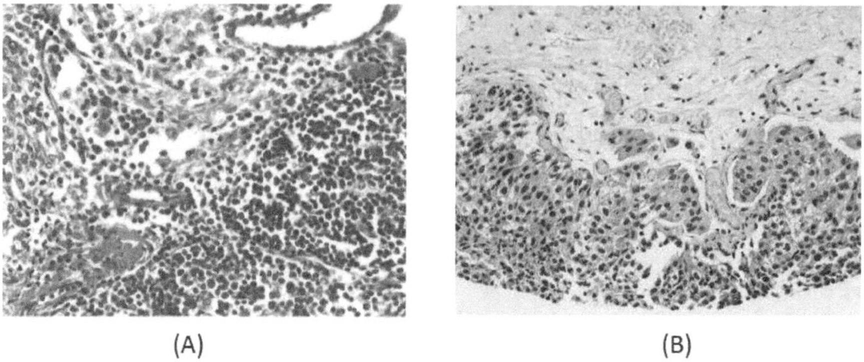

FIGURE 8.1: A high-power photomicrographs showing microinvasive solid tumors. (A) Single aggressive malignant cells of a type B2 thymoma detaching from the main malignant mass, located in the anterior mediastinum and myasthenia gravis. (B) pT1 urothelial carcinoma invading into the lamina propria. In both cases, it is possible to notice single isolated cells or irregular small clusters of cells invading the surrounding tissue. Image (A) courtesy of the Institute for Cancer Research and Treatment of Candiolo (I.R.C.C.), image (B) taken from www.visualhistology.com.

dynamics of the invasive phase of tumor development, in particular of the morphological instability resulting in the shed of aggressive single cells, represents therefore a fundamental issue in cancer research, as also commented in [94].

In the last decades some help has been provided by computational approaches, able to realistically reproduce selected features of the biological system and to test potential anti-tumoral strategies, as pointed out by the books [57, 240] and by the excellent reviews [15, 30, 59, 179, 239, 241, 321, 317, 392]. The variety of numerical methods now available for performing simulations of nonlinear solid tumor growth and invasion has been typically based on either continuous or discrete approaches. Among continuous techniques, single-phase models deal with tumor front invasion and surface growth, as reviewed in [48, 74, 239, 392], while multiphase mixtures present tumor interface instabilities, see [15, 245, 318]. However, all these approaches overlook the behavior and the mutual interactions of single cells, which, as seen, are fundamental in determining the invasiveness of cancers and the subsequent metastatization. Discrete techniques, which include cellular automata and agent-based models [5, 12, 202, 235, 262, 326, 395], are instead able to preserve the identity of each simulated individual, to capture more naturally their biophysical properties, such as shape change, adhesion and intrinsic motility, and to handle local dynamics. On the contrary, as already commented in the Preface of the book, purely discrete models translate complex microscopic processes into simple phenomenological rules, are difficult to study analytically and the associated computational cost rapidly increases with the number of cells modeled, which makes it difficult to simulate lesions greater than one millimeter.

Following the recent literature [13, 158, 159], we here use the multi-level modeling framework introduced in Chapter 4 to reproduce and quantitatively analyze the first phase of tumor invasion. In particular, the main focus is on the emergence of different morphologies at the front of tumor invasion, which result both from cell-based processes (such as cell elasticity, adhesive properties and motility) and from subcellular molecular dynamics (such as growth factor internalization, ECM protein digestion and MMP secretion). The resulting model is able to characterize the morphology of the invasive tumor and to quantify its malignancy in term of invasive depth in several different conditions. Furthermore, the proposed approach has the potential to make clear the relevance of the various mechanisms involved and to suggest possible intervention strategies able to reduce the aggressiveness of the lesion by controlling its morphological stability, i.e., by enforcing its compactness so that it could in principle be more easily resectable.

8.2 Mathematical Model

The malignant cells are represented at a mesoscopic level by compartmentalized objects, that locally interact with each other and with the microenvironment through their membranes. The molecular biology is instead incorporated with a macroscopic description of the evolution of nutrients, ECM proteins and tumor matrix metalloproteinasis (continuous CPM objects), see Figure 8.2 for a schematic representation. As a key feature of our discrete-continuum composite approach, the different scales affects each other, as the distribution of nutrients and ECM proteins in the extracellular environment influence cells' properties and phenomenology (with carefully-calibrated constitutive relations).

8.2.1 Cell-Level Model

The simulation domain is a planar square domain $\Omega \subset \mathbb{R}^2$. The cancer cells are bicompartmental individuals η of type $\theta = T$, composed of a central cell nucleus ($\tau = N$) and the surrounding cytosol, $\tau = C$, as depicted in Figure 4.3. The tumor mass lives on an extracellular matrix, $\tau = M$, which is assumed to be isotropically distributed throughout the simulation domain, forming no large-scale structures. It reproduces in fact the mixture of soluble components (among others, long carbohydrate polymers, and nonproteoglycan polysaccharides), which, together with the water solvent, compose the so-called interstitial medium. For any cell η, we define the state vector of each compartment:

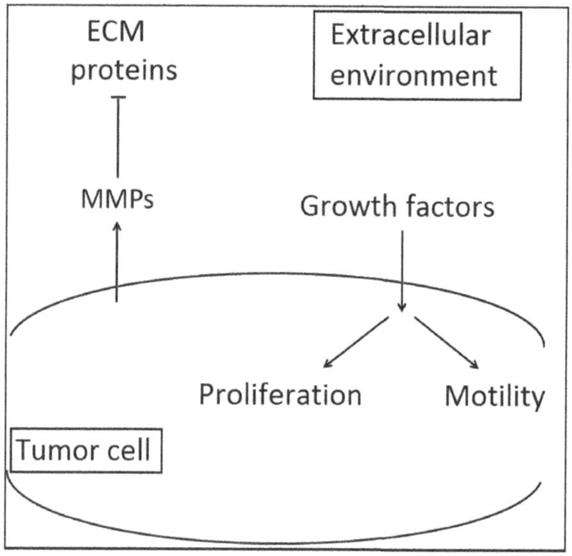

FIGURE 8.2: Schematic diagram of the key processes included in the mathematical model. The extracellular environment provides to tumor cells soluble growth factors, whose uptake enhances cell intrinsic motility and proliferation. Tumor cells, in turn, secrete matrix metalloproteinases (MMPs), which degrade ECM proteins.

- If Σ_σ is such that $\tau(\Sigma_\sigma) \in \{N, C\}$, $\mathbf{s}_{\eta,\Sigma_\sigma}(\mathbf{x}, t) = (n(\mathbf{x}, t)) \in \mathbb{R}_+$, where $n(\mathbf{x}, t)$ is the local intracellular concentration of growth factors, expressed in units of μM.

The system *Hamiltonian* is given by:

$$H(t) = H_{shape}(t) + H_{adhesion}(t) + H_{haptotaxis}(t). \tag{8.1}$$

H_{shape} includes the constraints that regulate cell volume and surface (which, in two dimensions, correspond to the cell area and perimeter), written as nondimensional relative deformations in the quadratic form of Equation (4.6). The cell target dimensions correspond to their initial measures and are given in the Appendix. Assuming that the cells do not significantly grow during invasion, the fluctuations of their volumes are kept negligible with high constant values of $\lambda^{surface}_{\eta,\Sigma_\sigma:\tau(\Sigma_\sigma)\in\{N,C\}}$. Moreover, because tumor cells are typically deformable, as they are able to significantly remodel to invade their surroundings more efficiently, for any η and for Σ_σ such that $\tau(\Sigma_\sigma) = C$, $\lambda^{perimeter}_{\eta,\Sigma_\sigma} = \lambda^{perimeter}_{\eta,C}$ needs to be sufficiently low. The relative rigidity of the nucleus (with respect to the cytoplasm) is instead modeled, as usual, by a relatively higher value of $\lambda^{perimeter}_{\eta,\Sigma_\sigma} = \lambda^{perimeter}_{\eta,N}$ when $\tau(\Sigma_\sigma) = N$.

$H_{adhesion}$ is straightforwardly differentiated into the internal and external contact energy contributions, cf. Equations (4.1), (4.2), and (4.3). In particular, the high negative value set for $J^{int}_{N,C}$ prevents cells from fragmenting (we refer to [298, 352] for comments). $J^{ext}_{T,T}$ represents instead the adhesive strength between the membranes of two nearby cells, a measure of the quantity of active and exposed cadherins . Finally, $J^{ext}_{T,M}$ evaluates the heterophilic adhesive bonds between the integrins on the cell surface and suitable ligands (such as laminin and fibronectin) in the extracellular matrix . By setting constant and homogeneous values for both $J^{ext}_{T,T}$ and $J^{ext}_{T,M}$, we assume a uniform distribution of adhesion molecules on the cell surface and a homogenous density of ligands in the external microenvironment. In particular, since malignant cells have been demonstrated to have reduced cell–cell adhesiveness but increased cell–ECM adhesiveness, due to a change in the relative expression of the corresponding adhesive molecules [193], we set $J^{ext}_{T,T} \approx 2\ J^{ext}_{T,M}$.

$H_{haptotaxis}$ reproduces the effect of cell preferential movement in the direction of zones with higher concentration of extracellular matrix proteins (such as fibrin, vitronectin, and some of the collagen family) and is implemented with a local linear-type relation of the form (1.9):

$$\Delta H_{haptotaxis} = \mu^{hapt}\left[p_t(\mathbf{x}_{target}, t) - p_t(\mathbf{x}_{source}, t)\right], \tag{8.2}$$

where $\mathbf{x}_{source}$ and $\mathbf{x}_{target}$ are, respectively, the source and the final lattice site randomly selected during a trial update in a MCS, and

$$p_t(\mathbf{x}, t) = p(\mathbf{x}, t) + \sum_{\mathbf{x}' \in \Omega'_{\mathbf{x}}} p(\mathbf{x}', t),$$

where $\mathbf{x} \in \{\mathbf{x}_{source}, \mathbf{x}_{target}\}$, evaluates the local level of ECM proteins sensed by the moving cell membrane site, as $p(\mathbf{x}, t)$ is their amount at site $\mathbf{x}$ (defined in Equation (8.6)). Finally, $\mu^{hapt} \in \mathbb{R}_+$ represents the local strength of haptotaxis, which is assumed constant for all malignant individuals.

Given the *Hamiltonian*, the transition probability of a spin flip has the form of Equation (4.24). In particular, as usual in this book, $p(T_{\eta,\Sigma_\sigma}(t)) = \tanh(T_{\eta,\Sigma_\sigma}(t))$. For any cell η, T_{η,Σ_σ}, with $\tau(\Sigma_\sigma) = N$, is a low constant positive value, $T_{\eta,N}$ that mimics the passive motion of the nucleus of η. If $\tau(\Sigma_\sigma) = C$, $T_{\eta,\Sigma_\sigma} = T_{\eta,C}$ represents instead, as seen, the intrinsic motility of the overall cell, which is a measure of the agitation of its cytoskeleton (i.e., by the random actin polymerization/depolarization dynamics) and is in this case assumed to be enhanced by the intracellular level of growth factors (i.e., motogen agents) in a dose-dependent manner. Adopting again the notation of Equations (4.9) and (4.10), for any malignant individual η and for $\tau(\Sigma_\sigma) = C$, we have therefore that:

$$\mathbf{s}^{T}_{\eta,\Sigma_\sigma}(\mathbf{x}, t) = \mathbf{s}^{T,A}_{\eta,\sigma}(\mathbf{x}, t) = (n(\mathbf{x}, t)); \tag{8.3}$$

and

$$T_{\eta,\Sigma_\sigma}(t) = T_{\eta,\Sigma_\sigma}(\mathbf{s}^{T,A}_{\eta,\Sigma_\sigma}(\mathbf{x}, t)) = f_T(\mathbf{s}^{T,A}_{\eta,\Sigma_\sigma}(\mathbf{x}, t)) = T_0\left[\frac{n_\eta(t)}{N_0 + h\,(n_\eta(t) - N_0)}\right], \tag{8.4}$$

where the quantity $n_\eta(t) = \sum_{\mathbf{x}\in\eta} n(\mathbf{x}, t)$ is the total amount of growth factors inside cell η, as $n(\mathbf{x}, t)$ is their local intracellular concentration, defined in Equation (8.5), while $N_0 = \sum_{\mathbf{x}\in\eta} n_0$ is instead their overall basal level. T_0 therefore corresponds to the basal motility of tumor cells, while T_0/h is their asymptotic motility for saturating concentrations of chemicals. The above hypothesis are in agreement with experimental observations providing, through classical wound healing experiments, that high concentrations of growth factors stimulate the migratory capacity of different tumor cell lines (see, for example, [101, 104] for the hepatocyte growth factor, [38, 383] for the vascular endothelial growth factor families, and [375] for the fibroblast growth factor).

8.2.2 Molecular-Level Model

Available growth factors are supplied to the medium, diffuse and decay at a constant rate, and are consumed by tumor cells. Their spatial profile, $n(\mathbf{x}, t)$, therefore satisfies the following equation:

$$\begin{aligned}\frac{\partial n}{\partial t} &= \underbrace{D_n \nabla^2 n}_{diffusion} - \underbrace{\lambda_n n \delta(\tau(\Sigma_{\sigma(\mathbf{x})}), M)}_{decay} + \\ &- \underbrace{\min\{n_{max}, \chi_n n\}(1 - \delta(\tau(\Sigma_{\sigma(\mathbf{x})}), M))}_{uptake} + \underbrace{S}_{production}, \qquad (8.5)\end{aligned}$$

where $\delta(\tau(\Sigma_{\sigma(\mathbf{x})}), M) = 1$ in the extracellular environment and 0 within cells. D_n is the characteristic diffusion coefficient, homogeneous throughout domain Ω, and λ_n is the decay rate in the ECM. The third term in (8.5) models the local uptake by tumor cells, which follows a piecewise-linear approximation of a Michaelis–Menten law. In particular, $\chi_n \gg \lambda_n$, as we assume that the nutrient natural decay is negligible compared to the uptake by tumor cells. S describes the production (or input) of chemical factors at a constant rate ϕ_n per unit of time by a planar source, whose location and extension will be discussed in the next section.

The substrate contains matrix soluble proteins with concentration $p(\mathbf{x}, t)$ (i.e., we neglect their production, assuming a uniform distribution at the beginning of each simulation, see next section for more details), that naturally decay and that are degraded by the metalloproteinases (MMPs) secreted by malignant cells, $m(\mathbf{x}, t)$. The change in the local amount of ECM proteins is therefore described by:

$$\frac{\partial p}{\partial t} = -\underbrace{\lambda_p p}_{decay} - \underbrace{\chi_p p m}_{degradation}, \qquad (8.6)$$

where λ_p and χ_p are, respectively, nonnegative decay and degradation rates, constant within the entire simulated substrate. In particular, we assume $\lambda_p \ll$

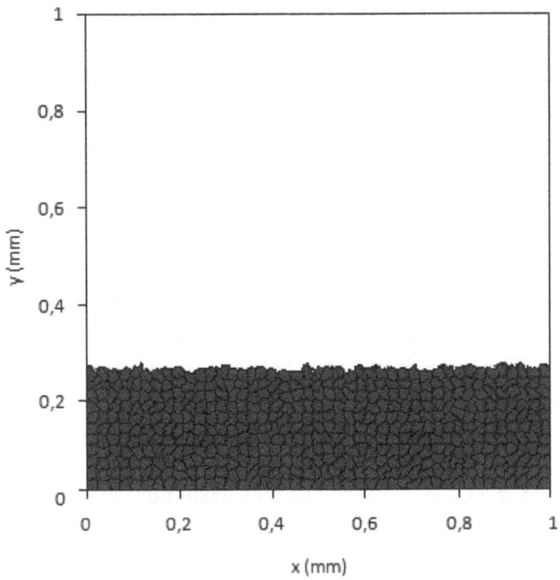

FIGURE 8.3: Initial condition (i.e., at $t = 0$) of the tumor mass for all sets of simulations. The dimensions of cells, as well as all the other model parameters, are given in Table C.8.

χ_p to indicate a much higher dissolution of ECM proteins due to the activity cell proteolytic enzymes than to the physiological decay. The local level of tumor-secreted MMPs is instead governed by:

$$\frac{\partial m}{\partial t} = \underbrace{D_m \nabla^2 m}_{diffusion} - \underbrace{\lambda_m m \delta(\tau(\Sigma_{\sigma(\mathbf{x})}), M)}_{decay} + \underbrace{\pi_m (1 - \delta(\tau(\Sigma_{\sigma(\mathbf{x})}), M))}_{production}, \quad (8.7)$$

where λ_m is the decay rate in the ECM and D_m the diffusion coefficient, whose low value models the fact that proteolysis is strongly localized in the regions close to the cell membrane, in agreement with experimental evidence in [286]. π_m is the constant production rate of degrading enzymes, active at each cell site.

8.3 Simulations in Standard Conditions

The simulation domain Ω is a square lattice of 500×500 sites. The characteristic length of each site is 2 μm, and therefore Ω has a physical size of 1 mm. One MCS is set to correspond to 20 sec: the overall simulations stop after 15000 MCS, so that they reproduce a time-lapse of nearly 4 days. As represented in Figure 8.3, the initial conditions of all simulations consist of a layer of tumor cells, whose initial dimensions correspond to their target measures given in the Appendix. This configuration is consistent with a spatially extended cancer mass which is invading from an epithelial cell lining down its basement membrane into the surrounding stroma.

The boundary conditions for both growth factors, n, and proteolytic enzymes, m, are periodic at the left and the right sides of the domain (i.e., at $x = 0$ and $x = 500$) and zero flux at the bottom (i.e., at $y = 0$). In particular, the lateral periodic conditions are set since the simulated malignant population is a section of a much larger lesion, that could not be reasonably modeled here. The zero fluxes at $y = 0$ are set assuming that this part of the tissue is far enough from the front of the tumor mass. For the same reason, we set a no flux condition for the MMP field at the top of the domain (i.e., at $y = 500$). The upper border of the grid represents instead an extended, planar source for the environmental growth factors, which are supplied by the host stroma throughout the basement membrane: consequently, we set $n(\partial(x, y = 500) \in \partial\Omega, t) = \phi_n$, where $\mathbf{n}_{\partial \mathbf{x}}$ is the outward normal to $\partial \mathbf{x}$.

The parameter describing the biophysical properties of the cells, such as their basal motility, elasticity and adhesive strengths, have been evaluated, consistently with the biological considerations provided in the previous section, through preliminary simulations, that showed the model consistency in a wide range of values. The intracellular basal level of chemical factors, n_0, is sufficiently high to avoid that cells irreversibly enter the apoptotic state. The extracellular environment is initially seeded with a saturating level of proteins, $p_{ext,0}$, while there are no growth factors or secreted MMPs in the extracellular medium.

To quantify the effects of the different model parameters on the tumor phenotype, we concentrate on the final depth of invasion of the mass at the end of the simulations (i.e., at $t = 15000$ MCS ≈ 4 days), given by:

$$d_f = \frac{1}{N} \sum_{i=1}^{N} (d_i - d_0), \tag{8.8}$$

where d_i represents the final distance between a tumor cell and the bottom border of the domain and d_0 is the initial width of malignant mass. The average of $N = 10$ randomly chosen values is used to avoid biases towards accounting for outlier individuals. d_f has a clear clinical relevance, since it quantitatively estimates the severity of the disease and characterizes the presence of metastasis delocalized with respect to the main body of the tumor.

We first simulate the evolution of the tumor with the standard parameter setting. The malignant mass is observed to have an overall movement biased toward the extracellular environment, which is significantly invaded, as $d_f \approx 200$ μm; see Figure 8.4(B). Consistently with the experimental evidences provided in the biological introduction (see also Figure 8.1), the invasive phenotype is largely mediated by the aggressive behavior of the more

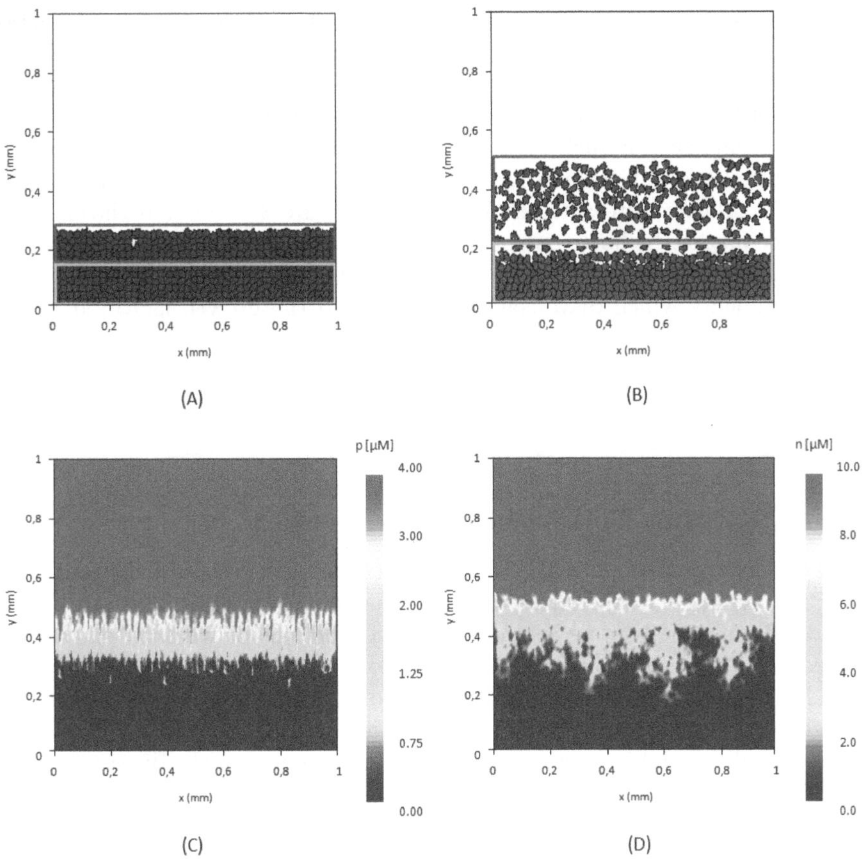

FIGURE 8.4: (See color insert.) Invasion of the tumor mass and relative modification of the microenvironment with the standard parameter setting, given in the Appendix. (A) Regions in which the tumor mass is divided form its front, i.e., external (dark border) and internal (light border). (B) Configuration at t = 15000 MCS ≈ 4 days. (C) ECM protein distribution and (D) growth factor concentration in the host tissue at the end of malignant invasion.

"external" cells (i.e., those placed near the front of the mass, see dark bordered region in Figure 8.4(A)), which loosen contacts, dissociate, move from their original site and start wandering in the close proximity, displaying an evident ability to spread in the surrounding tissue. The increased metabolism of these individuals is due to the availability of growth factors which, together with the interactions with the local microenvironment (both the haptotactic movement and adhesive preference for ECM components) induces a sort of mesenchymal transition able to overcome the physiological regulation of the mechanism of contact-inhibition of cell locomotion (firstly proposed by Abercrombie to describe the influence on cell crawling of the contact interactions between individuals [1]). The rest of the population remains instead relatively compact, with only some cells wade just outside the body of the tumor, see light-bordered region in panels (A) and (B) of Figure 8.4. The migratory and invasive capacity of such "internal" individuals is in fact inhibited by both the limited access to chemicals and by the high cellular density, which enforce cell–cell adhesive interactions. As shown in Figure 8.4(C-D), at the end of the invasive phase, the host tissue is significantly modified, as the matrix proteins have been dramatically degraded (notice in the same figure the paths of protein gradients which drive cell migration) and the growth factors have been significantly consumed.

8.4 Varying Cell Adhesive Properties

As widely commented in the previous sections, the values of the parameters J^{ext}s have a clear biological relevance, as they describe the relative preference of tumor cells to be in contact with other cells or with the extracellular medium. At a molecular level, they are therefore a measure of the expression and the engagement of the different types of cell adhesion molecules, cadherins and integrins, respectively. Indeed, the variation of the J^{ext}s is expected to have a substantial impact on the overall invasiveness of the disease.

We first change the cell–cell adhesive strength $J^{ext}_{T,T}$; see Figure 8.5. At low values of the parameter (i.e., $J^{ext}_{T,T} < J^{ext}_{T,M}$, which means higher cell–cell adhesiveness), the tumor remains compact, with the highly motile external cells that clump along the front of the mass, rather than invading significantly the surrounding tissue; see Figure 8.5(A). This phenomenology is due to the fact that the cell-cell adhesive interactions are too strong to be overcome by the other forces experienced by malignant individuals and therefore stabilize tumor morphology. At large values of $J^{ext}_{T,T}$, the tumor instead expands and invades deeper: the external cells quickly spread away from the rest of the mass creating a dispersed front (fewer of them are in contact with each other within 12 h) and a repulsion occurs also among individuals within the main body of the mass. The subsequent formation of islands of free matrix within the center of the tumor, which are then only partially filled in again, increases the overall bias toward invasion, as reproduced in Figure 8.5(B). From these results, we can confirm that an alteration in cell–cell adhesive properties is an important consideration in determining the severity of the disease. In particular, a biomedical intervention that is able to enforce cell–cell adhesive interactions allows to control the morphology of the lesion which, remaining smooth and densely packed, can be more easily treated surgically. The presented computational outcomes are consistent with a number of *in vitro* experiments: for

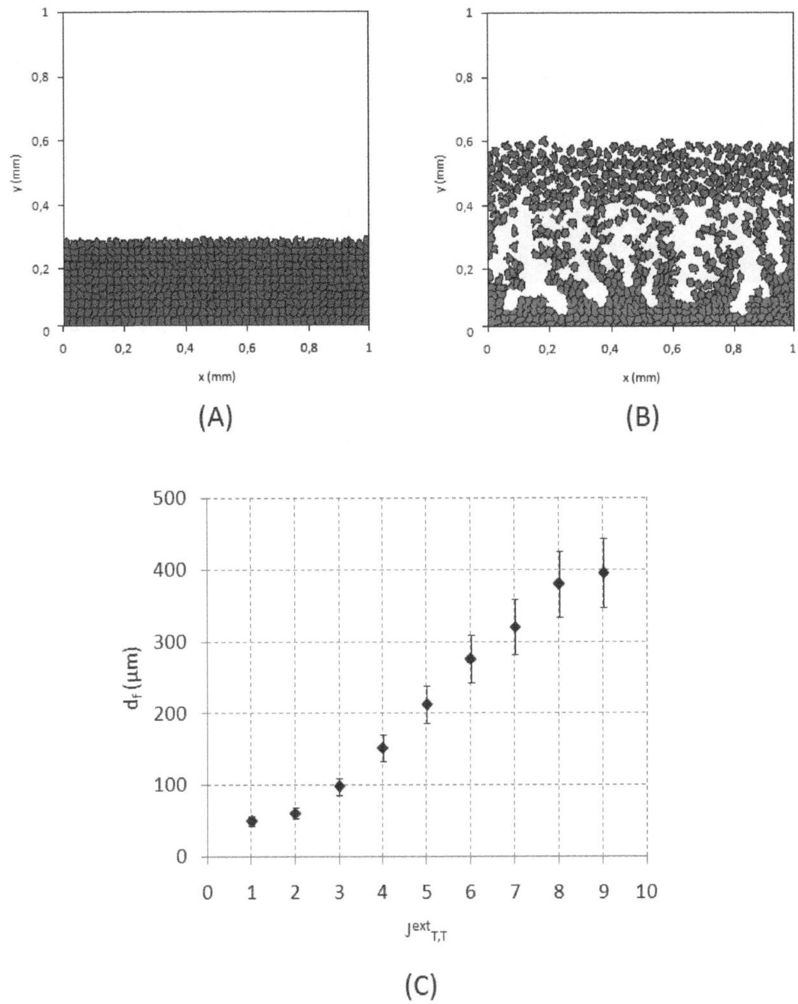

FIGURE 8.5: Invasiveness of the tumor mass in response to alterations of cell–cell adhesive strength, $J^{ext}_{T,T}$ (i.e., all the other model parameters are the same as in the simulation in Figure 8.4). Decrements in cell–cell adhesion lead to increments in the malignancy of the lesion. On the contrary, by enforcing cell–cell adhesion, the mass remains compact and smooth with a low invasive potential. Examples of final configurations for (A) $J^{ext}_{T,T} = 1$ and (B) $J^{ext}_{T,T} = 8$. In this last case, it is possible to see islands of free matrix formed in the main body of the mass, as repulsion occurs also among internal individuals. (C) Final depth of invasion d_f vs. $J^{ext}_{T,T}$. The error bars show standard deviation over 10 realizations.

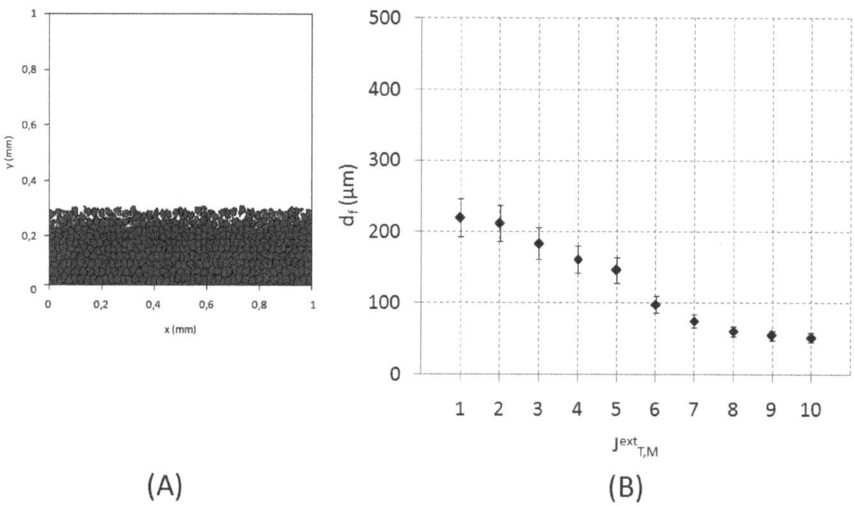

FIGURE 8.6: Invasiveness of the tumor mass in response to alterations of cell–matrix adhesive strength, $J^{ext}_{T,M}$ (i.e., all the other model parameters are the same as the simulation in Figure 8.4). Decrements in cell–matrix adhesive interactions discourage the aggressive behavior of the lesion. In particular, for high $J^{ext}_{T,M} > J^{ext}_{T,T}$, the malignant cells remain closely packed each other. (A) Final configuration for $J^{ext}_{T,M} = 8.5$. (B) Final depth of invasion d_f vs. $J^{ext}_{T,M}$. The error bars show standard deviation over 10 realizations.

example glioma cell lines with a high N-cadherin expression have been seen to invade less significantly different types of collagenous matrices [182].

We next vary $J^{ext}_{T,M}$. The decrement of $J^{ext}_{T,M}$ has a relatively subtle effect on the malignancy of the tumor, as its invasive distance does not dramatically change (see Figure 8.6(B)). This is due to the fact that tumor cells prefer to adhere to extracellular matrix elements already with the standard parameter setting, (i.e., $J^{ext}_{T,M} < J^{ext}_{T,T}$, see Appendix C). At high values of $J^{ext}_{T,M}$ (i.e., $J^{ext}_{T,M} > J^{ext}_{T,T}$, which means a lower cell–matrix adhesiveness) the invasion is instead strongly discouraged, as also the cells at the front remain close to the body of the mass, with occasional individuals shed in the extracellular environment (see Figure 8.6(A)). The cancerous cells are in fact simulated to have an underexpression of integrin molecules: the haptotactic force alone is unable to balance the cell–cell adhesive interactions and therefore to cause the scatter of individuals. These considerations suggest that the use of drugs that interfere with cell–matrix adhesive interactions, either by altering the amount of ligand in the matrix or by inhibiting integrin molecules (which, as seen, are overexpressed in tumors), has the potential to limit malignant invasion, and eventually, to render the cancer more partial to resection.

8.5 Varying Cell Elasticity

We now turn to address how the invasiveness of the tumor changes as a function of $\lambda_{\eta,C}^{perimeter}$, that describes the degree of cell membrane elasticity (see Figure 8.7). Small values of $\lambda_{\eta,C}^{perimeter}$ allow the cells to reorganize their shape more substantially and to traverse the extracellular matrix more readily. The consequent enhancement of invasiveness is due to a distribution of the external individuals, which, able to remodel their shape to adapt to the directions of ECM protein gradients, can venture further from the main body of the mass, as reproduced in Figure 8.7(A). At high values of $\lambda_{\eta,C}^{perimeter}$, a decrement of the invasive distance instead occurs, as the malignant cells cannot reorganize and expand efficiently in their environment, even if they have an increased motility and the preference to heterophilic (i.e., cell–matrix) contacts; see Figure 8.7(B). Indeed, we can conclude that the ability of tumor cells to undergo continuous and dramatic changes in their morphology during motion has a big impact on the aggressiveness of the disease, as also provided in [332]. A therapeutic approach that targets the dynamics of polarization/depolarization of the cytoskeletal of cancer cells, as the use of phalloidin-like compounds, may be therefore potentially effective.

8.6 Altering Cell-Substrate Interactions

We finally analyze how the invasiveness of the tumor is influenced by the haptotactic capacity related to the coefficient μ^{hapt}; see Figure 8.8(A). An increased sensitivity of cells to ECM protein gradients prompts in fact the tumor front to more significantly move in the extracellular environment, eventually promoting invasion. On the contrary, low values of μ^{hapt} results in an isotropic movement of malignant individuals, which are therefore unable to maintain directional persistence and to deeply penetrate in the host tissue.

The inhibition of cell haptotactic sensitivity can therefore represent a potential efficient intervention strategy. In particular, it can be reproduced *in vitro* by adding saturating amounts of matrix components, that extinguish substrate inhomogeneities. In this regard, experimental evidence have demonstrated that, consistently with our results, cells cultured in matrices with high collagen densities show a dramatic downregulation of their migratory capacity, driven by the fact that their integrin receptors are completely engaged without maximal cell spreading and movement [123, 156].

An analogous behavior is seen by varying the cell proteolytic activity, measured by the parameter χ_p; see Figure 8.8(B). In particular, low values of χ_p translate into a partial degradation of ECM proteins; see Figure 8.8(C), whose

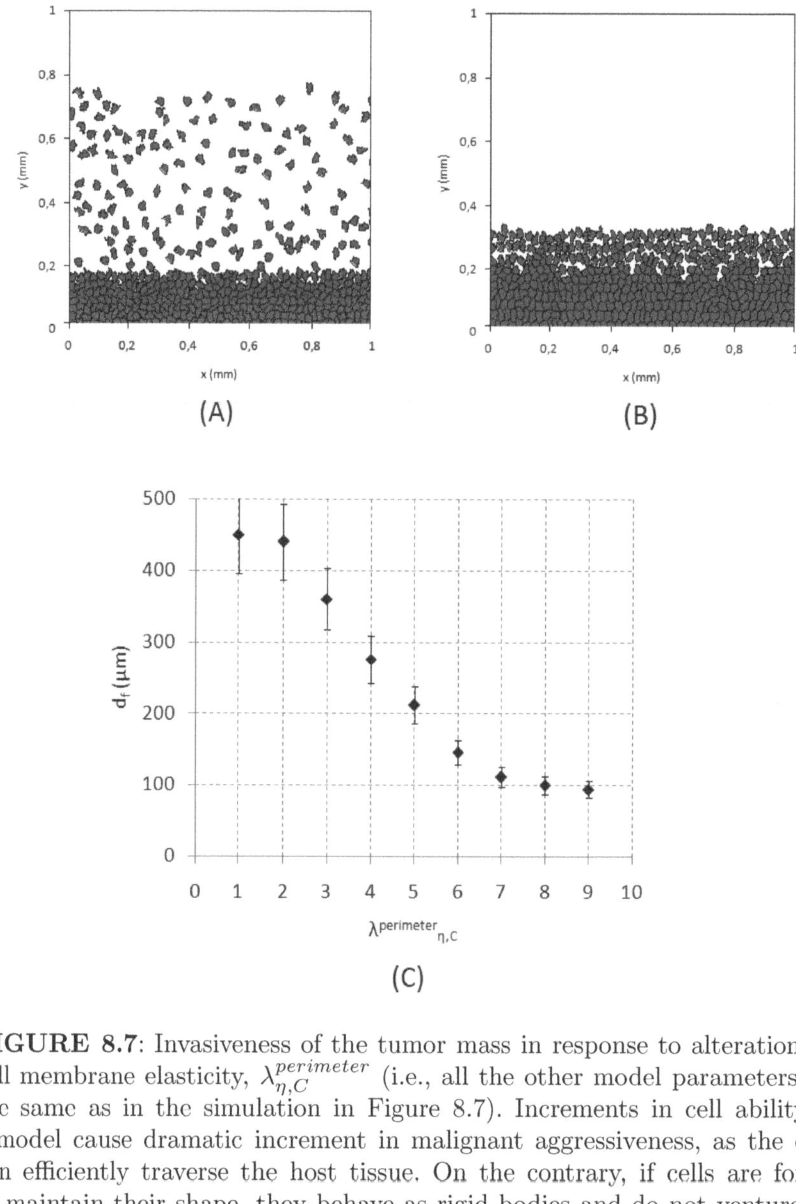

FIGURE 8.7: Invasiveness of the tumor mass in response to alterations in cell membrane elasticity, $\lambda_{\eta,C}^{perimeter}$ (i.e., all the other model parameters are the same as in the simulation in Figure 8.7). Increments in cell ability to remodel cause dramatic increment in malignant aggressiveness, as the cells can efficiently traverse the host tissue. On the contrary, if cells are forced to maintain their shape, they behave as rigid bodies and do not venture far from the front of the main mass. Examples of final configurations for (A) $\lambda_{\eta,C}^{perimeter} = 1$ and (B) $\lambda_{\eta,C}^{perimeter} = 8$. (C) Final depth of invasion d_f vs. $\lambda_{\eta,C}^{perimeter}$. The error bars show standard deviation over 10 realizations.

pattern remains almost homogeneous and significantly inhibits the haptotactic component of the migration of cancer cells (that is fundamental for tumor

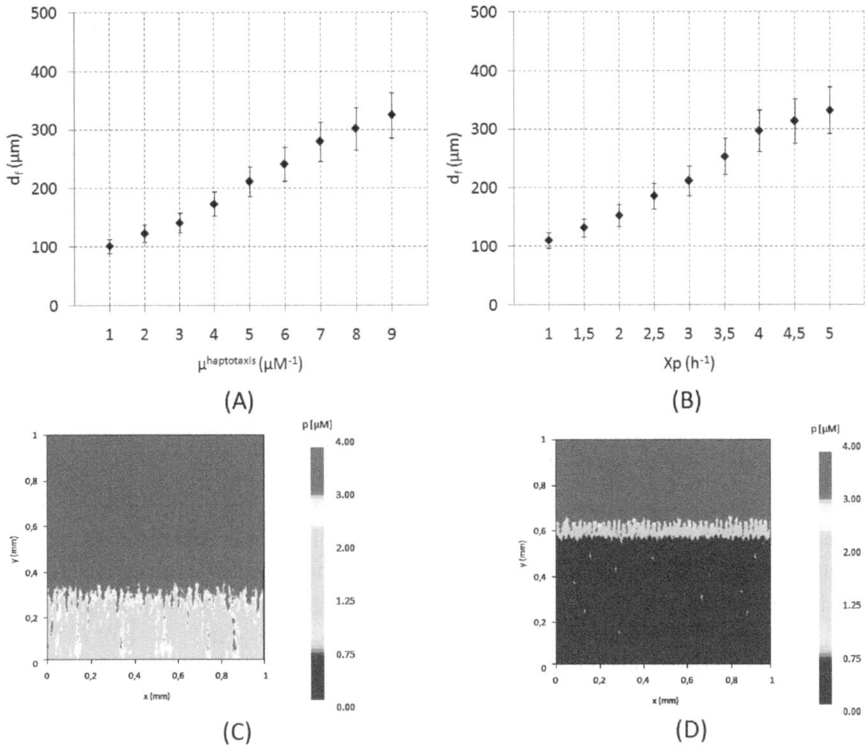

FIGURE 8.8: (See color insert.) Invasiveness of the tumor mass in response to alterations in the interactions between malignant cells and ECM substrate. (A) Final depth of invasion d_f vs haptotactic coefficient μ^{hapt} (all the other model parameters are the same as in the simulation in Figure 8.4), error bars show standard deviation over 10 realizations. Inhibitions of cell sensitivity translate in decrements of the aggressiveness of the lesion. (B) Final depth of invasion d_f vs. degradation rate of ECM proteins χ_p (all the other model parameters are the same as in the simulation in Figure 8.4): The error bars show standard deviation over 10 realizations. A downregulation of the activity of tumor MMPs results in a more homogeneous pattern of protein distribution: therefore malignant individuals have not a directional path to follow during their motion. Final pattern of ECM protein concentration for (C) $\chi_p = 1$ and (D) $\chi_p = 5$. In the last images, we do not explicitly represent the cell configuration: however the invasive depth of the shed malignant individuals is easily recognized by looking the interface between the zones with partially degraded matrix components and those with the initial amount of substrate.

aggressiveness). High values of χ_p result instead in a dramatic digestion of the matrix substrate and in the consequent formation of steep gradients, as reproduced in Figure 8.8(D), which enhance the directional movement of malignant individuals, eventually causing an unstable and dispersed morphology of the tumor. The scattered cells at the front of the mass in fact deeply penetrate into the host tissue to reach regions with sufficient amounts of ECM proteins.

These observations agree with the correlation experimentally found between the malignancy and the metastatization ability of most tumors and the activity of the secreted matrix metalloproteinases, see for example [207]. In this regard, it is useful to underline that the inhibition of MMP production alone has been demonstrated to be insufficient to decrease the overall malignancy of the disease, see [327].

8.7 Effect of Cell Proliferation

So far, we have correlated the invasive phenotype of a solid tumor with the biophysical properties of mutated cells, and with their interactions with the local microenvironment. However, malignant individuals are seen to have also a higher proliferation rate than their unmutated counterparts. In particular, as suggested by a number of experimental works as early as [369], the time between cell divisions has a stochastic distribution, which depends on both the internal state of each cell and on the time from its last mitotic process (i.e., except for extremely rapidly dividing cells, those who recently divided will be still growing and the likelihood of their reentering the S phase is extremely small, see [7]).

For each cell η, we therefore define its present probability P_η to undergo mitosis with the following functional form, which resembles that used in a similar approach [395]:

$$P_\eta(t) = \begin{cases} 0 & \text{if} \quad (t - t_\eta) \leq t_0; \\ \left[\dfrac{n_\eta(t) - N_0}{n_\eta(t)}\right]^+ \dfrac{(t - t_\eta)^2}{(t - t_\eta)^2 + 1} & \text{if} \quad (t - t_\eta) > t_0, \end{cases} \tag{8.9}$$

where $n_\eta(t)$ and N_0 are, respectively, the present and initial intracellular amount of growth factors in cell η and $[\,\cdot\,]^+$ is the positive part of the function. The last time that cell η underwent mitosis is denoted by t_η while the dormant period during which the cells are prohibited to proliferate (i.e., it corresponds to the G1 phase, during which the cells are metabolically active and grow) is denoted by t_0 and is set equal to 1300 MCS (i.e., $\approx$ 7 h). Since the mitotic rate greatly varies from tissue to tissue, and between different types of malignancies, this choice represents a compromise between very slow growing and very aggressive tumor types. Relation (8.9) takes into account that, once the time since last division exceeds t_0, the probability of another mitosis slowly increases, and approaches 1 as it becomes very long. Moreover, a cell needs

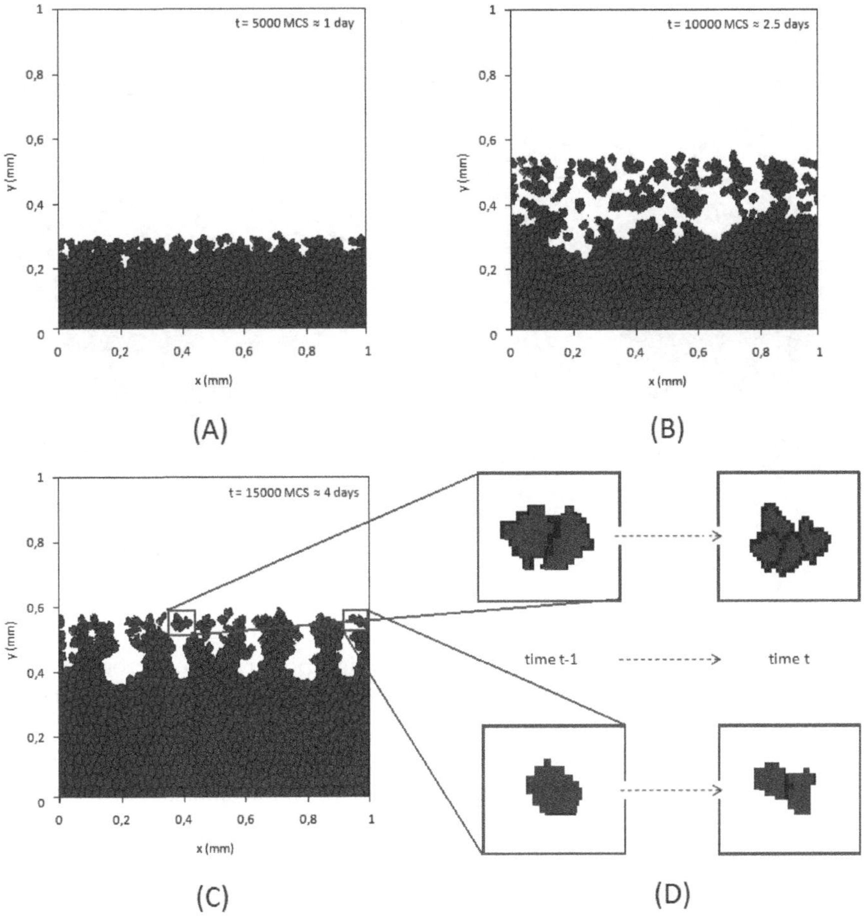

FIGURE 8.9: Tumor development with the inclusion of cell proliferation. External cells quickly undergo mitosis, constituting a front of little islands, (B), which, due to further cell proliferations, come in contact with the main tumor mass by short and thick (4–5 cell-wide) "fingers", (C). The high cellular density, in turn, enforces cell–cell adhesive interactions, which balance the effect of the haptotaxis and the increased cell motility, and do not permit the breakage of the formed fingers. (D) Zoom view of cell divisions.

a sufficient quantity of growth factors to proliferate, as widely demonstrated in literature with mitogenic assays (refer for example to [272, 375, 423]) and explained in details in [7]. Procedurally, the mechanism of cell division is implemented by dividing the proliferating individual into two identical daughter cells, with both the nucleus (which is randomly located) and the cytosolic region halved with respect to their parent; see Figure 8.9. However, due to the shape constraints in Equation (8.1), they will gradually "maturate" into

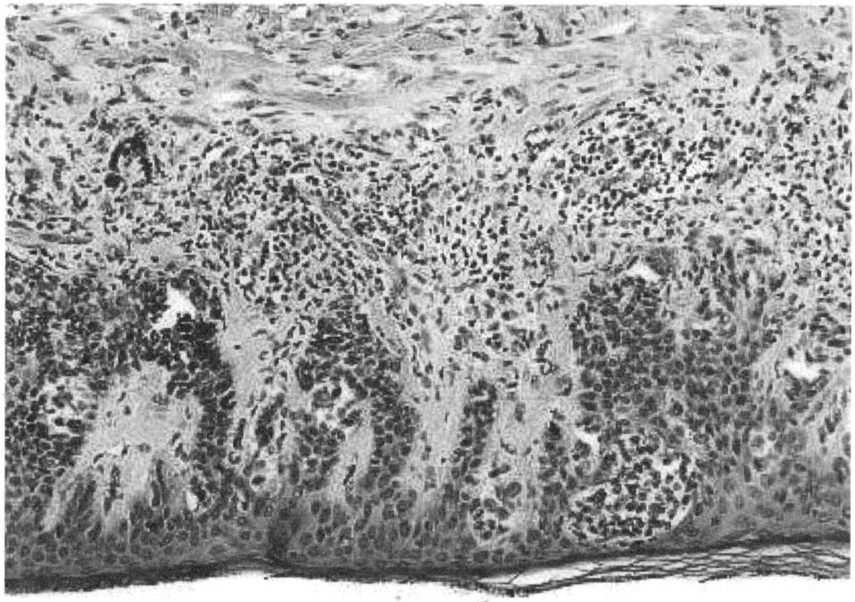

FIGURE 8.10: For comparison purpose, high-power photomicrograph (courtesy of I.R.C.C.) of a lentiginous and junctional moderate melanocytic dysplasia in the epidermis overlying the dermal component. The papillary dermis is widened by a mixture of tumor cells, inflammatory cells, and fibrous tissue. At the bottom of the image, invading fronts of the neoplasm feature tentacular or finger-like extensions, similar to those reproduce by the computational model.

full-size cells. We further assume that both daughter cells evenly inherit all the parent's biophysical properties (i.e., its motility and adhesive properties). Finally, the newly formed individuals are placed symmetrically about the parent cell center of mass with a random orientation.

Intuitively, cell proliferation is expected to be dramatically proinvasive, on the basis that an increment in the cell population will facilitate the invasion of the extracellular environment. However, tumor invasiveness only slightly increases, as $d_f \approx 260$ μm (see Figure 8.9). The explanation of this counterintuitive result is that the external cells, whose metabolism is accelerated by the high quantity of available nutrients, quickly divide, forming a front of little islands, as reproduced in Figure 8.9(B). Such cell clusters go on increasing in size, due to further cell proliferations, and come in contact with the main tumor mass by short and thick (4–5 cell-wide) "fingers" (see Figure 8.9(C)). The increased cellular density, in turn, enforces cell–cell adhesive interactions, which balance the effect of the haptotaxis and of the enhanced cell motility. The formed fingers therefore do not break and, consequently, the

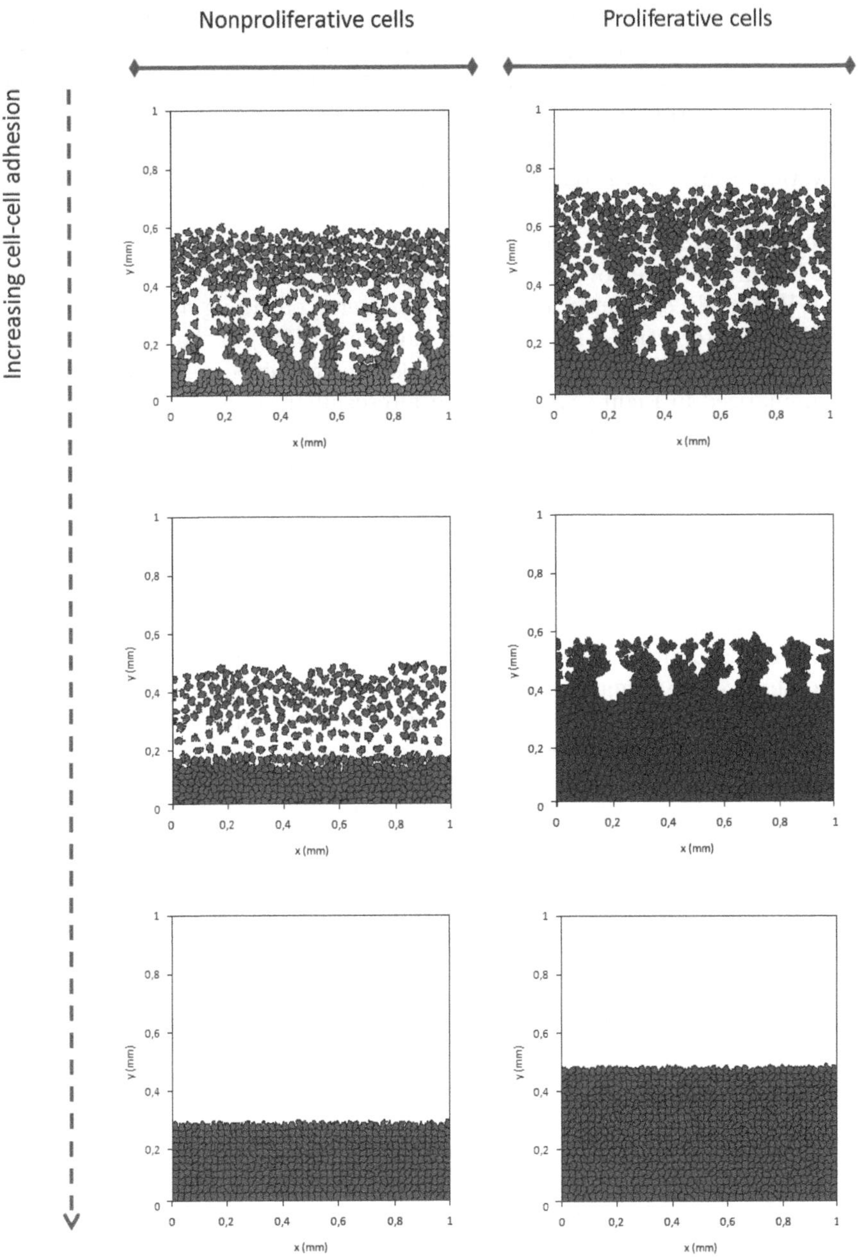

FIGURE 8.11: Resulting morphologies of the microinvasive tumor mass for different intercellular adhesive interactions (i.e., given by parameter $J^{ext}_{T,T}$) in the case of either nonproliferative or proliferative malignant individuals.

overall invasion is partially discouraged. The phenomenon of "fingering" has been captured by other discrete models [370, 395], and characterizes many malignancies (see for example Figure 8.10). In particular, recent studies of photo-micrograph have shown that the "ragged" surfaces of different tumors contains fractal components at the smaller, cellular scale, that relate to the overall severity of the disease [96, 227]. However, tumors with an unstable fingered morphology are typically more aggressive and hard to be treated than smoother ones: in fact, even if their invasive depth is limited, they are difficult to be surgically removed.

We finally characterize the tumor patterns emerging for different cell–cell adhesive interactions (i.e., given by parameter $J^{ext}_{T,T}$) in the case of either non-proliferative or proliferative individuals. The obtained results are graphically summarized in Figure 8.11. At high intercellular adhesiveness, the cancerous mass remains compact and smooth: indeed, its invasive distance is specifically determined by the mitotic rate of malignant individuals. On the opposite, at low cell–cell adhesion, the tumor invades the surrounding environments with a front of dispersed aggressive individuals, with islands of free tissue forming within the main mass. Also in this case, cell replication enhances the invasive potential of the cancer, leading to the most aggressive disease, which deeply penetrates in the surrounding tissue with an increasing number of scattered individuals. Finally, at intermediate values of intercellular adhesive connections, we observe the formation of a ragged tumor front only in the case of proliferative individuals: this suggests that fingers of malignant cells can emerge only with a specific balance between their mitotic potential and their adhesive strength. Reviewing the results in Figure 8.11, we can therefore conclude that tumor patters are primarily differentiated by the variation of the cell adhesive potential, while the invasive depth is definitively dependent from the cell mitotic rate. Of particular interest is the unstable fingered morphology characterizing a small region of the cell biophysical parameter space.

8.8 Early Stages of Tumor Spheroid Growth

During the first period of avascular growth (or when cultured *in vitro* in a three-dimensional gel, as studied in Chapter 3), a primary tumor may also organize in a sphere-like hyper-proliferative colony, which has to rely upon the host for crucial substances, such as oxygen, nutrients, and growth factors [7, 275, 401]. Substrate gradients therefore form within the growing cancer mass, causing a well-localized phenotypic specification of malignant individuals, which typically differentiate in an outer viable rim of highly metabolic and proliferative cells, an interior band of quiescent individuals and a central necrotic core, formed by cells dead for deprivation of vital chemicals [47]. In particular, in this stage, the increment in cell mass due to the aberrant mitosis of external cells is only able to balance the loss of volume of the interior necrotic region. This limits the size of the tumor spheroid, which reaches a well-characterized steady state [88, 316]. Indeed, in order to grow further, the cancer mass will have to coopt the existing vasculature, or to stimulate the formation of new capillaries. .

The above-described diffusional limitations of growth can be overcome by the carcinoma with morphological instabilities which, as a result of many factors, including the downregulation of intercellular adhesion, specific mechanical stresses, or the enhancement in cell motility, increment the surface area of its interface with the host, thereby allowing internal individuals to have greater access to vital molecules.

8.9 Mathematical Model

The previous model is indeed generalized to three-dimensions, focusing on the growth of a multicellular spheroid and on the detachment of invading cells from it. In this case, the cancer cells are uncompartmentalized standard individuals Σ_σ of type $\theta = C$. This choice is done to avoid unnecessary computational costs given the fact that intracellular units do not play a fundamental role in the early stage of spheroid tumor growth. This means that most of the parameters previously used are formally transferred to this model dropping the subscript η or substituting it with Σ_σ, e.g., in Equation (8.3) and in the similar ones $\mathbf{s}_{\eta,\Sigma_\sigma}$ becomes simply $\mathbf{s}_{\Sigma_\sigma}$, as now the internal state vector characterizes the entire cell.

With respect to the two-dimensional case treated in the previous section, an essential feature of three-dimensional growth is volume change because of growth and death. Growth was already introduced in Section 8.8, where a probability of undergoing mitosis (now P_{Σ_σ}) was introduced in Equation (8.9). Necrosis is here introduced as a consequence of lack of suitable growth factors. In fact, in the presence of an adequate intracellular level of growth factors, malignant cells keep almost the same volume during tumor growth: therefore, for any Σ_σ, $A^{volume}_{\Sigma_\sigma} = A^{volume}_C$ is their initial volume and $\lambda^{volume}_{\Sigma_\sigma} = \lambda^{volume}_C$ is a high constant value. On the contrary, when the internal level of growth factors drops below a certain threshold, here defined as n_l, a cell enters in a irreversible necrotic state, and starts to lose volume at a constant rate. This process, the model counterpart of the biological lysing of death cells, is modelled by setting:

$$\frac{dA^{volume}_C}{dt} = -k \tag{8.10}$$

if $n(\Sigma_\sigma, t) < n_l$. Zero volume cells are finally removed.

Moreover, given that $A^{surface}_{\Sigma_\sigma} = A^{surface}_C$ is the initial cell surface, because tumor cells are typically deformable, as they are able to significantly remodel to invade their surroundings more efficiently, for any individual Σ_σ, we use a sufficiently low $\lambda_{\Sigma_\sigma}^{surface} = \lambda_C^{surface}$.

Since, the cancer cells are uncompartmentalized objects, $H_{adhesion}$ has the classical form given in Equation (1.5). In particular, $J_{C,C}$ represents the adhesive strength between the membranes of two nearby cells, as usual a measure of the quantity of active and exposed cadherins, while $J_{C,M}$ evaluates the heterophilic adhesive bonds between the integrins on the cell surface and suitable ligands in the extracellular matrix. As in the bidimensional set-up, we set constant and homogeneous values for both $J_{C,C}$ and $J_{C,M}$. However, tumor cells within spheroid masses prefer to adhere to one another rather than to the host (the strong homotypic interactions within solid tumor masses are typically regulated by the over-expression of E-cadherins and by other intercellular mechanisms): therefore $J_{C,C}$ has to be lower than $J_{C,M}$. In particular, as given in Table C.9, we set $J_{C,C} = 1/2\ J_{C,M}$: this is an arbitrary choice, however a complete screening of the role played by the cell–cell adhesive strength in determining the tumor phenotype will be performed in the Results section (cf. Figures 8.16 and 8.17).

The other major change is due to the fact that, in the three-dimensional set-up, the production of diffusing growth factors, $n(\mathbf{x}, t)$, is set (at a constant rate per unit of time) at the entire border of the domain. Again the matrix soluble proteins, $p(\mathbf{x}, t)$, are assumed to have a uniform distribution at the beginning of each simulation and are degraded upon contact by the metalloproteinases (MMPs) secreted by malignant cells $m(\mathbf{x}, t)$, that, in turn, diffuse throughout the tissue and undergo some form of decay (passive, active, or due to the neutralization by endogenous matrix inhibitors).

8.10 Simulations

The simulation domain Ω is a cubic lattice of $350 \times 350 \times 350$ sites. The characteristic length of each site is 1 μm, and therefore Ω represents a tissue with a volume of 0.04 mm^3. One MCS is set to correspond to 20 sec: the overall simulations stop after 56250 MCS, so that they reproduce a time-lapse of nearly 15 days. The PDE for the evolution of nutrients is numerically solved with a finite difference scheme on a grid with the same spatial resolution as Ω, characterized by 30 diffusion steps per MCS. This temporal scale is sufficiently small to guarantee the stability of the numerical method.

As represented in Figure 8.12, we start all simulations with a cluster of 24 cells in the center of the lattice. Each cell is initially a sphere whose volume is consistent with the average dimensions of human glioma cells [50]. The specific initial configuration reproduces an avascular solid tumor spheroid, which is invading the surrounding host tissue.

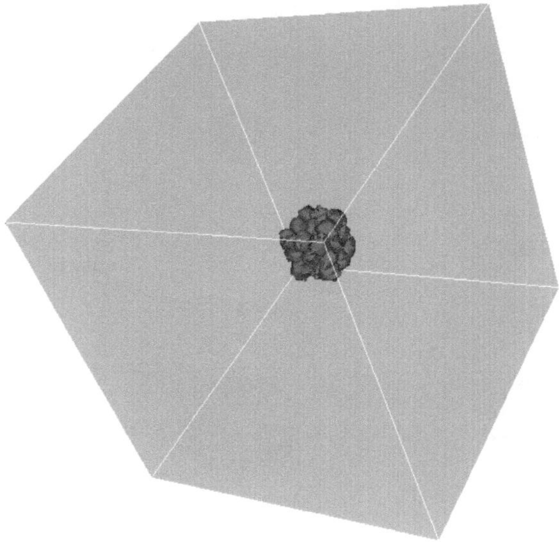

FIGURE 8.12: Initial condition (i.e., at $t = 0$) of the tumor mass for all sets of simulations. We start with a cluster of 24 cells placed in the center of the lattice. The dimensions of cells, as well as all the other model parameters, are given in Table C.9.

The boundary conditions for the proteolytic enzymes, m, are zero flux at the border of the entire domain (i.e., $\mathbf{n}_{\partial \mathbf{x}} \cdot \nabla m(\partial \mathbf{x} \in \partial\Omega, t) = 0$, where $\mathbf{n}_{\partial \mathbf{x}}$ is the outward normal to $\partial \mathbf{x}$), assuming that these parts of the tissue are far enough from the cancer mass. The borders of the domain represent instead an extended source for the environmental growth factors, which are supposed to be supplied by the host: consequently, we set $n(\partial \mathbf{x} \in \partial\Omega, t) = \phi_n$. The intracellular basal level of chemical factors, n_0, and its threshold for the necrotic transition have been experimentally measured for glioma spheroids grown from glioblastoma multiforme (GBM) cells [29, 67, 138]. The extracellular environment is initially seeded with a low level of growth factors $n_{ext,0}$, and with a saturating level of ECM, $p_{ext,0}$. Finally, there are no secreted MMPs in the extracellular medium at the beginning of the simulations.

To quantify the effects of the different model parameters on the tumor phenotype, we concentrate on the evolution of invasive depth of the mass, defined as $d(t)$ and measured as the radius of the sphere circumscribing the 95% of malignant cells. This choice is done to avoid biases toward accounting for few shed individuals. d has a clear clinical relevance, since it quantitatively estimates the severity of the disease and characterizes the presence of a significant number of metastasis delocalized with respect to the main body of the tumor.

We first simulate the growth of the tumor in the standard parameter set-

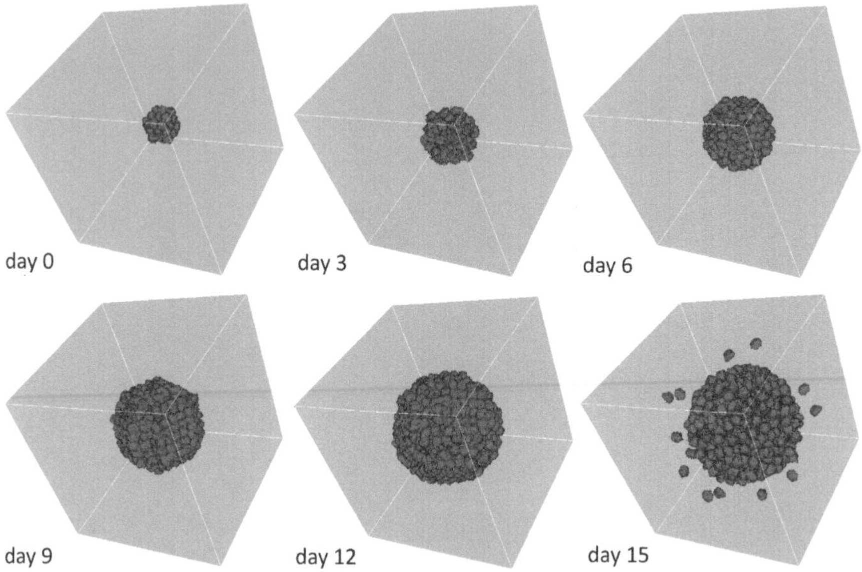

FIGURE 8.13: Time-series of the simulated tumor evolution. For the first 9–10 days, the malignant spheroid exponentially grows. It then stabilizes in a diffusion limited steady-state, where the proliferation of external viable cells is only enough to exactly balance the lost of mass due to the necrosis of the inner region, due to nutrient deprivation. In the final configuration, few single individuals are separated from the bulk of the tumor, as their motility has overcome intercellular adhesive interactions.

ting. As reproduced in the time series of Figure 8.13, the malignant mass grows nearly exponentially for the first 9–10 days (see also Figure 8.14). Due to the abundance of chemical supply, external cells (i.e., those located near the tumor/host interface) can in fact freely proliferate, while, in this stage of development, the basal intracellular level of growth factors is sufficiently high for the survival of more internal individuals. As the evolution proceeds, the cancer cells near the center of the tumor begin the irreversible necrotic transition. Without exposition to the external environment, their internal amount of nutrients in fact drops below the quantity needed for viability. Indeed, a core of death individuals, which loose volume, forms at the center of the mass. In this phase, the proliferation of external cells is only sufficient to balance the loss of mass in the central region of the tumor, whose growth therefore slows until reaching a diffusion-limited steady state, characterized by $d(t = 15$ days$) \approx 90$ μm.

In the steady configuration, a small fraction of cells is escaped from the bulk of the spheroid; see again Figure 8.13. The increased metabolism of these individuals, due to the availability of environmental growth factors, has in fact resulted in a dramatic increment in their motility, which has overcome the

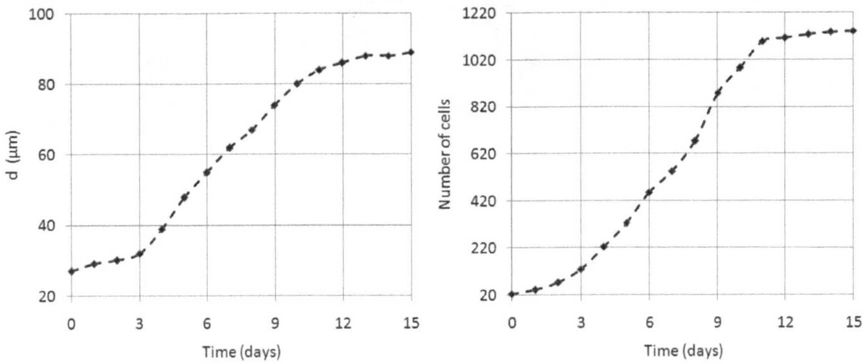

FIGURE 8.14: Regression of the time-evolution of both the invasive radius d (left panel) and the number of cells (right panel) for the tumor spheroid, developing under the standard parameter settings, as depicted in Figure 8.13.

intercellular adhesive forces and the physiological regulation of the contact-inhibition of movement, inducing a sort of mesenchymal transition. These shed individuals, which are difficult to be clinically detected, have the greatest potential to invade the host and further metastasize. As seen in the previous sections, they display in fact an evident ability to wonder in the close proximity and to spread in the surrounding tissue, mainly driven by haptotactic mechanisms via local ECM gradients created by the activity of their proteolytic enzymes. Evading destruction by the immune system, they may subsequently establish secondary colonies with devastating consequences for the wellbeing of the patient, as the likelihood of success of therapeutic interventions strongly decreases [162, 324].

A specific description of the cell phenotypic differences within the spheroid is obtained by considering the distribution of the intracellular levels of growth factors, given in Figure 8.15 (left panel). Cells in the more external ring of the tumor have an abundant supply of chemicals and, as seen, can freely proliferate and eventually shed in the extracellular environment, as a consequence of their increased metabolism and motility. As the distance between the spheroid border increases, the local availability of growth factors progressively falls. Indeed, a sort of equilibrium point is reached at which cells have an internal amount of vital factors too low to undergo mitosis, but sufficiently high to survive. These individuals are therefore in a so-called quiescent state. Finally, in the core of the mass, the supply of chemicals is negligible, so that inner cells, once consuming their basal level of growth factors, are unable to stay alive. They therefore die due to nutrient deprivation forming, as seen, a region of necrotic debris, which loose volume and are eventually removed. Summing up, malignant individuals within a solid, avascular mass display three different phenotypic state: *proliferating*, *quiescent*, and *necrotic*.

Figure 8.15 represents also the final distribution of tumor MMPs and of

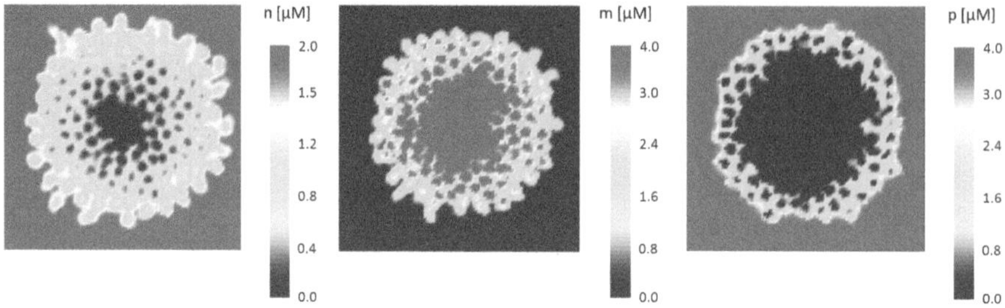

FIGURE 8.15: (See color insert.) Final patterns (i.e., at $t = 15$ days) of the microscopic variables included in the model, taken in a representative section of the domain. From left to right panel: concentration of growth factors (n), tumor proteolytic enzymes (m), and matrix proteins (p). It is straightforward to notice that cells in the center of the spheroid have a negligible amount of chemicals and therefore undergo the necrotic transition. The more external individuals have instead an abundant supply of vital nutrients, which increase their metabolism. Indeed, the extracellular environment is completely modified by the tumor, as the matrix components are degraded by the localized activity of malignant MMPs.

matrix proteins. In particular, it can be clearly observed that the host tissue is significantly modified, as the matrix components have been dramatically degraded by the activity of tumor proteolytic enzymes. Indeed, steep protein gradients have formed, which eventually drive, via haptotactic mechanisms, a further invasion of the single, aggressive individuals.

The evolution of the malignant mass captured in the model reproduces the first stages of the growth of several tumors grown as spheroids in spinner cultures, such as ovarian [53, 363] or breast [155] carcinomas. Moreover, the model results are consistent with the development of avascular gliomas , both embedded *in vitro* in collagenous gels [208, 376] and implanted *in vivo* in mice [2, 69]. It is useful to underline that such a qualitative agreement occurs even without an exact bookkeeping of diffusing growth factors and a detailed inclusion of the cell cycle. Indeed, a similar growth of primary solid tumors have been also predicted by different types of theoretical models, see again the reviews [15, 58, 74, 321] and the comprehensive books [240, 316].

As briefly sketched in the previous section, the values of the parameters Js have a clear biological relevance, as they describe the relative preference of tumor cells to be in contact with other cells or with matrix components. At a molecular level, they are in fact a measure of the expression and the engagement of the different types of cell adhesion molecules, cadherins and integrins, respectively. As the variation of the Js may be expected to have a substantial impact on the overall development of the tumor, we analyze the model outcomes obtained by varying the cell–cell adhesive strength, defined by $J_{C,C}$. As reproduced in Figure 8.16, at low values of $J_{C,C}$, which mean

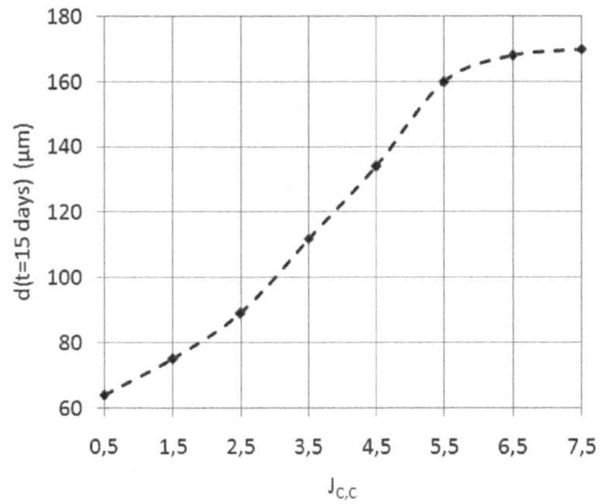

FIGURE 8.16: Invasiveness of the tumor mass in response to alterations of cell–cell adhesive strength, $J_{C,C}$ (i.e., all the other model parameters are the same as in the simulation in Figure 8.13). Regression of the final depth of invasion $d(t = 15 \text{ days})$ vs $J_{C,C}$. Downregulations of intercellular contact interactions enhance the invasive potential of the tumor.

overexpressions of intercellular adhesive molecules, the final invasive radius of the spheroid slightly decreases. In particular, the tumor remains compact and strictly packed, without single individuals shed in the surrounding tissue, see left panel of Figure 8.17. An enhanced activity of cadherins in fact is able to stabilize the morphology of the disease, reducing its aggressiveness: a smoother carcinoma mass can be in fact more efficiently treated surgically.

As $J_{C,C}$ starts to increase, the tumor expands and invades deeper. The progressive reduction of cell-cell adhesive interactions results in fact in an increasing number of single individuals able to escape from the main mass into the external environment (see also central panel of Figure 8.17).

Finally, at even large values of $J_{C,C}$ (i.e., $> J_{C,M}$), not only the external cells quickly loose contact, dissociate and spread away from the rest of the mass creating a dispersed ring (fewer of them adhere with each other within 2 days, not shown), but a repulsion occurs also among individuals within the core of the mass. The surface area of the tumor/host interface therefore increases, thereby allowing internal cells greater access to vital growth factors needed to survive and eventually to proliferate. The overall tumor is therefore able to overcome the diffusional limitations and continues to expand both in the number of cells and in their dispersion, which further increases the overall bias toward invasion, as reproduced in the right panel of Figure 8.17.

Interestingly, there is not a critical value of the cell homotypic adhesion that clearly separates a compact phenotype of the tumor from a scattered configuration: rather the metastatic potential of the disease increases progressively with the downregulation of cell–cell contact strength.

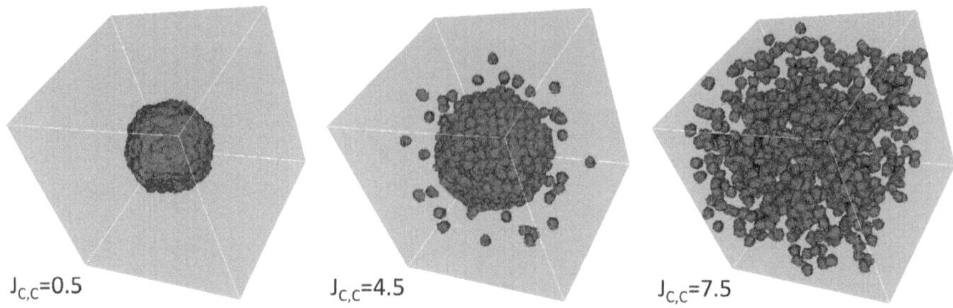

FIGURE 8.17: Invasiveness of the tumor mass in response to alterations of cell–cell adhesive strength, $J_{C,C}$ (i.e., all the other model parameters are the same as in the simulation in Figure 8.13). By enforcing cell–cell adhesion, the spheroid remains densely compact and smooth with a low invasive potential. On the contrary, decrements in cell–cell adhesion lead to increments in the malignancy of the lesion, as external cells quickly shed in the host tissue and, consequently, also internal individuals have access to vital growth factors, surviving and proliferating. Examples of final configurations for $J_{C,C} = 0.5$ (left panel), $J_{C,C} = 4.5$ (central panel), and $J_{C,C} = 7.5$ (right panel).

These results represent a definitive confirmation that an alteration in cell–cell adhesive properties is an important consideration in regulating the morphology of the disease and, eventually, in determining its severity. As a support of the model outcomes, a number of experimental studies have recently demonstrated that the downregulation of cadherin molecules is implicated in a variety of metastatic cancers [71, 72, 83, 385]. In particular, glioma cell lines with low N-cadherin expressions have been observed to aggressively invade matrix gels, whereas the same populations with a high N-cadherin activity have been instead shown to grow slower and to expand less significantly in the host tissues [205, 182].

Chapter 9

Cell Migration in Extracellular Matrices

9.1 Biological Introduction

As already seen in Chapter 5, cell migration on and within tissues plays a critical role in a diverse array of processes, such as in developing embryos, where the coordinated movement of cells of different origin along ECM layers is crucial for organogenesis, and migratory defects at all stages lead to severe embryonic malformations [220]. In adult organisms, cells are normally quiescent, except in immune surveillance or inflammation, where leukocytes actively migrate from blood vessels into infected tissues and then into the lymph node for effector functions [145], and in wound healing, where migration contributes to the repair of both basement membrane-underlaid epithelium and connective tissues. In pathological conditions, cell migration is involved in chronic inflammatory diseases such as atherosclerosis, or in cancer cell invasion into ECM and metastatization [337]. The process of cell migration is finally exploited in biomedical engineering applications for the regeneration of various tissues, such as cartilage, skin, or peripheral nerves *in vivo* or *in vitro* [63, 178, 373, 418].

Cell motile behavior is modulated by a spatiotemporally integrated set of multilevel mechanisms that are influenced not only by the biochemistry of extracellular and intracellular signaling, but also by the biophysics of the surrounding environment, whose basic component is the extracellular matrix (ECM). The ECM forms, in fact, the architecture of a number of structures, i.e., two-dimensional (2D) flat basement membranes, or three-dimensional (3D) connective tissues and can be described as a complex network of insoluble structural fibrous proteins such as collagen type I, and soluble glycosaminoglycans and glycoproteins, that together provide microstructural guidance cues and biochemical stimuli for moving individuals.

How does a cell migrate then on and in ECMs? For the basic program of cell migration over flat ECM substrates, four requirements have been identified. Migrating cells

1. Are morphologically polarized in the direction of motion.

2. Adhere dynamically to their environment via adhesive molecules, i.e., integrins.

3. Generate the force necessary for propulsion by contraction of cytoskeletal elements.

4. Retract their rear ends [7, 229].

For migration within three-dimensional (3D) porous environments, in addition to the basic principles above, the cell needs to find its way throughout steric obstacles [110, 332, 410]. This can be achieved by

1. Passing through constricted openings of the ECM by significant cell deformation and cytoskeletal force generation.

2. Activating a cell-derived proteolytic machinery able to degrade matrix components and to open space for cell movement [141, 142, 219, 336, 411].

This basic motile behavior is further modulated by a number of mechanisms that include determinants from both the surrounding ECM and the cell itself (refer to [229, 410] and references therein) that we aim to systematically analyze with our multi-scale modeling approach. Also in this case, each cell is modeled as a discrete compartmentalized object, differentiated into nucleus and cytosol. On the other hand, unlike what we have seen in most of the previous chapters, the environments are constituted of two components, an inhomogeneous fibrous collagen-like network, and a homogeneous interstitial medium.

The model is highly flexible, being capable of characterizing the migratory behavior of cells in several conditions, both on 2D substrata and in 3D ECMs. In the simulations, characteristics like cell shape and directionality are not imposed *a priori*, but are a result of the interaction with the matrix fibrous component. As an outcome, we focus on experimentally addressable characteristics of cell locomotion, i.e., cell overall displacement, velocity and persistence time, and cell shape, predicting how these quantities are influenced by manipulations of properties of either the matrix (i.e., adhesive ligands, fiber distribution, pore size, elastic modulus), or the cell (i.e., adhesive strength, deformability, and proteolysis).

Consistent with experimental observations, our findings provide evidence for a biphasic cell migratory behavior for planar substrate in response to variations of the number of matrix ligands or adhesion strength, with maximal movements at intermediate values. In 3D matrix environments, the geometrical distribution of the collagenous network, such as matrix alignment or pore size, or the matrix elasticity will be demonstrated to affect cell behavior in a similar way. Further, the cell compartmentalization allows one to discern the effect of the mechanical rigidity of the nucleus that, being higher than the cytosol, limits the migration capacity of the entire cell. Finally, we will include ECM-directed proteolysis, resulting in enhanced migration in restricted environments.

9.2 Mathematical Model

The simulation domain is a d-dimensional regular lattice $\Omega \subset \mathbb{R}^d$. In particular, $d = 2$ describes the case of cells moving on a gel and $d = 3$ the case of cells moving in a 3D ECM. In both cases, the simulated cells are defined as bi-compartmentalized units η of type $\theta = E$, composed, as usual, of the nucleus of type $\tau = N$ and the surrounding cytosol, $\tau = C$. It is useful to emphasize that in this model we do not consider the molecular state of moving cells: therefore, it is not necessary to define their internal state vector.

The cell population resides in an extracellular matrix, which is differentiated in a medium-like state, $\tau = M$, and a collagen-like state, $\tau = F$. The medium-like state reproduces the mixture of soluble components (among others, proteoglycans and glycoproteins in water), which compose the interstitial fluid. It is assumed to be isotropically distributed throughout the simulation domain, forming no large-scale structures.

The collagen state represents instead a network of insoluble macromolecules, such as collagen, that associates into first-order fibrils and second-order fibers and displays the most abundant structure in mammalian tissues. Each fibrous component is treated as CPM standard and non-compartmentalized CPM objects Σ_σ. The dimensions, density, and distribution of the fibrous structures will be specified in next sections and will reproduce 2D and 3D matrix types, respectively, typically employed for *in vitro* assays.

The inclusion of an explicit two-component matrix environment, already present in some other CPM applications [25, 162, 246, 334] and in Chapter 3, is a fundamental aspect of this work: it allows an accurate analysis of how cells migratory behavior is influenced by the heterogeneous fibrillar extracellular environment and therefore by the ECMs' specific biophysical and biomechanical properties while they glide in medium of constant and homogeneous physical properties.

The system *Hamiltonian* is

$$H(t) = H_{shape}(t) + H_{adhesion}(t). \tag{9.1}$$

In this case, H_{shape} models the geometrical attributes of both the cells and the matrix threads, which are written as nondimensional relative deformations in the standard quadratic form of Equation (4.6). Trivially, in the 2D case, we have constraints on the cell surface and perimeter, while in the 3D case, we have constraints on the cell volume and surface. Indeed, assuming that cells do not significantly grow during migration, the fluctuations of their volumes are kept negligible with high constant values $\lambda_{\eta,N}^{volume} = \lambda_{\eta,C}^{volume} \gg 1$, for any individual η and for Σ_σ such that $\tau(\Sigma_\sigma) \in \{N, C\}$ (respectively, in 2D $\lambda_{\eta,N}^{surface} = \lambda_{\eta,C}^{surface} \gg 1$).

On the other hand, when moving in matrix environments, cells are typically deformable, while their nuclei show a rigidity higher than the cytoplasm region: therefore, for any η and for Σ_σ such that $\tau(\Sigma_\sigma) = C$, we set $\lambda_{\eta,C}^{surface} \ll 1$, while and for Σ_σ such that $\tau(\Sigma_\sigma) = N$, we set $\lambda_{\eta,N}^{surface} \gg 1$ (respectively, in 2D $\lambda_{\eta,C}^{perimeter} \ll 1$ and $\lambda_{\eta,N}^{perimeter} \gg 1$). The extracellular environment is instead assumed to have homogeneous mechanical and microstructural properties: in particular, the matrix fibers are assumed to be rigid by setting $\lambda_F^{surface} = \lambda_F^{perimeter} \gg 1$ when $\tau(\Sigma_\sigma) = F$ in both two and in three dimensions. However, it is useful to emphasize that in the following we will analyze how the explicit variation of fiber and nucleus stiffness will affect cell migratory phenotypes within three-dimensional matrices.

$H_{adhesion}$ is, as usual, differentiated in the contributions due to either the generalized contact tension between the nucleus and the cytoplasm within the same cell, or the effective adhesion between a cell and both the medium and the fibrillar matrix component, and, in the case of collision, between cells. In particular, $J_{N,C}^{int} \ll 0$ implicitly models the forces exerted by intermediate actin filaments and microtubules to anchor the nucleus to the cell cytoskeleton, preventing cells from fragmenting.

$J_{E,M}^{ext}$ and $J_{E,F}^{ext}$ evaluate instead the heterophilic contact interactions between cells and matrix components: specifically, they are a measure of the affinity between cell surface adhesion complexes (i.e., sugar-binding receptors or integrins) to either nonsolid (i.e., glycosaminoglycans in medium) or solid (i.e., fibrillar collagen) extracellular ligands, respectively [348]. In particular, we assume $J_{E,F}^{ext} < J_{E,M}^{ext}$ since, as widely demonstrated in the literature, most cell lines in standard conditions adhere more strongly with the fibrous part of the extracellular matrix rather than with its soluble component (see [382] and the references therein).

$J_{E,E}^{ext}$ represents the local adhesive strength between neighboring cells, a measure of the local quantity of active and exposed cadherin molecules. It is kept high to avoid cell–cell adhesive interactions upon accidental cell collisions that may affect cell movement.

Setting in general constant and homogeneous values for the bond energies Js corresponds to assuming a uniform distribution of adhesion molecules on cell surfaces and of ligands in the external environment, without any change during the observation time. A summary of values of all the model parameters used in the simulations is given in Table C.10 in Appendix C.

Given the *Hamiltonian*, the transition probability of a spin flip has the form of Equation (4.24). In particular, we use $p(T_{\Sigma_\sigma}(t)) = \tanh(T_{\Sigma_\sigma}(t))$. Indeed, for each cell η and for $\tau(\Sigma_\sigma) = N$, $T_{\eta,\Sigma_\sigma} = T_{\eta,N}$ gives the agitation rate of its nucleus, while, for $\tau(\Sigma_\sigma) = C$, $T_{\eta,\Sigma_\sigma} = T_{\eta,C}$ is a measure of the intrinsic motility of the overall individual, as it gives the frequency of the ruffles of its cytosol (we again recall that on a molecular level, it is determined by polarization/depolarization processes of the actin cytoskeleton [267, 311, 328] and the references therein).

Finally, for $\tau(\Sigma_\sigma) = F$, $T_{\Sigma_\sigma} = T_F$ determines the vibration degree of matrix fibers. Indeed, for each cell, $T_{\eta,N}$ is a low value (< 1), resulting in the passive motion of the nucleus, which, unable to move autonomously, is dragged by the surrounding cytosol, characterized instead by a high $T_C \gg 1$ (see Chapter 6). In most simulations, the matrix fibers are instead assumed to be fixed by setting $T_F = 0$.

9.2.1 Simulation Details

To describe cell migration on and in ECM matrices, we start with default cell-ECM conditions and subsequently adapt them to specific conditions, such as matrix orientation, density, adhesiveness, etc., whose role will be of interest to examine.

The basic CPM in both 2D and 3D conditions contains certain common spatial and temporal characteristics. The spatial simulation domain Ω has periodic boundary conditions and a basic grid size of 1.3 μm. In all the 2D simulations, $\Omega \subset \mathbb{R}^2$ represents a 3.5 cm (i.e., $2.69 \cdot 10^4$-site-width) side-length squared section of an experimental dish, which is commonly used for planar migration assays [106]. In the 3D case, $\Omega \subset \mathbb{R}^3$ reproduces instead an experimental scaffold with a volumetric extension of 1 cm^3 (formed by $4.55 \cdot 10^{11}$ cubic voxels). The temporal resolution of the model is the MCS, which is set to correspond to 2 s to compare cellular dynamics with experimental observations. All the performed simulations last 12 h ($\approx$ 21600 MCS) to ensure the development of sufficiently long migration paths.

The basic cell-matrix model to be simulated consists of a heterogeneous ECM of fibrillar and amorph ("medium") components hosting a cell population of low density to allow for isolated motions with very rare cell encounters.

In all 2D simulations, we plate $1 \cdot 10^3$ cells/cm^2, as done in [106], while in all 3D simulations, we embed $2 \cdot 10^3$ cells/cm^3, reproducing the cellular density of the experimental migration assays performed in [177]. The cells that interact with collagen-like fibers, i.e., fibroblasts or cancer cells of epithelial or mesenchymal origin, display initially a round nonmigratory unpolarized morphology: therefore, as default conditions, we start with round flat disks with a central round nucleus in 2D, and as spheres with a spherical nuclear compartment in 3D. In both cases, their overall diameter is 10 grid sites ($\approx$ 14 μm), while the nucleus is 5 grid sites ($\approx$ 7 μm) in diameter. For the reader's convenience, we emphasize that the entire volume (respectively, the area in 2D) of a cell is the sum of the volume (respectively, the area in 2D) of the nucleus and of the cytosolic region, while its external surface (respectively, the perimeter in 2D) is instead the difference between the surface (respectively, the perimeter) of the cytosol and the surface (respectively, the perimeter) of the nucleus. These dimensions, given in Table C.10, reflect the mean measures of typical eukaryotic cells except white blood cells [7].

In our model, we set the length of a collagen-like fiber equal to 15 lattice sites ($\approx 20\ \mu$m). Its thickness would generally range between 100 nm and 0.5 μm [44, 325, 331], and therefore it would be substantially smaller than the grid resolution. Modeling them with the real size would considerably increase the computational cost of the simulations. Indeed, we here follow most CPM applications (see, for example [162, 334]), where a fiber is modeled as a single site thick. Each simulated fiber therefore is assumed to contain nearly 10^6 collagen-like molecules, given that a single matrix protein is approximately 300–600 nm long and 1.5–5 nm wide [7].

For the sake of simplicity, we will use the term *fiber* for both the basic short ECM structure ($\approx 20\ \mu$m long threads) simulated for the 2D condition, and the long structure crossing the entire spatial domain of the 3D cubic network.

The migratory properties of cells are quantified by evaluating their *average velocity* ($\overline{\mathbf{v}}_\eta$ defined as in Equation (1.13) where $t_{final} = 12$ h, 21600 MCS), and their *mean square displacement* (d^2_η defined as in (1.14)).

In particular, as demonstrated in a number of experimental [106, 177] and computational [106, 421] studies, at sufficiently long times the mean square displacements vary approximately linearly with the number of time steps. It can therefore be related to the cell *instantaneous velocity* ($\mathbf{v}_\eta$, defined in (1.12)) and its *persistence time* (p_η, which quantifies the directional productive motion) with the so-called *persistence-random-walk* (PRW) law:

$$< d^2_\eta(t) >= 2\mathbf{v}^2_\eta(t)p_\eta(t)[t - p_\eta(t)(1 - e^{-t/p_\eta(t)})]. \tag{9.2}$$

In particular, at still longer observation periods, (9.2) reduces to

$$< d^2_\eta(t) >\approx 2\mathbf{v}^2_\eta(t)p_\eta(t)t, \tag{9.3}$$

and the persistence time of a moving individual can be directly calculated as

$$p_\eta(t) \approx \frac{< d^2_\eta(t) >}{2\mathbf{v}^2_\eta(t)t}. \tag{9.4}$$

The PRW relation has been demonstrated to characterize the cells' migratory behavior more properly than other common methods, which calculate the average distance migrated by biological individuals in an arbitrary time interval, as commented in [121].

The quantitative analysis of cell morphological changes is carried out by evaluating the evolution of the cell *aspect ratio*, given by the ratio between the actual cell surface (respectively, perimeter in 2D) and the surface of the sphere having the same volume (respectively, the perimeter of the circle having the same area in 2D). It is useful to underline that in our model cell volume (respectively, area in 2D) is kept nearly fixed by high values of $\lambda^{volume}_{\eta,N} = \lambda^{volume}_{\eta,C}$ (respectively $\lambda^{surface}_{\eta,N} = \lambda^{surface}_{\eta,C}$ in 2D) Therefore, the aspect ratio gives a quantitative measure of cell membrane ruffling.

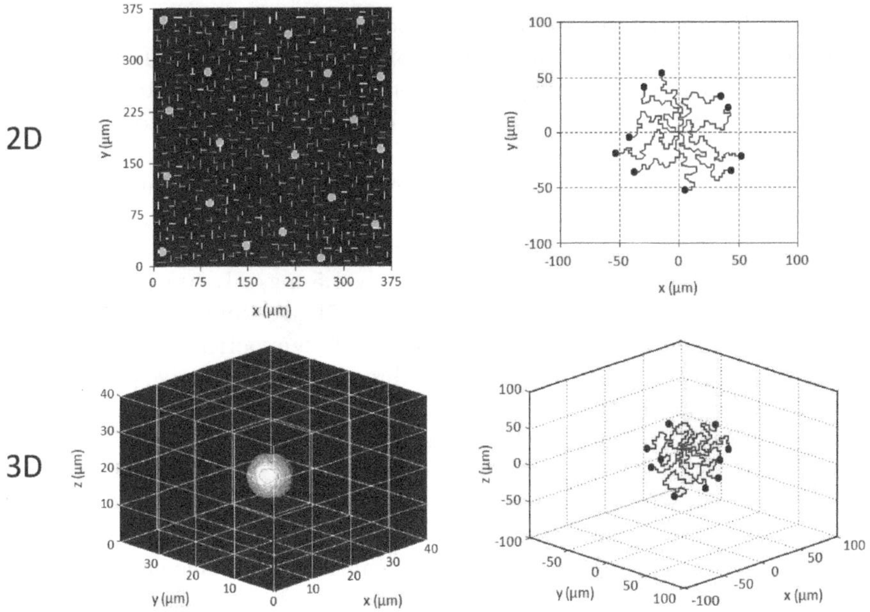

FIGURE 9.1: Standard cell–ECM models and migration in two and three dimensions. Left panels: For representation purposes, sections of the simulation domain Ω of both the 2D and the 3D case. Depicted are standard two-component substrates containing both an isotropic fibrous ECM of moderate density (light stripes) and the medium (black), and cells. In the 2D model, 500 fibers/mm^2 fibers are distributed equally and in both $x-$ and $y-$directions. In the 3D scaffold, matrix fibers are assembled into a regular cubic mesh, with a uniform distribution of pores of 10 μm side length. Right panels: Cell migration on or within the above-represented isotropic ECMs. Wind-rose graphs showing 10 randomly chosen cell tracks over 12 h. Black circles represent the ending location of each cell center of mass. In both conditions, cells display a Brownian random movement with net final displacement ca. 50 μm, MSD ca. $9 \cdot 10^4 \pm 0.5 \cdot 10^3$ μm^2, and velocity ca. 10 ± 0.6 μm/h. As reproduced from selected cell paths, the persistence time is low (ca. 1.5 ± 0.2 h). Here and in the following, all values are given as means $\pm$ s.d. over 50 randomly chosen individuals. The cell migratory behavior is consistent with the extracellular environment isotropy, and the absence of chemical gradients or other directional biases.

9.3 Isotropic Matrices

We first test the model for matrices containing an isotropic, moderately dense, fibrous network in both two and three dimensions. To form a planar substrate, we distribute $3 \cdot 10^5$ flat collagen-like fibers in each x and y-direction of the 3.5 cm-side length dish, yielding a density of 500 fibers/mm^2; see left top panel of Figure 9.1. This corresponds to a concentration of nearly $5 \cdot 10^3$ collagen-like molecules/μm^2, since, as seen, each thread is here assumed to be formed by 10^6 collagen-like molecules [7].

The analogous isotropic 3D scaffold consists of a regular cubic mesh of collagen fibers creating a uniform pore distribution 10 μm wide (i.e., the same order of magnitude of the initial cell diameter; see Figure 9.1, left bottom panel). This gives a fibrous concentration of nearly 1.5 mg/mL, as each collagen-like molecule has a molecular weight of 285 kDa [7]. We simulate a regular fibrous network to avoid the minor heterogeneities often experienced in experimental matrices, where the distribution of the threads and the relative pore diameters is only roughly constant [242, 294, 409].

As shown in the wind-rose graphs (right panels of Figure 9.1), when cells migrate on both 2D and in 3D matrices, the selected cell paths display a random walk, without any preferred direction, in the absence of biasing chemical gradients or matrix anisotropies.

Such migratory path structures and quantitative parameters are consistent with experimental results for both 2D and 3D porous ECMs, such as for human adult vascular smooth muscle cells (HSMCs) plated on flat type IV collagen (CnIV) substrates of similar concentrations [106] or human glioma cells plated on polyacrylamide ECMs [397], and for different fibroblastic and cancerous cell lines migrating within 3D fibrous matrices of similar geometrical and structural properties, i.e., NR6 mouse fibroblasts in collagen-glycosaminoglycan matrices [177], or human melanoma cells in collagen lattices [143]. Indeed, these comparisons provide confidence in the choice of parameters describing the biophysical and mechanical properties of the simulated cell–ECM system.

9.4 Anisotropic 2D and 3D Matrices

Next, we analyze the migratory characteristics of a cell population in the case of anisotropic matrices. In particular, we keep fixed the quantity of fibers as displayed in Figure 9.1, but progressively change their distribution by increasing their number along the same x-direction, leaving the remaining fibers disposed in their standard direction. The alignment of the matrix is quantified by evaluating a proper index that can be called *alignment index*, given by

$$N_{align} = \frac{1}{d-1}\left[d\frac{n_x}{n_{tot}} - 1\right], \qquad (9.5)$$

where d is the dimension of the domain, n_x the number of threads along the x-direction, and n_{tot} their overall number. This quantity scales the percentage of fibers aligned along the x-direction, so that it is zero in the case of isotropic

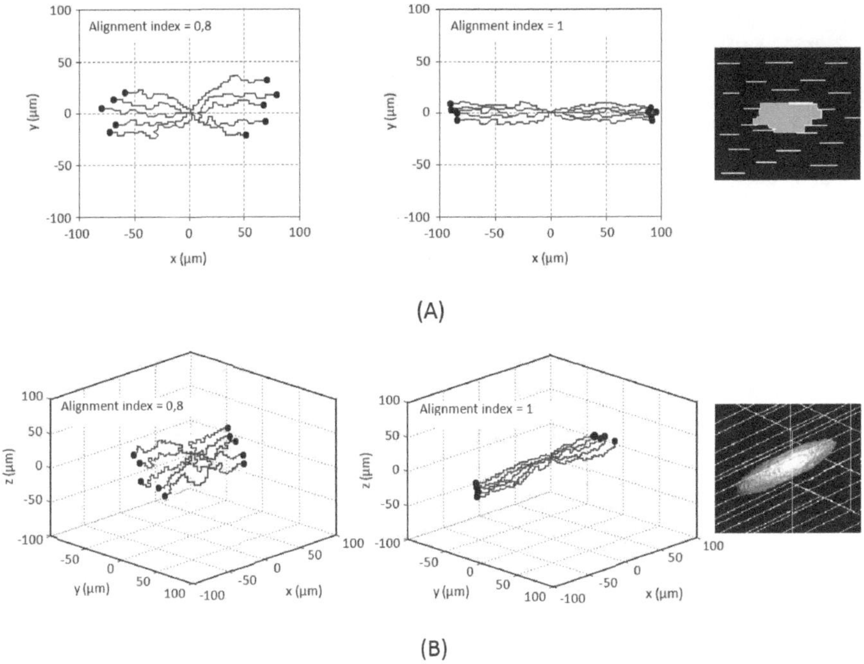

FIGURE 9.2: Orientation of cell migration along matrix fiber topology. Both on 2D (A) and within 3D matrix (B), the overall number of basic fibers remains fixed with respect to the standard simulations of Figure 9.1, while their percentage along the x-axis is increased. The alignment of the matrices is quantified by the alignment index N_{align}, defined in (9.5), which is 0 in the case of isotropic networks (already analyzed in the previous section) and 1 for fully aligned ECMs. The wind-rose graphs show 10 randomly chosen cell tracks over 12 h. The simulation-based images show instead cell morphologies in the case of completely aligned matrices.

networks (already discussed in the previous section) and 1 in the case of fully aligned matrices.

As a result, for both 2D and 3D migration, the paths gradually adapt toward anisotropic random walks, in particular, the directional cell motion increases toward the principal direction of alignment; see Figure 9.2(A-B). Interestingly, the cells' final average velocity and MSD remain constant despite increasing substrate orientation, with very similar values for both 2D and 3D conditions. However, the cells' 2D and 3D directed motile behavior in response to fiber distribution directly correlates with a strong increase in time (up to 5 h) that cells are able to perform persistent movement (i.e., without back-and-forth movements), as captured by the plots in Figure 9.3. Therefore, ECM geometry and architecture directly impact the migration pattern of individual cells. The directionality of cell movement is here not introduced *a priori*, but is

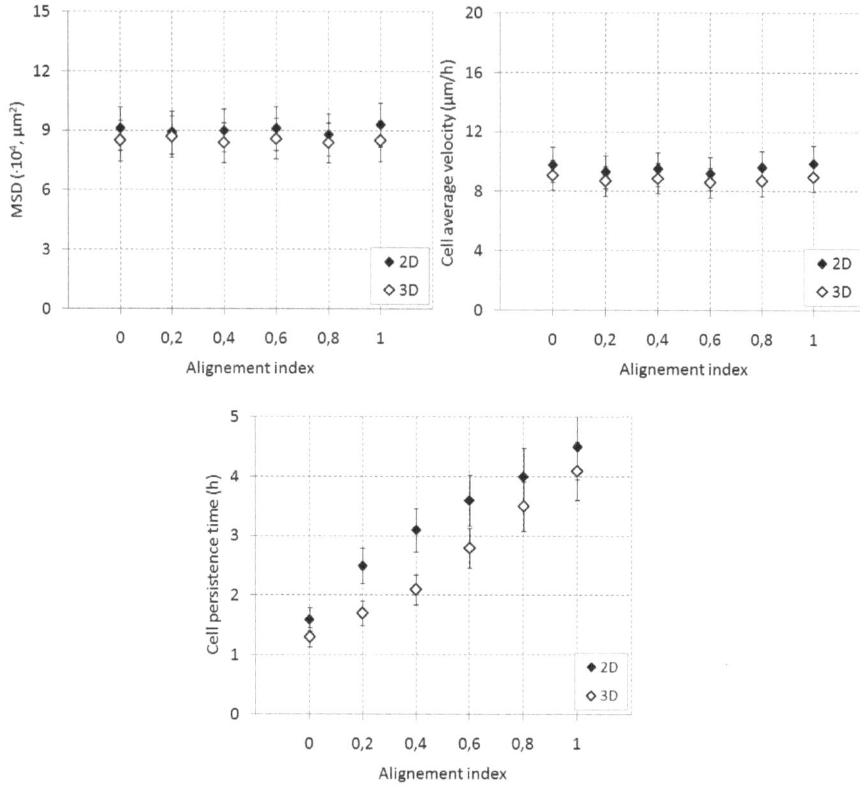

FIGURE 9.3: Orientation of cell migration along the matrix fiber topology. Cell mean square displacement (MSD), average velocity, and persistence time as functions of the alignment index N_{align}, defined in (9.5), for both 2D and 3D matrices. The values are given as means $\pm$ s.d. over 50 randomly chosen individuals. The directional component of cell motion increases as all fibers align, with no change in cell speed.

a direct result of well-defined *directional guidance cues* provided by the specific matrices. In fact, the anisotropy of the matrices induces a reorientation of moving cells in the direction of the threads, i.e., with the formation of clearly distinguishable leading and trailing edges; see the simulation-based images in Figure 9.2(A-B). The consequent motion along them is no longer an isotropic Brownian movement but is highly biased.

The influence of the ECM geometry and architecture on the migration of individual cells has been widely treated in the literature with *in vitro* and *in vivo* studies. In particular, several experimental models have demonstrated the cell preference to migrate along aligned matrix fibers, as in the case of fibroblasts in collagen-like gels [105] or neuronal cells on fibrin substrates [119]. Moreover, manipulations of the topology of two-dimensional matrices using lithographic and microprinting techniques have shown that the microstructural contact guidance not only affects cell migration, but also intracellular functions [46, 120, 232]. Lastly, *in vivo* intravital imaging studies of carcinoma cells in the mammary fat pad have pointed out the preferential chemotactic movement of invasive malignant cells along thick bundles of collagen fibers offering a 2D surface toward blood vessels [93], while in the lymph node paracortex, the aligned microarchitecture of collagen and fibronectin fibers ensheathed by fibroblastic reticular cells significantly influenced the migratory behavior of T-cells [19] .

9.5 Varying Fiber Density

Turning to the question of how the matrix topological structure affects cell migration in both two and three dimensions, we consider the effect of varying the density of the fibrous component of the substrate.

The cell population is therefore planted in lattices with an increasing number of collagen-like fibers, isotropically disposed as in the standard cases of Figure 9.1. In particular, in 2D, it means simulating migration over a surface containing an increasing amount of matrix fibers distributed equally and isotropically along the $x-$ and $y-$directions. Indeed, migration efficiencies develop a bell-shaped distribution from low toward high fiber numbers with a maximum at intermediate fiber numbers (Figures 9.4, 9.5, and 9.6). At low numbers of threads, the planar matrix is unsaturated and unpercolated and cells are found to have a short-range movement. Without finding enough collagen-like sites to attach at and to use for traction and significant displacement, they in fact fluctuate in the interstitial medium around their initial position. Indeed, the body of cells remains mainly round, regardless of their deformation ability. The distance from the nearest matrix fiber is in fact too high to for adhesive interactions; see Figures 9.5, and 9.6, bottom–right panels.

Increments in the number of bundles cause a progressive enhancement in cell migratory behavior. In particular, an optimal topology of the collagenous network (i.e., interfiber measure in 2D and pore dimensions in 3D) allows cells to efficiently move in the matrix environments. The specific distribution and distance of matrix threads results, in fact, in an effective *contact guidance* for moving individuals, which are forced to undergo a change in their shape toward a more motile mesenchymal morphology, exerting the maximal intensity of adhesive and traction forces needed for their motion. In particular, cell spreading is isotropic, as the homogenous geometry of the fiber networks does not lead to cell elongation in a preferential direction, as reproduced again in Figures 9.5 and 9.6, bottom–right panels and in the insets therein.

At the higher number of bundles, the extent of cell movement substantially decreases again in both 2D and 3D. However, the underlying biomechanical

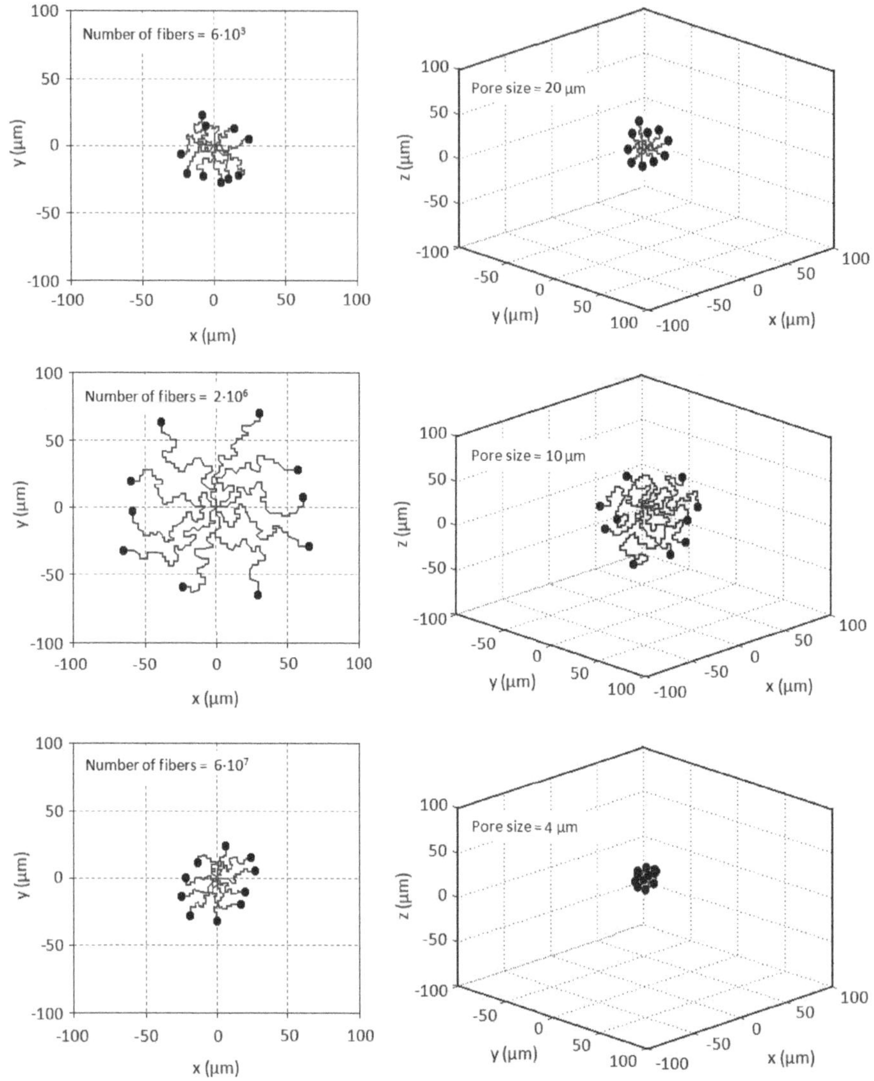

FIGURE 9.4: A bimodal relationship is observed also between cell migratory capacities and density of the fibrous component of the matrices, in both 2D (left panels) and 3D (right panels). In both cases, an increasing number of collagen-like threads is isotropically distributed as in the simulations referring to Figure 9.1. In particular, in three dimensions, increments in fiber number lead to decrements in pore sizes. Wind-rose graphs show 10 randomly chosen cell tracks over 12 h in the case of different densities of matrix fibers.

reasons and mechanisms are different. In the planar substrate, the overabundance of matrix threads causes cells to lose their preference toward collagen contacts, inhibiting their spreading and preventing a significant dispersion; see the inset in the bottom–right panel of Figure 9.5.

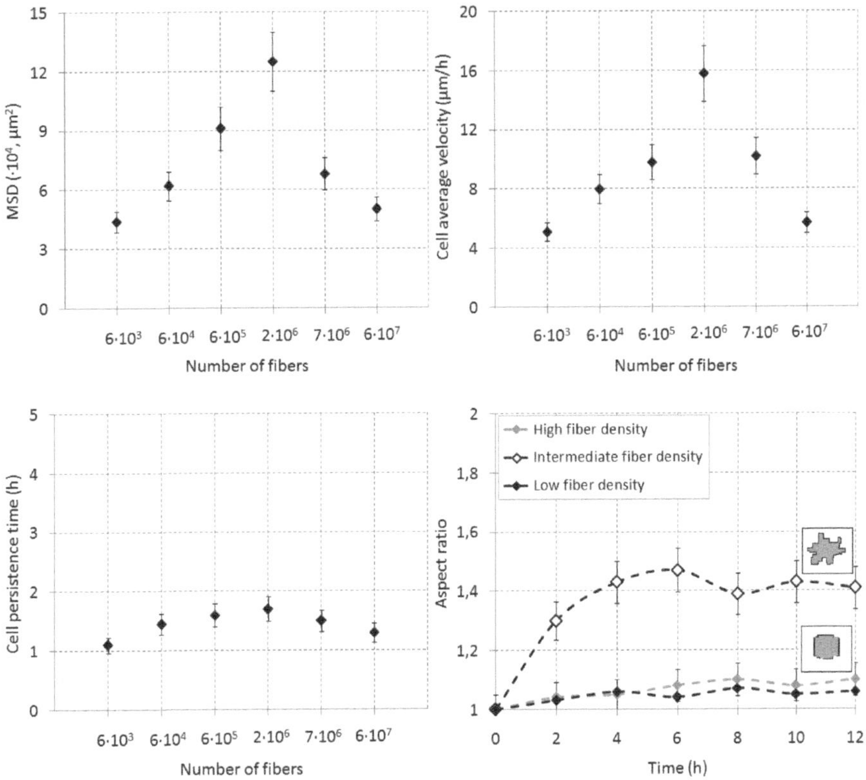

FIGURE 9.5: A bimodal relationship is observed between cell migratory capacities and the density of the fibrous component of the planar matrix. The other model parameters are the same as in the simulations of Figure 9.1. Cell mean square displacement (MSD), average velocity, and persistence time are given as functions of the number of fibers. In the bottom–right panel, the cell aspect ratio (see Section 9.2.1) is evaluated over time at low (i.e., $6 \cdot 10^3$), intermediate (i.e., $2 \cdot 10^6$), and high ($6 \cdot 10^7$) number of fibers. A biphasic behavior is indeed observed: at low fiber concentrations, cell morphology remains stationary, while it undergoes significant fluctuations at intermediate fiber densities (top black line). Finally, cell spreading decreases again for higher numbers of threads. Each value in the plots is shown as mean ± s.d. over 50 randomly chosen individuals.

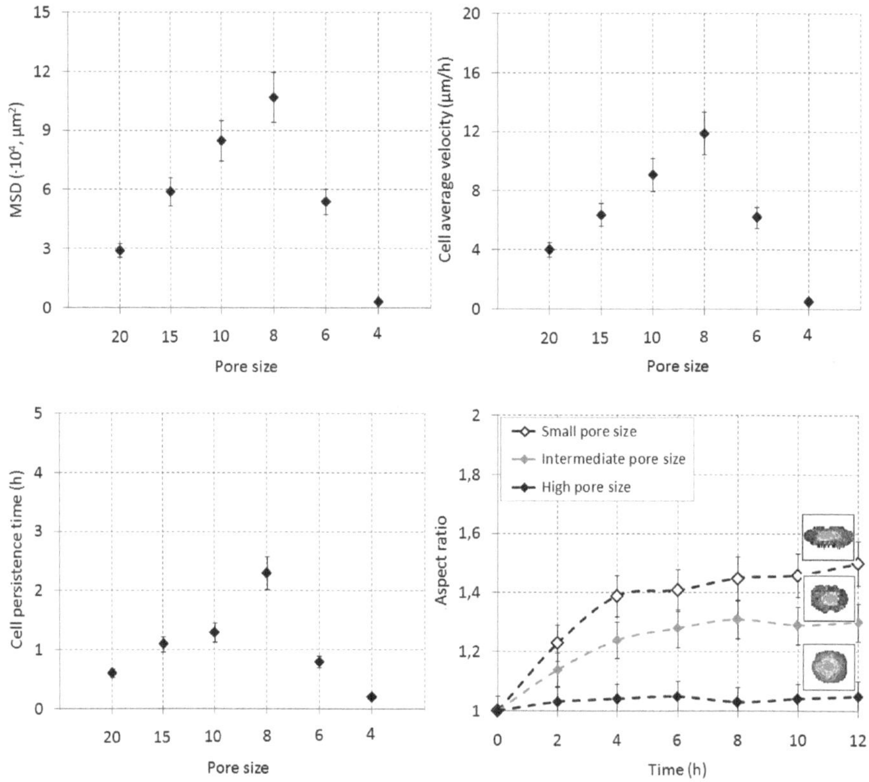

FIGURE 9.6: A bimodal relationship is observed between cell migratory capacities and the density of the fibrous component of the 3D scaffold (i.e., its characteristic pore size). The other model parameters are the same as in the simulations of Figure 9.1. Cell mean square displacement (MSD), average velocity, and persistence time are given as functions of the pore dimension. In the bottom–right panel, the cell aspect ratio (see Section 9.2.1) is evaluated over time at high (i.e., 20 μm), intermediate (i.e., 10 μm), and small (i.e., 5 μm) pore size. Cell elongation increases with decrements of pore dimension (i.e., increments in the number of fibers) until a sort of threshold value, defined by the measure of the nucleus. The rigidity of the nuclear region does not allow the entire individual to completely deform and squeeze through small pores. Each value in the plots is shown as mean ± s.d. over 50 randomly chosen individuals.

This interesting result, also found in [156] in the case of BALB/c 3T3 fibroblast cultured on a 2D type I collagen matrix, seems paradoxical. In general, it would, in fact, be expected that improved opportunities of cell-fiber contacts in the case of overabundant fibrous matrices would promote cell morphological transitions. A hint about the biophysical bases of this unexpected outcome is given by the fact that if the number of threads is above a certain threshold, moving individuals, in their search of cell-fiber adhesive interactions, do not need to spread in the surroundings. This explanation can be interpreted as the model counterpart of the so-called *integrin receptor saturation*, originally proposed and thermodynamically analyzed in [102], and observed in the phagocytosis of latex beads [171] and in the immune synapse between the antigen presenting cell and the T cell [168].

Finally, when the matrix is formed by a continuous carpet of fibers (i.e., $> 6 \cdot 10^7$, corresponding to more than $5 \cdot 10^5$ collagen-like molecules/μm^2), the cells' migratory capacity is clearly comparable to the case of the unpercolated substrate.

In the three-dimensional case, the abundance of matrix threads results in the formation of a scaffold characterized by small pores (i.e., half of a cell diameter or less), which prohibit any cell migration over long distances. An even complete stretch of cell cytosol, captured in the insets of the bottom–right panel of Figure 9.6, is in fact not sufficient to pass through such steric hindrances, as the nucleus cannot significantly deform ($\lambda_{\eta,N}^{surface}$ is high), causing the overall individual to be confined in a small area. In this regard, in the following we will examine how the migration ability of cells depends on the deformability of their nucleus.

The outcomes of the models are then consistent with the relative observations provided in the experimental literature. Different studies have, in fact, coherently shown that migration on planar substrates is limited for low fiber densities by the inability of the cells to form sufficient attachments to generate traction and to move forward [167, 228]. Optimal ligand densities, in contrast, preclude the formation of stable focal adhesions [17, 70, 224, 302] and require rapid focal adhesion turnovers, which result in maximal cell movements characterized by frequent membrane protrusions. Eventually, at high densities migration is blocked because integrin receptors engage into stable focal adhesions that exclude coordinated attachment-detachment for cell movement [123, 156]. Blocked migration due to stable focal contact formation is usually accompanied by an increased spreaded area (again, refer to [70]), which we, however, did not capture with our approach.

In 3D environments, neutrophil migration (both velocity and directional coefficient) has been reported to vary in a biphasic manner with the gel pore size [219], while mouse fibroblasts have been observed to migrate more significantly in collagenglycosaminoglycan (CG) scaffolds featuring small pore sizes, whereas they have exhibited less dispersion in matrices with larger pores [177].

9.6 Varying Cell-Fiber Adhesiveness

Cell-matrix adhesion is mainly mediated by integrins on the cell surface that form a link to connect ECM to the cytoskeleton and allow for the activation of signaling pathways. Adhesion can be modulated by a number of parameters, such as

1. The number of substratum ligands.

2. The expression and activation levels of integrins.

3. The resulting integrin–ligand binding affinity, which can be reduced by $\beta 1$ integrin antibodies that block integrin binding epitopes to ECM or by soluble ligands that compete with ligand binding, or can be enhanced by integrin–activating agents.

From the mathematical point of view, (1) has been just modeled by varying the fiber density of the substrate, whereas (2) and (3) by modulating cell-fiber adhesion parameter $J^{ext}_{E,F}$. Indeed, the number and the distribution of the collagenous threads remain the same as in the cases of Figure 9.1.

As a basic migration–adhesion relationship, the migratory capability of moving individuals can be sorted into the three regimes of high, intermediate, or low adhesion strength and in principle is valid for movements both over a surface or within a 3D matrix (see Figure 9.7). At high integrin engagement (say, $J^{ext}_{E,F} < 3$), cells display barely no detectable movement within the observation period, by being unable to detach from fibers. From an energetic viewpoint, cells minimize the Hamiltonian H by maintaining such an adhesive contact. Given the great difference between $J^{ext}_{E,F}$ and $J^{ext}_{E,M}$, moving individuals have in fact no benefit from further movements, meaning that an overly adhesive substrate causes the formation of integrin clusters on the cell surface that strongly bind to the substrate and do not allow the detachment needed for further migration.

Intermediate values of $J^{ext}_{E,F}$ (say, in the interval [3,6]) yield moderately high adhesive forces associated with a balance of attachment and detachment, allowing cells to efficiently move along 2D surfaces or within the fibrous network with a maximal distance covered.

Finally, above a certain value of $J^{ext}_{E,F}$ (say, > 6.5), adhesion is lacking, and consequently cells display barely detectable movement within the observation period. Given that $J^{ext}_{E,M} < J^{ext}_{E,F}$, cells actually prefer to fluctuate around the initial position in the interstitial fluid, avoiding contacts with the collagenous threads. Indeed, if a passive contact occurs, cells soon detach from the fiber without exerting the traction needed for further movements.

The biphasic distribution of MSD is associated with a similar corresponding distribution in velocity, but only a flat curve in persistence; see Figure 9.7. Therefore, the adhesion-dependent overall motility is mediated mostly by

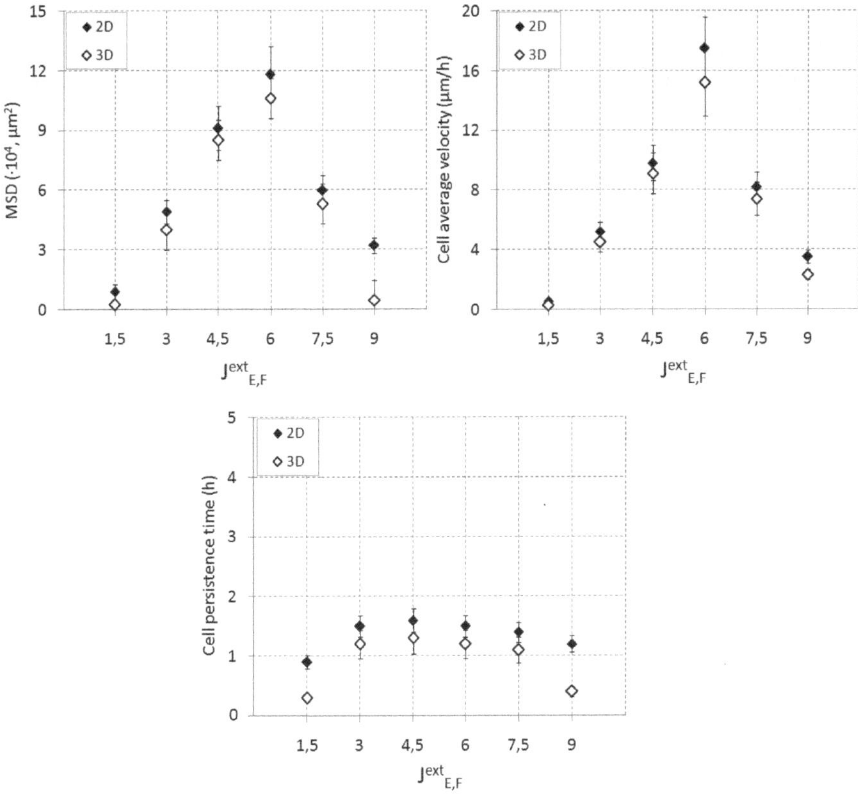

FIGURE 9.7: Biphasic relationship of cell motility and cell-fiber adhesion strength, given by the model parameter $J^{ext}_{E,F}$, for both 2D or 3D matrices. All other parameters remain unchanged, such as in the standard case of Figure 9.1. The values of cell MSD, average velocity, and persistence time are given as means ± s.d. over 50 randomly selected cells. The maximal migratory capacity is seen at intermediate values of cell-fiber adhesiveness.

cell velocity, whereas the persistent component of cell motion remains almost unaltered and refers to random movement (recall Figure 9.1). In particular, persistence levels smaller than 1 at the lowest and highest $J^{ext}_{E,F}$ levels correspond with a running on the spot phenotype observed in experimental assays for cells embedded in matrix [411].

Bimodal relations between cell migratory phenotype and cell adhesiveness with planar substrates has been captured in [106], where cell–matrix attachment strengths have been quantified by the critical shear stress measured using a modified radial-flow detachment assay, and in [167], where maximal locomotion of murine skeletal muscle myoblast over polystyrene surfaces coated with laminin (Ln) and fibronectin (Fn) has been demonstrated to occur at approximately one-third to one-half of the maximum adhesive strength. In [118], an analogous relationship has been instead found in the case of embryonic neural crest cells moving on bidimensional dishes coated with fibronectin molecules, where cell-substrate adhesiveness has been varied with the use of antibodies possessing different affinities for the $\beta 1$ subunit of fibronectin receptors (FnRs).

On the other hand, motile biphasic trends have been provided in recent 3D assays in the case of

- Human prostrate carcinoma cells, both parental and EGF-receptor overexpressed, whose speed has been plotted as a function of an adhesiveness parameter incorporating changes in functional ligands and receptor density at given Matrigel concentrations [422].
- Melanoma cells, cultured in collagen scaffolds, and stimulated with different concentrations of integrin-binding peptide RGD [50].

Finally, in [242] the authors have found that cell speed varied nonlinearly with modulations of the amount of ligand concentration, as it first increases, reaches a maximum, and then decreases again.

9.7 Varying Fiber Elasticity of 3D Matrix Scaffold

In the body, extracellular tissues display a range of elastic characteristics that are modulated by

1. The collagen content.
2. The amount of cross-links between collagenous molecules.
3. The presence of elastic fibers.

Rigid tissues are then dense and elastic tissues are often loose, and increasing the matrix density will increase their rigidity. In experimental studies using 3D ECM that were either modulated in density, i.e., fiber concentration [417], or rigidity [295, 296, 362, 367], the other component becomes influenced as well. However, to separate the related effects, we here simulate both varying scaffold stiffness (regulated by $\lambda_F^{perimeter}$) and the geometrical microstructure. To quantify such convoluting factors, we provide contour plots as joint functions of pore size and fiber elasticity that illustrate cell motile parameters as differently colored "landscapes" (see Figure 9.10). It is useful to emphasize that elastic fibers are also characterized by a low constant $T_F = 0.2$, as they are no longer rigid but can deform.

In Figure 9.10(A; top panels) at high pore size (i.e., 20 μm), cells display a reduced motile behavior, regardless of the fiber stiffness, as already shown in

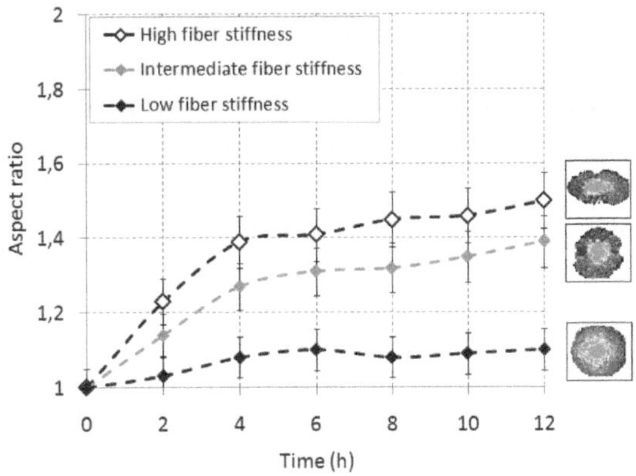

FIGURE 9.8: Dependency of cell deformation on fiber stiffness. Variations of cells' aspect ratio (defined as in Figure 9.5) during migration at different fiber elasticities: from bottom to top $\lambda_F^{perimeter} = 1, 5, 11$. In particular, the pore size is kept at 5 μm. The values are given as means $\pm$ s.d. over 50 randomly selected cells.

Figure 9.5(B). The rationale of this is that, in very loose tissues, cell migration is not supported by fibers from all sides of the cell. Consequently, the cell only migrates along single fibers that, however, when stiffened, again promote to some extent traction and therefore migration.

Next, at intermediate mesh dimensions, both cell velocity and persistence (and, consequently, the overall displacement) biphasically depend on matrix elasticity. If the collagenous threads are too elastic (e.g., $\lambda_F^{perimeter} < 3$), they can be easily deformed, without representing a sufficient anchor for the pulling force generation required for cell motion. With a moderate stiffness (e.g, $5 < \lambda_F^{perimeter} < 9$), the matrix fibers can be slightly arranged to form contact guidance cues, thereby facilitating cell migration. On the contrary, a too rigid network (e.g, $\lambda_F^{perimeter} > 9$) forms steric obstacles that can be somewhat less efficiently overcome by moving individuals.

Finally, small pore sizes allow motility only within elastic matrices, whereas migration is negligible for intermediate or high rigidities of the fibers. Migrating cells are able to move within small pores, in fact, only by significantly deforming the matrix network, creating an open space to pass through. Therefore, if the pore size is much smaller than the cell dimensions, the mechanical matrix characteristics exert an increasing influence. When evaluating the plots at constant rigidity, cell migration displays the same bimodal dependence on pore size previously captured in Figure 9.5(B).

Such variations in fiber rigidity induce a suite of cell morphological changes (see Figure 9.8). Cells plated within rigid scaffolds are typically well elongated and exhibit multiple fiber-associated constrictions along their bodies. They indeed adapt their shape and squeeze through rigid matrix arrangements by concomitant elongation of their body. Cells cultured in progressively soft matrices show instead decreasing elongation, and cells within complete compliant ECMs remain uniformly rounded, as they easily deform the collagen-like threads and lack cytoskeletal traction [397]. In summary, cells migrate in a biphasic manner at theoretical conditions of either increasing density or increasing stiffness alone, but, however, also at experimental conditions, upon the combined increase of density and stiffness together (imagine a decreasing curve in the plots of Figure 9.10 from left top to right bottom).

Bimodal relationships between cell migratory ability and the deformability of matrix scaffolds have been observed in experimental models of smooth muscle cells [310] and mouse fibroblasts, cultured in EDAC-cross-linked collagenglycosaminoglycan (CG) matrices with constant pore size [177]. A biphasic dependence on matrix rigidity has been previously reported in isotropic homogeneous networks, as in the case of prostate cancer cells coated in Matrigels with a fixed fibronectin level and variables stiffness [422]. Finally, the inhibition of cell motility in rigid ECMs has been also demonstrated in [397] for glioma cell lines.

Indeed, our results may be of particular relevance for the design of efficient synthetic biomaterials, used in tissue engineering applications and physiology. Matrix scaffolds with optimal values of pore size and stiffness may in fact accelerate cell in-growth into an initially acellular structure, which is a critical requirement for the development of implants for regenerative biomedical therapies.

9.8 Effect of Varying Nucleus Compressibility in 3D

Analysis of the results provided in the previous sections suggests that cells fail to neck down to micrometer dimensions and to migrate through steric barriers posed by matrix scaffolds in which the pore size is smaller than nuclear diameters and the component fibers are too rigid to be deformed. However, migration over significant distances in such highly constrained environments may be achieved with drastic deformations of their nucleus (whose rigidity is, as seen, the main reason of the halt in cell locomotion due to steric obstacles). Therefore, the degree of nuclear deformability may contribute to the migration efficiency of a cell. The nucleus elasticity is mainly regulated by both the chromatin structure and the lamin intermediate filaments that form a part of the nuclear envelope [144, 160]. The softness of a nucleus can be modeled by lowering the values of the nuclear rigidity $\lambda_{\eta,N}^{surface}$ from 8.5 (see Table C.10) to 0.5 (compare Figure 9.10(A) and (B)).

At high pore sizes of 10 μm or higher and lower fiber rigidity (i.e., at left–upper corner), migration remains unaltered regardless of nuclear elastic-

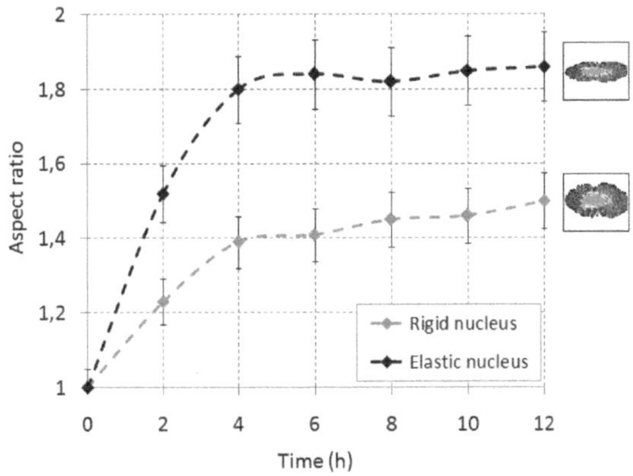

FIGURE 9.9: Dependence of migration-associated cell morphology over time on nuclear deformability. Evolution of cells aspect ratio during migration through small pore sizes (i.e., 5 μm) and high fiber stiffness (i.e., $\lambda_F^{perimeter} = 11$). The light line represents cells with a standard rigid nucleus ($\lambda_{\eta,N}^{surface} = 8.5$) and the dark line represents cells with an elastic nucleus ($\lambda_{\eta,N}^{surface} = 0.5$). All the other model parameters are the same as the standard simulation of Figure 9.1. Note that a high deformability of the nucleus allows cells to undergo dramatic morphological transitions fundamental for their migration in highly constrained environments. The values are given as means $\pm$ s.d. over 50 randomly selected cells.

ity, as moving cells do not experience steric hindrance. However, once mesh dimensions and scaffold rigidity move to intermediate ranges, nuclear elasticity somewhat facilitates cell movement, measured as MSD. The persistence time does not contribute to such an increase; i.e., cells with rigid nuclei were already able to migrate through the matrix networks characterized by intermediate dimensions without changing direction. On the contrary, the velocity does; i.e., they can quickly and efficiently remodel.

Finally, as pore size and matrix elasticity further decrease toward a highly constrained environment (lower–right corner), the simulations demonstrate that enhanced nuclear deformability is associated with enhanced cell migration. Such a facilitated locomotion is mediated by an elongated and deformable nuclear configuration allowing the entire cell to squeeze and stretch more easily and thereby pass through the steric obstacles of a dense and rigid matrix (Figure 9.9).

Our simulations relate to a number of experimental works, such as [408], where cell migration efficiency decreases with matrix density and is associated with nuclear deformation, or [27], where glioma cell lines significantly deform their nucleus upon recruitment of nonmuscle myosin II (NMMII) for squeezing through narrow locations in a brain model *in vivo*, thereby increasing their metastatic potential.

9.9 Effect of Matrix Degradation in 3D

In the previous sections, we have demonstrated that cells move within matrix fibers of varying density and stiffness that act as constraints, deforming both body and nucleus. As an additional mechanism to overcome limited space, cells may upregulate proteolytic enzymes that degrade ECM structure (i.e., matrix metalloproteinases, MMPs). They act either bound to the cell surface or when secreted into the extracellular space. Accordingly, cells degrade steric fiber obstacles either in a cell contact-dependent manner targeting locally confining fibers, or, in a diffusive manner leading to gradient formation and, consequently, a more overall weakening of the surrounding tissue structure [410]. As a result, barrier-free matrix spaces will be created, resulting in longer traveling distance.

The local concentration of the net proteolytic activity (both surface-bound or diffusive) is defined as $m(\mathbf{x}, t)$, and is assumed to evolve following a standard reaction-diffusion equation:

$$\frac{\partial m}{\partial t} = \underbrace{D_m \nabla^2 m}_{diffusion} - \underbrace{\lambda_m m \delta(\tau(\Sigma_{\sigma(\mathbf{x})}), M)}_{decay} + \underbrace{P}_{production}, \tag{9.6}$$

where $\delta(\tau(\Sigma_{\sigma(\mathbf{x})}), M) = 1$ in the interstitial medium M and 0 elsewhere. λ_m and D_m are, respectively, the decay rate and the effective diffusion coefficient of proteolytic enzymes. A low value of D_m models proteolysis being strongly localized in regions close to cell membranes, in agreement with experimental evidence in [336, 408]. $P(\mathbf{x}, t)$ models instead the local production of proteases either at the cell surface or secreted away from the external cell surface, at a constant rate π_m collagenous component:

$$P(\mathbf{x}, t) = \begin{cases} \pi_m & \text{if } \mathbf{x} : \tau(\Sigma_{\sigma(\mathbf{x})}) = C \text{ and } \exists \mathbf{x}' \in \Omega'_{\mathbf{x}} : \tau(\Sigma_{\sigma(\mathbf{x}')}) = M; \\ 0 & \text{else,} \end{cases} \tag{9.7}$$

where we recall that C stands for the cell cytosolic region.

The MMPs are capable of degrading the fibrous component of the matrix: to reproduce this biological effect, a lattice grid site $\mathbf{x}$ belonging to a degraded collagenous fiber becomes a generalized medium (fluid) site when its local level of MMPs, $m(\mathbf{x}, t)$, is sufficiently high (in our simulations above 2.5 μM). This change is implemented by changing its type τ from F (fiber) to M (medium), as in Chapter 3.

Referring to Figure 9.10(A) and (C), the comparison of cell migration of either MMP-active and MMP-inactive individuals reveals that at high and intermediate pore size and/or low matrix rigidity, the proteolytic machinery does not appreciably affect cell motion. The loose fiber network does not represent a significant obstacle for cell migration, which therefore is not enhanced further by MMP activity. In the case of small pores formed by rigid collagenous fibers (lower right), MMP activity promotes instead appreciable cell migration. This suggests that proteases, by degrading matrix fibers, are able to break steric obstacles in the close proximity of moving individuals, opening spaces for them to sample greater distances without turning back.

The role of MMP activity in cell migratory behavior captured by the model is in good agreement with the experimental results provided in [322] for dermal fibroblasts embedded in molecularly engineered polyethylene glycol (PEG) hydrogels, where a significant increment in the number of migrating individuals was observed upon upregulation of proteolytic enzymes.

In conclusion, summarizing all the examined parameters, cell migration is greatly influenced by a number of complex ECM- and cell-derived characteristics that, in addition, display a number of interdependencies [146] and together determine the net outcome on migration.

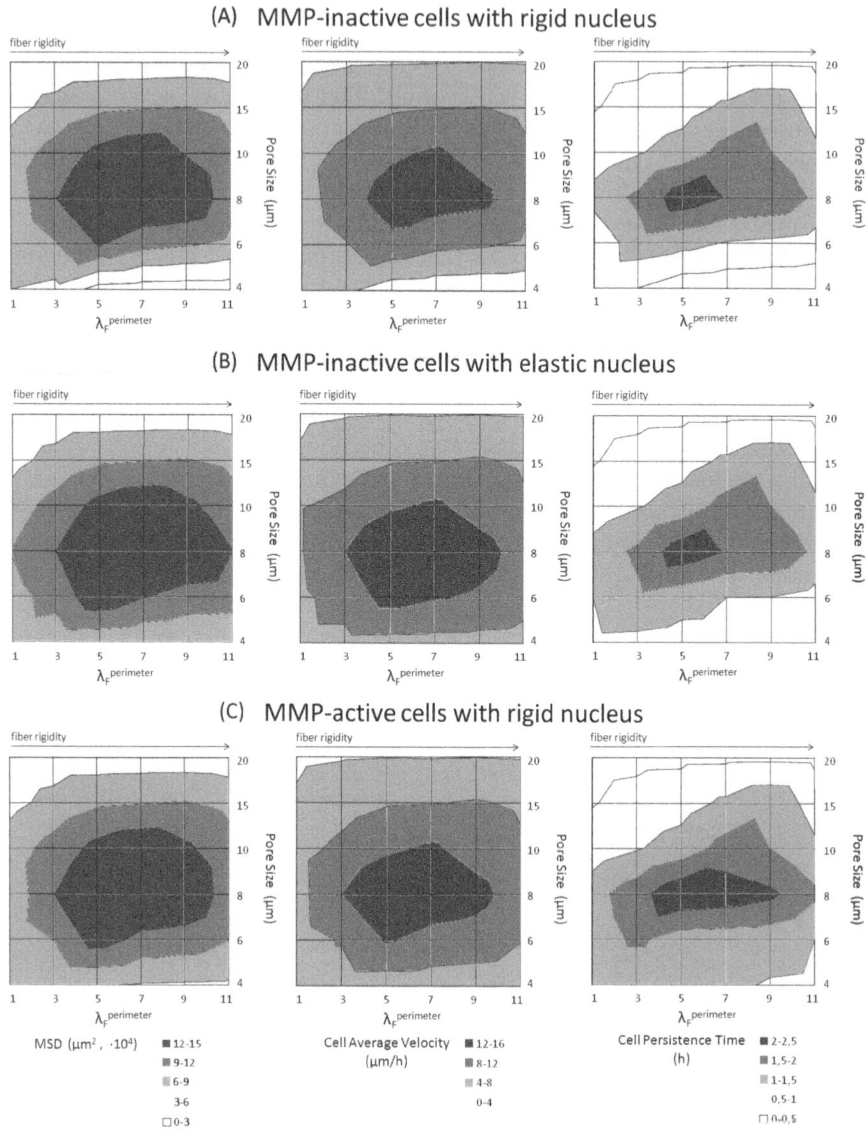

FIGURE 9.10: (See color insert.) 3D cell migration as a function of fiber rigidity $\lambda_F^{perimeter}$ and pore size. Contour plots of cell migratory parameters for (A) standard conditions (see Figure 9.1) that further include the absence of proteolytic activity and a stiff nucleus ($\lambda_{\eta,N}^{surface} = 8.5$), (B) cells as in (A) but with an elastic nucleus ($\lambda_{\eta,N}^{surface} = 0.5$), and (C) cells as in (A) but expressing matrix degrading enzymes, regulated by Equation (9.6). Each value is given as the mean over 50 randomly chosen individuals (see Section 9.2.1).

Chapter 10

Cancer Cell Migration in Matrix Microchannels

10.1 Biological Introduction

Cell migration on and within matrix environments plays a critical role in a huge number of physio-pathological phenomena, both during embryo development and in the adult organism, as observed in the introduction of the previous chapter.

In particular, tumor cells are able to abandon their primary site and migrate through the surrounding parenchyma, in order to enter the circulatory system and invade other healthy tissues. On this journey, cancer individuals need to continuously regulate their migratory and invasive behavior, as they are exposed to a variety of biochemical and biomechanical interactions, modulated both by the biophysics and by the microstructure of the environment [110, 141, 218]. *In vivo* connective tissues can, in fact, provide interstitial extracellular matrices (ECMs) with heterogeneous composition, density, and organization: they can comprise both loose regions formed by sparse collagenous fibers and areas with tightly packed threads.

In the first case, analyzed in Chapter 9 and in several experimental [122, 124, 146, 409] and theoretical [353, 421, 422] works, important physical determinants of cell movement include the degree of ECM alignment, the width of resulting pores, and the ECM stiffness.

In the second case, matrix environments behave instead as physical barriers, adjacent to channel/track-like spaces, which, depending on their dimensionality, may either guide, hinder, or completely prevent cell movement [189, 408]. Remarkably, recent studies have reported that cells achieve significant movement in highly constrained physical spaces, determined by dense matrices, by drastic morphological deformations, in addition (or even in substitution [411]) of the proteolytical degradation of the local environment [408]. The repeated adaptation of cell shape requires substantial reorganization both of the cytoskeleton and of other organelles, and in particular of the nucleus, the most voluminous and rigid intracellular compartment. This type of migration, called *amoeboid movement* since it is associated with flexible amoeba-like shape changes, proceeds in a poorly adhesive mode and is mostly independent from specific molecular and biochemical interactions with the surrounding system [172]. Notable, more aggressive and metastatic cancer cells have been recently shown to have a significant compliance with respect to nonmetastatic individuals, which remain stuck at the vessel walls or when crossing the endothelium and therefore are forced to be confined in the primary site.

For these reasons, a quantitative assessment of cell deformability has the potential to be of significant value for diagnostic purposes, such as screening and cancer grading, and for a more detailed prediction of the course of the disease in individual patients [347]: indeed, it has given rise to a number of *in vitro* models. For instance, 3D lattices consisting of reconstituted fibrillar collagen are typically used to study cell migration in structures mimicking highly confined *in vivo* connective tissues [409]. However, in most cases, these bioengineered scaffolds lack well-controlled spatial characteristics, because small and large pores results from the stochastic fiber polymerization processes, therefore failing to recreate defined trails and barriers [122, 319, 409]. Indeed, specific mechanisms of reassembly of fibrillar matrices have been recently combined with microlaser procedures able to generate predefined tracks that create spatially defined patterns of connective tissue organizations [196]. In this regard, geometrical characteristics of 3D matrix environments can be easily controlled and modulated also with migration assays whose key features are micro-sized channel structures [197, 332].

Such different types of experimental systems are here reproduced and simulated by the extended compartmentalized CPM. The method is used to extract the main features of tumor cell invasiveness by working with a 3D channel environment. As an outcome, we focus on the experimentally addressable characteristics of cell locomotion, i.e., overall displacement and velocity, predicting how these quantities are influenced by manipulations either of the geometrical features of the channels, or of the biophysical properties (i.e., elasticity) of the cells themselves. We then use the simulated migration chip to compare the migration of cells inside the microchannels and their movement on a 2D flat surface, like the one located before the entrance of the channels. Consistently with experimental observations on different tumor cell lines, our approach allows one to discern the effect of the mechanical rigidity of each cell compartment (i.e., the nucleus and the cytosolic region) in the migration capacity of the entire individual. Moreover, our findings provide evidence of the facts that migration characteristics of cells are very different in 2D and in highly constrained 3D environments and that even the underlying dynamics change.

10.2 Mathematical Model

The model of the microchannel device is analogous to the one presented in the previous chapter.

The simulated cancer cells, called η of type $\theta = E$, are defined as bicompartmentalized units, composed of the nucleus, $\tau = N$, and the surrounding cytosol, $\tau = C$. Also, in this case, they are not assigned an internal vector: the purpose of the model is to analyze the influence on their movement of geometrical and elastic characteristics and not of their molecular state. The extracellular environment is differentiated into a medium-like state, $\tau = M$, and a polymeric-like state, $\tau = P$. The medium-like state represents the standard mixture of soluble components, which, together with the water solvent, compose the interstitial fluid. The polymeric state reproduces instead the structured wafer, which, after subsequent replica molding processes, is typically covered with fibronectin-based solutions and used in the standard microfabricated channel migration chip; see [332] and the references therein.

The system *Hamiltonian* is defined as

$$H(t) = H_{shape}(t) + H_{adhesion}(t) + H_{persistence}(t). \tag{10.1}$$

For each cell η, H_{shape} models the geometrical attributes of its subcellular compartments, which are written as nondimensional relative deformations in the quadratic form of Equation (4.6). Assuming that cells do not significantly grow during migration (which is consistent with the time-scale of the phenomenon of our interest), the fluctuations of their volumes are again kept negligible with high constant values

$$\lambda_{\eta,N}^{volume} = \lambda_{\eta,C}^{volume} \gg 1,$$

for any individual η and for Σ_σ such that $\tau(\Sigma_\sigma) \in \{N, C\}$. The stiffness of the intracellular compartments (given, respectively, by $\lambda_{\eta,\Sigma_\sigma}^{surface} = \lambda_{\eta,C}^{surface}$, for $\tau(\Sigma_\sigma) = C$ and $\lambda_{\eta,\Sigma_\sigma}^{surface} = \lambda_{\eta,N}^{surface}$, for $\tau(\Sigma_\sigma) = N$) will be instead discussed in the results section.

$H_{adhesion}$ is differentiated in the contributions due to either the generalized contact tension between the nucleus and the cytoplasm within the same cell, or to the effective adhesion between cells or between a cell and a channel wall. In particular, $J_{N,C}^{int} \ll 0$ implicitly models the forces exerted by intermediate actin filaments and microtubules to anchor the nucleus to the cell cytoskeleton, preventing cells from fragmenting. A null contribution is instead given to the adhesive interactions between a moving cell and an extracellular component (i.e., we assume $J_{E,M}^{ext} = J_{E,P}^{ext} = 0$). This choice, successfully used in another similar model of *in vitro* cell migration [350], is done to analyze the direct influence of cell deformability on its motile behavior, and is consistent with the experimental literature, which widely demonstrates that most cell lines display sustained amoeboid motility in confined environments in a poorly adhesive mode [172, 225, 332]. $J_{E,E}^{ext}$ is finally kept high to avoid cell–cell adhesive interactions that may affect the early phases of their movement (i.e., on the flat surface outside the channels).

Finally, if moving cells are able to polarize, i.e., to differentiate in a leading and a trailing surface, they display a directional movement dictated by their longer axis, implemented with a term analogous to Equation (2.4). In particular, $\Delta t = 60$ MCS, whereas μ^{pers}, which controls the persistence time, is again a function of the length of the cell:

$$\mu_{pers}^{\eta}(t) = \mu_{pers,0}\left[\frac{L_{\eta}(t)}{L_0} - 1\right], \tag{10.2}$$

As in Equation (6.4), L_η is the current length of the longer axis of the cell, which is approximated with an ellipsoid [260], and L_0 is the initial cell diameter. Obviously $L_\eta \geq L_0$, since we have assumed that the cell deforms but does not grow during migration.

Given the *Hamiltonian*, the transition probability of a spin flip has the form of Equation (4.24). In particular, we use $p(T_{\Sigma_\sigma}(t)) = \tanh(T_{\Sigma_\sigma}(t))$. Indeed, for each cell η and for $\tau(\Sigma_\sigma) = N$, $T_{\eta,\Sigma_\sigma} = T_{\eta,N}$ gives the agitation rate of its nucleus, while, for $\tau(\Sigma_\sigma) = C$, $T_{\eta,\Sigma_\sigma} = T_{\eta,C}$ is as usual a measure of the intrinsic motility of the overall individual. Indeed, for each cell, $T_{\eta,N}$ is a low value (< 1), resulting in the passive motion of the nucleus, which, unable to move autonomously, is dragged by the surrounding cytosol, characterized instead by a high $T_C \gg 1$ (see Chapter 6). The matrix structure is instead fixed (i.e., $T_P = 0$). A summary of the parameters used in the model is given in Table C.11.

10.3 Simulations

The simulation domain $\Omega \subset \mathbb{R}^3$ is a $480 \times 288 \times 48$ regular grid, with periodic boundary conditions in the y direction and no flux in the others. The characteristic size of each grid site is 1 μm. The lattice reproduces a micro-fabricated device with channel structures of various widths and a planar surface just outside their entrances; see Figure 10.1. This architecture is typically used in the literature to analyze cell migration both on open spaces and through precisely confined environments; see [332] and the references therein. The temporal resolution of the model is an MCS, which is set to correspond to 2 s in order to compare the simulated cellular dynamics with the relative experimental observations. The overall observation time is set equal to 8 h ($\approx$ 14400 MCS) to ensure the development of sufficiently long migration paths, as done again in [332]. Initially, the cells are seeded on the planar substrate in close proximity of the channel entrances and display an unpolarized morphology. Indeed, they are hemispheres with a diameter of 30 μm, while the nucleus, whose location and geometry is estimated from experimental images, is a central sphere with a diameter of 10 μm. These dimensions, given in Table C.11, reflect the mean measures of human pancreatic epithelial cancer cells (Panc-1) [28, 332].

Following the definition used in [332], we here distinguish cell migratory behavior as follows:

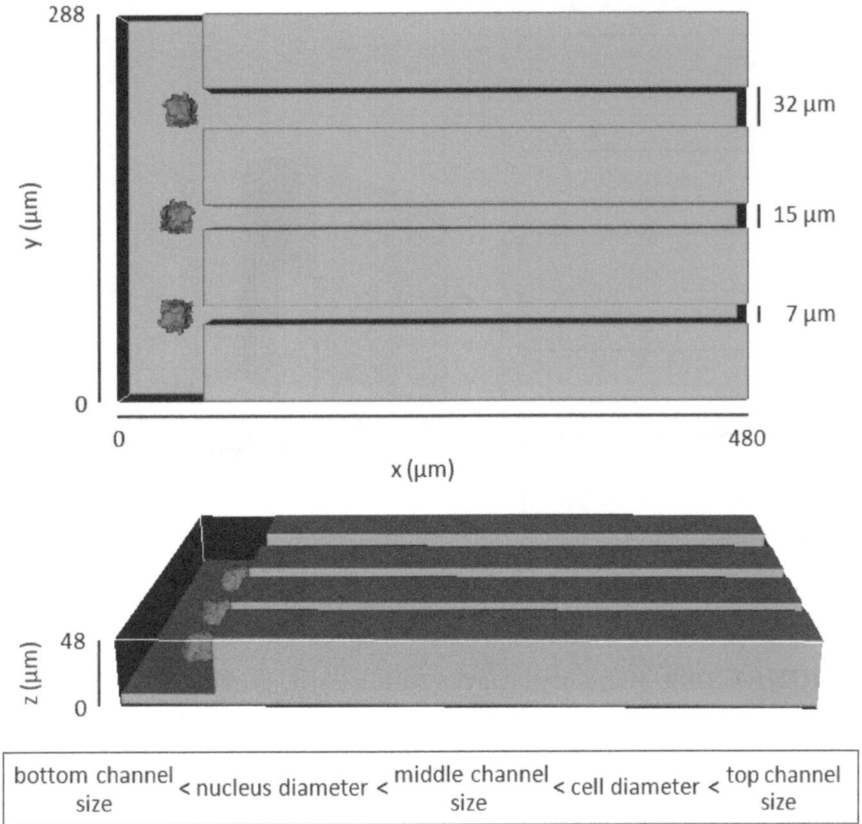

FIGURE 10.1: Sections of the simulation domain Ω. The lattice reproduces a micro-fabricated device with channels of various widths and a planar surface just outside their entrance. In particular, the top channel is larger than the overall cell diameter, while the bottom channel is smaller than nuclear dimensions. Initially, the cells are seeded on the flat substrate near the channel walls.

- Cells that only penetrate a channel with a part of their cytoplasm but not with the nucleus are called (partially) *penetrating*.
- Cells that completely enter in a channel structure but are not able to migrate to the other side within the observation period are called *invasive*.
- Cells that reach the opposite border of a channel are finally called *permeative*.

In the first set of simulations, we force cancer cells to maintain their hemispherical shape by setting high $\lambda_{\eta,C}^{surface} = \lambda_{\eta,N}^{surface} = 15$. This is the model

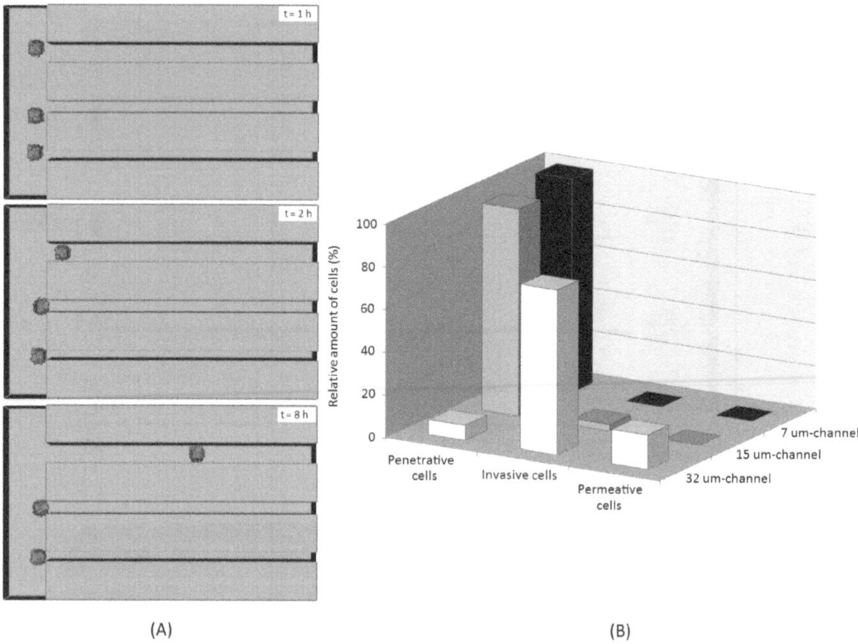

FIGURE 10.2: Migratory behavior within the microchannel structure of cancer cells unable to remodel at all (i.e., characterized by a high $\lambda^{surface}_{\eta,\Sigma_\sigma} = 15$, for $\tau(\Sigma_\sigma) = \{C, N\}$). (A) Images of a time-lapse simulation taken at $t = 1$ (top panel), 2 (middle panel), and 8 (bottom panel) h. Only the largest channel can be entered by moving individuals, which, forced to remain in their initial hemispheric morphology, are prohibited from squeezing through smaller structures. (B) Summary of cell migratory behavior within the matrix device. In the case of middle and small channels, cells display a penetrative non-invasive phenotype, whereas in the case of the biggest channel, they are typically invasive. The quantitative evaluation of specific cell motile phenotypes, represented in the histogram plot, is obtained by performing 100 simulations.

counterpart of the activity of phalloidin-like compounds, which block the reorganization of the cell cytoskeleton by inhibiting actin–myosin interactions. As reproduced in Figure 10.2, all individuals initially show a random migration on the flat surface (i.e., until nearly 2 h). Then, when approaching the microchannels, they start walking along them. In this regard, it is useful to emphasize that this preferred cell movement is completely autonomous, as we do not include in the model any chemical gradient or bias, or any *a priori* direction for moving individuals.

In the case of the channel with the largest cross–section, the cells typically display invasive behavior, as they are able to enter and migrate within the structure, but cannot reach the opposite border, as shown in Figure 10.2.

Migrating individuals in fact fluctuate and "rebound" between channel internal walls, whose width does not represent a significant geometrical *contact guidance*. Indeed, cells do not reach an appreciable directional velocity, remaining almost in the middle of the structure. Cell migration both in the middle and in the small channel (whose widths are smaller than cellular dimensions) is instead characterized by a penetrative phenotype. Without the possibility to deform in a substantial way, the individuals are in fact prohibited from squeezing through the confined environments. Therefore, they continue to isotropically wander in the close proximity of the channel entrances.

We then analyze how the motile behavior of cells is affected by the remodeling of their cytoskeleton. Indeed, the elasticity of the cell cytosolic region is allowed by a lower value of $\lambda_{\eta,C}^{surface} = 0.5$, whereas the nuclear cluster is maintained rigid by retaining the previous high value for $\lambda_{\eta,N}^{surface} = 15$. With respect to the previous set of simulations, the initial nondirected movement of cells on the planar substrate is accompanied by a significant spreading (i.e., until a 1.3–fold increase in surface area; see Figure 10.3(A, top panel)). As reproduced in the same image, the cell migratory phenotype in the largest channel remains unaltered, regardless of cytosolic deformation abilities. As already observed, moving individuals in fact do not experience significant steric hindrances that require substantial morphological changes and almost display the same invasive phenotype.

At the intermediate channel width (i.e., smaller than cellular dimensions and bigger than nuclear dimensions), cells with a deformable cytoplasm are instead able to squeeze into and move within the microstructure. In particular, they remodel toward an elongated mesenchymal shape and typically migrate to the other end of the device, displaying a common permeative phenotype; see again Figure 10.3. The transition from a stationary cell morphology to a polarized shape, which is completely self–generating and due to the geometry of the matrix environment, is fundamental in determining the persistent component in cell movement. From a modeling viewpoint, cell elongation in fact increases the relative magnitude of the persistence term (10.2) in the overall hamiltonian, given by coefficient μ_{pers}^{η}. Therefore, once a cell has established the direction of movement within a channel, it is energetically disadvantageous to change direction and is forced to maintain the direction of locomotion.

Finally, cells are still not able to enter the smallest channel. The front end of their cytoplasm quickly extends into the structure, while the voluminous nuclear region cannot deform and passes through a highly constrained space, therefore inhibiting the individual from pulling inside its entire body, as clearly reproduced in Figure 10.3(C).

Our results are consistent with the experimental observations provided in [28], where the authors used a Boyden chamber assay to correlate an increment in the ability of pancreatic cancer cells (Panc-1) to squeeze and migrate through microporous membranes to a drop in their elastic modulus, measured by a micro-plate based single-cell stretcher.

We next address the question of to what extent a variation in cell nucleus

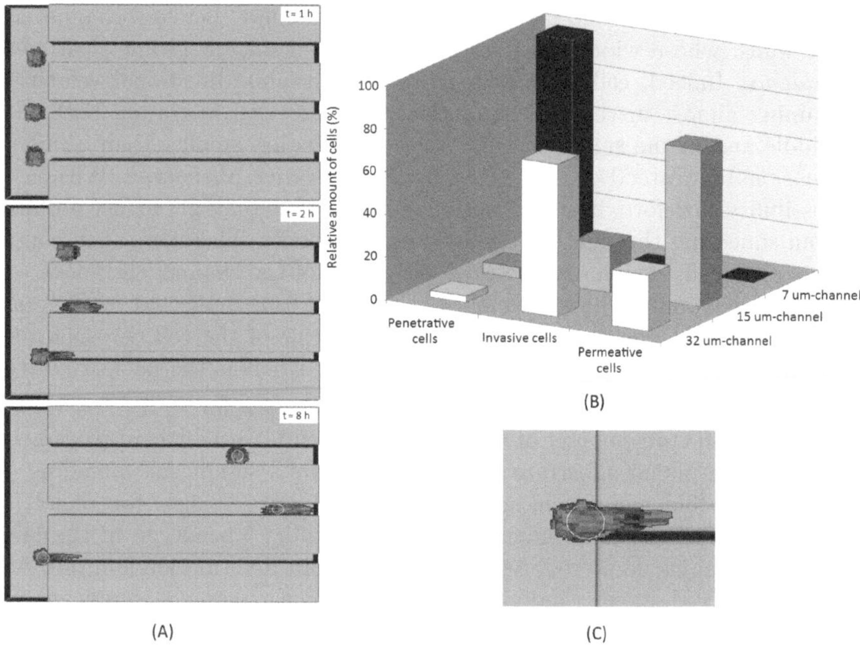

FIGURE 10.3: Migratory behavior within the microchannel structure of cancer cells with an elastic cytosol and a still rigid nucleus (i.e., $\lambda_{\eta,C}^{surface} \ll \lambda_{\eta,N}^{surface}$). (A) Images of a time-lapse simulation taken at $t = 1$ (top panel), 2 (middle panel), and 8 (bottom panel) h. By dramatically remodeling their cytoskeleton, cells are now able to also enter the intermediate channel. However, due to the presence of a stiff nucleus, they cannot penetrate the smallest structure. For better visualization, in the bottom panel, the nucleus is encircled manually. (B) Summary of cell migratory behavior within the matrix device. In the case of the widest channel, cells display the same invasive behavior as in Figure 10.2. In the case of the middle channels, they instead acquire a permeative phenotype. Finally, in the smallest channel, cells are still penetrative but uninvasive. The quantitative evaluation of specific cell motile phenotypes, represented in the histogram plot, is obtained by performing 100 simulations. (C) Magnification of moving cell in the close proximity of the smallest channel entrance. It is possible to see how the stiff voluminous nucleus is not able to pass through the confined space, allowing only the penetration of part of the cytosol.

elasticity enables cell invasion and movement into highly confined spaces. Indeed, we lower the model parameter $\lambda_{\eta,N}^{surface}$, fixing it equal to 0.9 (it however remains higher than $\lambda_{\eta,C}^{surface} = 0.5$, since the cytosolic region is typically softer). This mimics either induced reorganizations of the chromatin structure and/or of lamin intermediate filaments (whose assemblies form a part of the

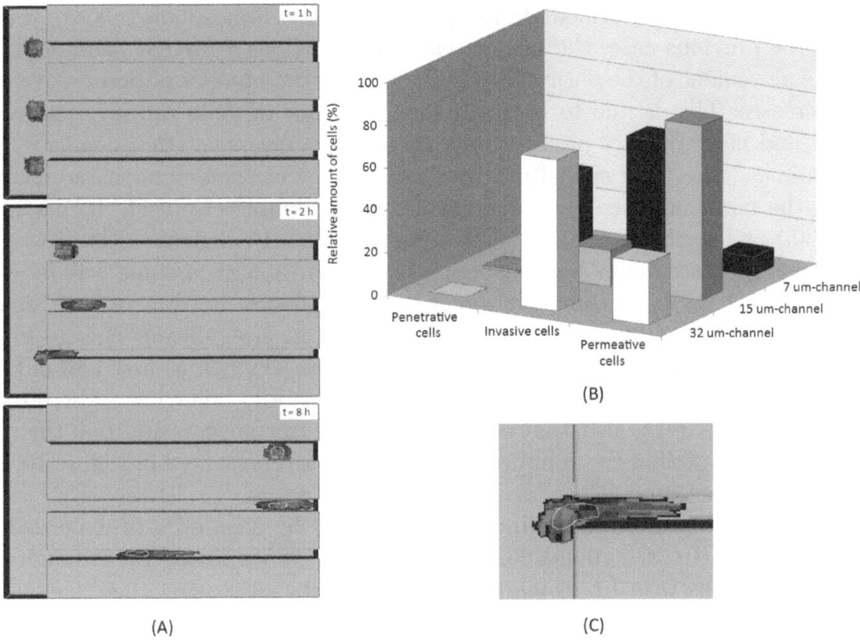

FIGURE 10.4: Migratory behavior within the microchannel structure of cells with an elastic cytosol and a deformable nucleus (i.e., $\lambda_{\eta,N}^{surface} = 0.9$). (A) Images of a time-lapse simulation taken at $t = 1$ (top panel), 2 (middle panel), and 8 (bottom panel) hours. The enhancement in nucleus elasticity enables cells to enter also the smallest channel. For better visualization, in the bottom panel, the nucleus is encircled manually. (B) Summary of cell migratory behavior within the matrix device. In the case of the widest channel, cells display the same invasive behavior as in Figure 10.2. In the case of the middle channels, the permeative phenotype is enforced. Finally, in the smaller channel, cells acquire an invasive ability. The quantitative evaluation of specific cell motile phenotypes, represented in the histogram plot, is obtained by performing 100 simulations. (C) Magnification of moving cell in close proximity to the smaller channel entrance. It is obvious that the nucleus can now squeeze to pass within the channel entrance, allowing the entire individual to invade the structure.

nuclear envelope mediating its rigidity [144, 160]) or experimental treatments with micromolar concentrations of bioactive lipid sphingosylphosphorylcholine (SPC), whose activity leads to a substantial modification of the keratin network toward a perinuclear rearrangement [28, 361].

The analysis of the model outcomes, summarized in Figure 10.4, reveals that, in the case of the largest channel, the deformability of the nucleus does not appreciably affect cell migratory behavior. When the channel dimension decreases to the intermediate range, though the cell morphology looks similar to the previous case, the nuclear elasticity provides a further facilitation for the movement of the entire individual. In fact, the number of permeative cells increases. This is due to both the fact that moving cells can now more easily and more quickly pass through the channel entrance (the so-called *entry time*, i.e., the time span from the first contact of the cell with the channel to the moment when the whole cell body is completely inside it, decreases [230]) and to the fact that they can more efficiently migrate within its walls; also their voluminous nuclear region is, in fact, able to assume a more elongated and "mobile" configuration; see Figure 10.4(A, bottom panel). Finally, the enhancement in nucleus elasticity results in a significant change in the migratory phenotype of the cells in the smallest channel, as now half of them can invade the structure. This change in the motile behavior is the obvious consequence of the fact that cells are now able to compress and pull their nuclear region within the confined environment, as represented in Figure 10.4(A, bottom panel) and (C).

The computational findings are supported by a number of experimental approaches. In [27], glioma cell lines have been shown to squeeze through narrow locations in a brain model *in vivo*, thereby increasing their metastatic potential, by significantly compressing their nucleus upon recruitment of non-muscle myosin II (NMMII). Moreover, very recently, other authors have been shown that the directional persistence of cancer cells in microsized structures is mainly regulated by the steric hindrance due to the presence of a rigid and voluminous nucleus [197, 198]. In particular, in [28, 332], Panc-1 cells have been shown to overcome size exclusion in microchannels architectures upon treatments with bioactive SPC. Finally, in [230], the authors have demonstrated that the number of acute promyelocytic leukemia (APL) cells able to migrate through filters of 5 μm (i.e., smaller than cell diameters) significantly reduced upon exposition to paclitaxel, which stabilizes the intracellular microtubule network. Analogous results have been provided therein also with primary human neutrophils moving in 3–μm-size pores.

10.4 Migration Velocities

In Figure 10.5, we analyze the *average velocity* ($\overline{\mathbf{v}}_\eta$, defined as in Equation (1.13)), both on the flat surface and within the different channel structures. This comparison allows one to elucidate and to further quantify the differences between cell migratory phenotypes in specific geometric conditions. The random movement of rigid cells on the planar substrate is characterized by a low 0.4 μm/min-speed, which slightly increases for individuals with a deformable cytoplasm, reaching almost 0.6 μm/min. Indeed, there is no further variation upon the enhancement of the nuclear compressibility. An analogous behavior,

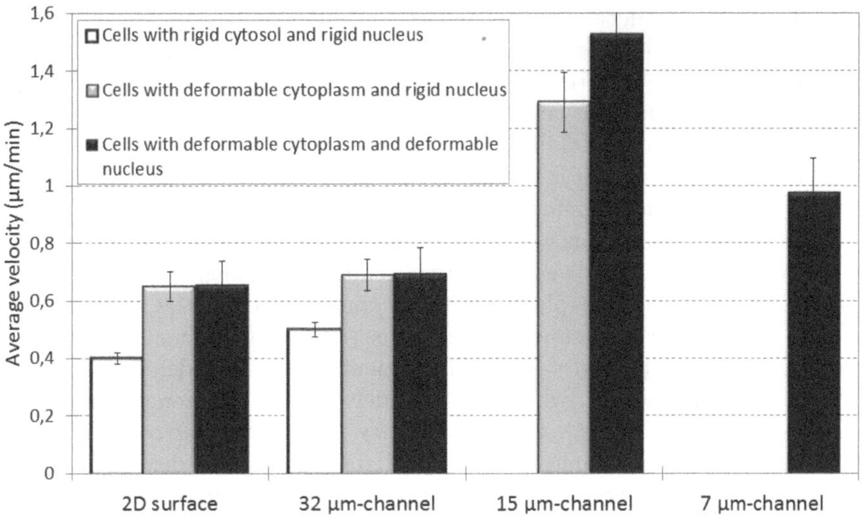

FIGURE 10.5: Comparison of the average migration speed of cells moving either on the flat surface or within the different microchannels. Cell velocity on both the bidimensional substrate and in the larger channel is substantially low, and does not significantly depend on the cytosol and nucleus rigidities. On the contrary, cell speed within the more confined structures is generally two-fold higher and is further enhanced by the possibility of nucleus deformability. The values of the cell average velocity are represented as means ± s.d. over 50 "effective" realizations. This means that, for this statistical analysis, we only consider cells that completely enter within the channel of interest (i.e., with an invasive or permeative phenotype). The others are classified as nonmotile, assigned an undefined velocity, and are not taken into account. Finally, if in a given case the number of cells entering a channel structure is not significant, we do not evaluate the relative average speed. In particular, statistical significance ($p < 0.05$) is determined via both the Students' t-test for motile fraction data and via the Kolmogorov–Smirnov test for non-normally distributed data sets. The specific values regulating the elastic properties of the cell subcompartments are those used in the relative sets of simulations.

with almost similar values, is observed for cell migration in the largest channel, confirming that extracellular environments whose dimensions are greater than cellular measures do not represent guidance cues, but rather behave as open spaces.

In the case of the intermediate channel, cells with an elastic cytosol display instead an approximately two-fold increment in cell migration speed (≈ 1.1 μm/min), which is significantly enhanced when allowing nuclear deformability (≈ 1.5 μm/min). The different migration speeds of cells within either the big or the middle structure indeed reflect the different overall phenotypes, i.e., invasive and permeative, respectively.

Finally, in the case of the smallest channel, the cell average velocity (obviously evaluated only for cells with an elastic nucleus) slightly decreases again to less than 1 μm/min. The explanation resides in the fact that although the nucleus is deformable, it is however stiffer (and less motile) than the surrounding cytoplasm and therefore takes more time to remodel and move, slowing the overall individual, as we will described below in greater detail.

These results first reveal two distinct migratory phenotypes that are proposed to occur for cells placed either in open structures (i.e., 2D surfaces or large channels) or in confined architectures (i.e., channels smaller than cellular dimensions). In the first case, the movement of cells is widely independent of their elastic properties, whereas it is widely known to strongly rely on their adhesive strengths [225]. On the contrary, the efficacy of cell migration in 3D constrained environments is mainly determined by the deformation ability of the moving individuals (in particular, of their voluminous nucleus), which adapt to the geometrical characteristic of the environment. In this regard, it is indeed possible to identify an optimal dimension of an extracellular structure that results in sustained cell locomotion: smaller than the cellular measures but higher than the nuclear diameter.

The drastic differences in cell migration speeds due to the specific environment dimensionality captured by the model are consistent with recent studies performed on NIH-3T3 fibroblasts [110], leukocytes [225], and pancreatic cancer cells [332]. In these cases, the authors have in fact demonstrated that cell movement in 3D confined structures is more rapid, uniaxial, and closely dependent on cell morphological transitions. In striking contrast, the basic program of cell migration over flat ECM substrates strongly requires a dynamical adhesion to their environment via adhesive molecules, i.e., integrins, which generate the force necessary for propulsion and movement [7, 229]. Analogous conclusions have been found also by a theoretical model that has reported that the migration of cells in a microsized channel strongly depends on partial pressure differences formed between the channel walls and either the leading or the rear edge of the individual, without the necessity of specific cell-surface adhesion molecules [181].

10.5 Migration Modes

We finally investigate whether in the different cases cells display a specific phenomenological/mechanical movement. In particular, in order to elucidate if cell distinct parts show coordinated movement, we separately tracked the leading edge, the nucleus, and the rear edge of the cells, and we plot their absolute position inside a channel versus time in a 2-h span.

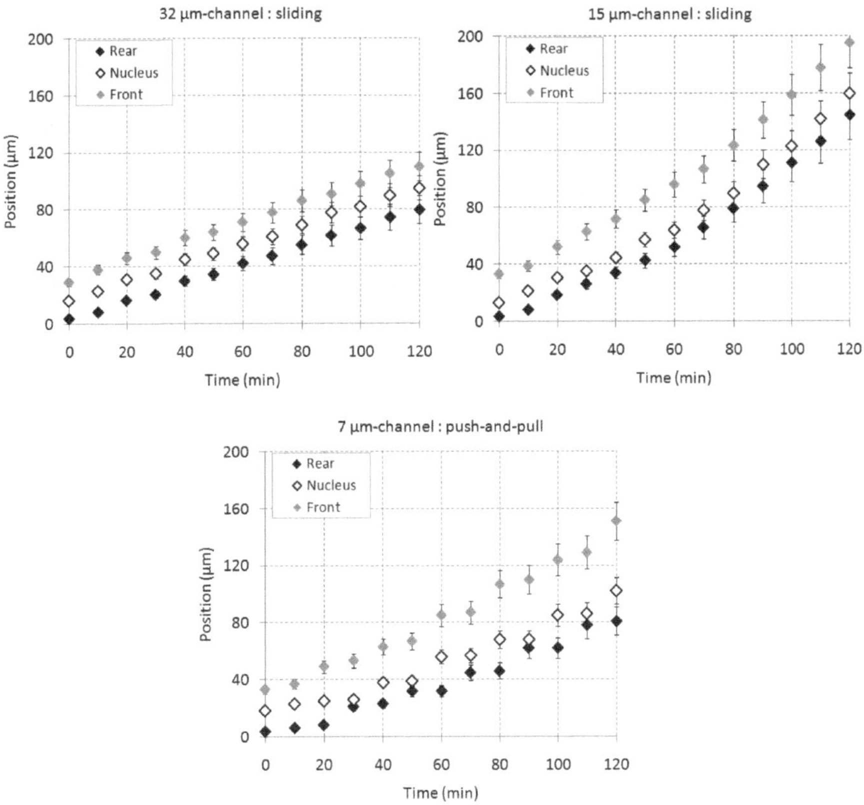

FIGURE 10.6: Cell migration dynamics inside the different channels in the case of cells able to remodel both their cytoplasm and their nucleus. Typical profile of a sliding-like migration pattern, where the cell front, nucleus and rear are moving in a synchronous way, is observed for large and intermediate channel structures (top and middle panels). Typical profile of a push-and-pull-like migration behavior, where the cell front smoothly moves while rear moves in an oscillating manner (i.e., depending of the time taken by nucleus to deform and displace) is found in the case of the smallest channel (bottom panel). The relative position of the different cell parts is plotted as the mean $\pm$ s.d. over 50 "effective" (see definition in Figure 10.5) realizations.

Indeed, we consider the case of completely deformable individuals (i.e., with elastic subcompartments). As reproduced in Figure 10.6 (top and middle panels), for large and medium channels, we observe a smooth *sliding* motion, characterized by an equidistant movement of the cell internal regions, that maintains the overall cell length. In particular, in the intermediate structure, the cell front is significantly distant from the central and the back parts of the individual: this is due to the elongation process (captured also in Figure 10.6) caused by the confined geometry of the channel.

Instead, in the smallest channel, cells display a stepwise *push-and-pull* behavior, characterized by oscillating variations of cell length; see Figure 10.6 (bottom panel), which has been also observed in [332]. This migratory phenomenology has a possible interesting mechanical explanation. The geometry of the channel causes the cytosolic front of the cell to protrude. Pushed by the leading front, the overall cytosolic region, characterized by high elasticity and mobility, then deforms and moves forward, while pulling onto the nucleus with the force exerted through the contact tension $J^{int}_{N,C}$. However, as a consequence of its lower elasticity and motility (i.e., $T_{\eta,N} < T_{\eta,C}$ and $\lambda^{surface}_{\eta,N} > \lambda^{surface}_{\eta,C}$), the nuclear cluster takes more time to remodel toward a very elongated morphology and to displace, and therefore lags behind (as reproduced in Figure 10.4 and explained in more detail for the chemotactic migration of an endothelial cell in Chapter 6), limiting the elongation of the overall cell (which has to maintain its volume). Indeed, the displacement of the trailing part of the cytosol decreases the length of the individual. However, when such a rear part has moved far enough to the right, the nucleus, in order to avoid the split of the cell, is forced to move. Therefore, it pushes the leading part of the cell, which can go on elongating, increasing again the overall cell length.

It is noteworthy that this dramatic and continuous reorganization of the nuclear cluster is not necessary for cell movement in the middle and large channels. Indeed, in those cases, the nucleus undergoes only slightly deformations and is therefore more easily dragged by the rest of the cell, whose whole locomotion is eventually characterized by the above-described sliding dynamics.

The push-and-pull pattern is similar to the classical migration of fibroblasts on flats surfaces [229], and is consistent with the results obtained again in [332] for pancreatic cancer cells moving in microchannel structures.

Part III

Appendix

Appendix A

Computational Implementation

The implementation of the CPM core algorithm (i.e., the definition of the *Hamiltonian* and the modified Metropolis iteration) requires a simple code, which has never created problems for most researchers. Originally, CPM users have therefore usually written their own software programs, which were specifically built according to the specific application they were developing. However, as commented in [20], these proprietary versions were usually incompatible, making the exact replication of published results hard and the integration of new CPM extensions considerably difficult.

In order to obtain a common CPM-based program that would allow researchers to communicate and combine efforts and to compare and validate their results, several groups have released open-source, extensible CPM modeling packages over the last few years. Among others, the CompuCell3D[1] environment has recently become a standard program for the implementation of basic CPM applications. CC3D was jointly set up by groups at the University of Notre Dame[2] and the Biocomplexity Institute at Indiana University at Bloomington[3] and it is still under full and active development.

Among its C++ core modules, CC3D uses an XML-based markup language and reasonably simple Python scripting (i.e., plug-ins and steppables) to define the simulated system and to control the model parameters. The resulting simulations are then run from a flexible graphical player.

Our extended CPM has been implemented by a modified version of CC3D. Although the compartmentalization approach was already a standard feature of the program, the other developments have required structural changes. The introduction of the new probability function in Eq. (4.24) has led to a modification of the core of the code (i.e., of the part that deals with the basic iteration of the Metropolis algorithm). A group of Python routines, which are set to run within the CC3D framework itself, has instead been implemented for the numerical solution both of individual internal dynamics and of external fields, as well as for the interface of their outcomes with the basic CPM. In principle, these separated microscopic models could also have been approached by adopting external connections between the standard CC3D and existing packages that deal with biochemical pathways and reaction-diffusion systems

[1] http://www.compucell3d.org
[2] http://www.nd.edu
[3] http://www.biocomplexity.indiana.edu

(such as BioSpice[4] or Systems Biology Workbench[5], see [20]). Our choice was made to avoid possible compatibility problems that usually arise due to the interoperability between different software programs.

However, in the freeware spirit of CC3D and of the other programs that implement the CPM, any interested readers can contact the authors of this book to obtain further information on the program modifications and, if necessary, get either the entire software or parts of it.

Procedurally, the grid for the numerical solution of each PDE is matched with the CPM lattice and, at every time step, each computational module is used as the initial condition for the others. Indeed, the main lines of the computational algorithm are as follows:

- The discrete CPM evolves through an MCS, following the standard Metropolis rule.

- The local quantities of the chemicals are computed, based on the new cell configuration. In particular, after the *spin flip*, the target site, $\mathbf{x}_{target}$, is assigned the same concentrations of the microscopic variables as the moving source site, $\mathbf{x}_{source}$.

- The continuous equations of the fields are rederived, according to the new distribution of the continuous variables themselves and to the new boundaries of the simulated discrete objects, and solved using a finite element scheme, characterized by 10 diffusion time steps per MCS (this temporal step is sufficiently small to guarantee numerical stability).

- The biophysical properties of each discrete individual (given by its Potts coefficients) are updated, given its new intracellular state and the new configuration of the lattice.

- The *Hamiltonian* functional of the system is updated, and the system is ready to evolve again.

Finally, it is evident that the improvements in the CPM presented so far require increasing computational power, and therefore cause a significant slowdown of the speed of the relative simulations. For example, a realization of the multilevel wound healing model presented in Chapter 5 takes almost 4 h when using a normal notebook, while a simulation of the analogous phenomenon approached with the basic CPM as in Chapter 2 lasts less than 30 min. To mitigate this issue, we strongly advise researchers to make a preliminary study of the problem of interest and to decide what is the level of detail really needed.

Typically, if the biological phenomenon requires an accurate description of the microscopic scale of individuals, it should also allows one to take into account a small number of individuals, while if it requires the simulation of a

---

[4] http://www.biospice.org
[5] http://www.sbw.sourceforge.net

huge population (for example, a full model of organ morphogenesis requires the simulation of millions of cells), the level of detail for each element can be significantly decreased. In other words, researchers should always follow this rough principle:

$$\text{number of simulated individuals} \cdot \text{details of each individual} = \text{constant}.$$

However, it is not always possible to either simplify the internal structure of simulated individuals or to reduce their number. For instance, a full extended CPM of the morphogenesis of a complete organ or of an entire embryo would require the representation of 10^6-10^8 compartmentalized cells, with the relative subcellular dynamics. Given that CPM realizations are usually run hundreds of times to average out the results determined by the same values of parameters, the first outcomes might be obtained after months.

Efficient computational techniques that able to significantly increase the simulation speed would be therefore needed. As commented in [20, 81], one of the main issues in CPM implementations is that too low acceptance probabilities of spin flips (i.e., 10^{-4}-10^{-6}) often waste much computational time. In this regard, among the non-Metropolis Monte Carlo algorithms, *rejection free* dynamics such as *N-fold way* and *kinetic Monte Carlo* are particularly productive [41, 131, 231]. For each time step, they do not consider a trial index copy, which may or may not be accepted, but choose only from among the set of allowed lattice updates (i.e., those that decrease the system energy). Obviously, the net computational gain will depend on the balance between the average number of possible spin flips and the average acceptation rate.

The *Random-Walker* (RW) algorithm instead reduces, but does not eliminate, the rejection rate by selecting as target sites only those belonging to an object boundary [73]. The automatic rejection of non-boundary sites, characteristic of a normal algorithm, is therefore eliminated.

However, the increment of the computational speed is appreciable only in the presence of large individuals. The most attractive area of improvement in CPM computational implementation is represented by the use of distributive computing, where the overall simulation domain is divided in equivalent subdomains, which are in turn assigned to different nodes, as addressed again in [20]. A first parallel version of the original Potts approach has been implemented in [412] on a model of grain growth, where the effective energy consisted only of local grain boundary interactions, so that each spin flip changed only the energies of its neighbors.

Moreover, a recent RW distributing implementation of the CPM ran significantly faster [174]. However, the proposed parallel scheme required shared memory with all processors sharing the same subdomains. This issue therefore limited the total domain extension to the memory size of a single computer.

Indeed, following the useful and detailed explanations provided in [20, 81], the main difficulty in all forms of CPM parallelization is that the effective energy is nonlocal: when a given object crosses between nodes, any modification to it requires efficient parameter passing between nodes, so that the overall computation does not use stale values. Therefore, naive parallelizations, where the activity of each processor is restricted to a predefined subdomain of the total lattice, will increase the frequency of interprocessor communication for synchronization and, eventually, the waiting time of each node will be much greater than its calculating time. An interesting attempt to overcome this issue has been made with the *checkerboard algorithm*, presented in [81]. Its authors have used an improved data structure to describe simulated individuals and have further decomposed the subdomain assigned to each node into smaller subgrids (i.e., sublattices), chosen so that the corresponding ones on different processors do not interact. In this way, an update in one sublattice affects only the nearest-neighbor sublattices. Each node is therefore able to determine the spin flips affecting neighbor nodes, accumulate them, and pass them synchronously. In this case, the speed gain increases with the size of the subgrids per processor and decreases with the interaction range. Such a basic checkerboard parallelization can also implement rejection-free or RW methods, by using either equivalent or master-slave computations [20].

Appendix B

Glossary

Arachidonic acid (AA, sometimes ARA): A polyunsaturated omega-6 fatty acid 20:4(ω-6). It is the counterpart to the saturated arachidic acid found in peanut oil. In chemical structures, arachidonic acid is a carboxylic acid with a 20-carbon chain and four cis-double bonds; the first double bond is located at the sixth carbon from the omega end. Some chemistry sources define "arachidonic acid" to designate any of the eicosatetraenoic acids. However, almost all writings in biology, medicine, and nutrition limit the term to all-cis-5,8,11,14-eicosatetraenoic acid. Arachidonic acid is a polyunsaturated fatty acid present in the phospholipids (especially phosphatidylethanolamine, phosphatidylcholine, and phosphatidylinositides) of membranes of the body's cells, and is abundant in the brain, muscles, and liver. In addition to being involved in cellular signaling as a lipid second messenger involved in the regulation of signaling enzymes, such as PLC-γ, PLC-δ, and PKC-α, -β, and -γ isoforms, arachidonic acid is a key inflammatory intermediate and can also act as a vasodilator.

ARO: A poorly differentiated thyroid carcinoma-derived cell line.

CD44: A cell-surface glycoprotein involved in cellcell interactions, cell adhesion, and migration. In humans, the CD44 antigen is encoded by the CD44 gene on Chromosome 11. CD44 is a receptor for hyaluronic acid and can also interact with other ligands, such as osteopontin, collagens, and matrix metalloproteinases (MMPs). CD44 function is controlled by its post-translational modifications. This protein participates in a wide variety of cellular functions including lymphocyte activation, recirculation and homing, hematopoiesis, and tumor metastasis. Transcripts for this gene undergo complex alternative splicing that results in many functionally distinct isoforms.

Bonferroni correction: In statistics, is a method used to counteract the problem of multiple comparisons. The correction is based on the idea that if an experimentalist is testing n dependent or independent hypotheses on a set of data, then one way of maintaining the family-wise error rate is to test each individual hypothesis at a statistical significance level of $1/n$ times what it would be if only one hypothesis were tested. So, if it is desired that the significance level for the whole family of tests should be (at most) β, then the Bonferroni correction would be to conduct each of the individual tests at a significance level of β/n. "Statistically significant" simply means that a given result is unlikely to have occurred by chance assuming the null hypothesis is actually correct (i.e., no difference among groups, no effect of treatment, no relation among variables). The Bonferroni correction is derived by observing Boole's inequality. If n tests are performed, each of them significant with probability p (where p is unknown), then the probability that at least one of them comes out significant is (by Boole's inequality) $\leq np$. We have then to equate this probability to β, which is the significance level for the entire series of tests. By solving for p, we get $p = \beta/n$. This result does not require that the tests be independent.

Carboxyamidotriazole (CAI): An anti-invasive and anti-angiogenic agent that alters calcium-mediated signal transduction in ECs by blocking agonist-activated calcium entry in a dose-dependent manner. In particular, it is currently under investigation as an orally administered tumoristatic agent in Phase II and III clinical trials for different tumors.

CD157: An ADP-ribosyl cyclaserelated cell surface molecule regulating leukocyte diapedesis during inflammation. CD157 is expressed by ovarian cancer cells and mesothelium, and it potentiates the adhesion, migration, and invasion of serous ovarian cancer cells through different extracellular matrices. Indeed, it may be clinically useful as a prognostic tool and therapeutic target.

Cdc42: A small GTPase of the Rho-subfamily that regulates signaling pathways that control diverse cellular functions including cell morphology, migration, endocytosis and cell cycle progression.

Carboxyfluorescein succinimidyl ester (CFSE): A fluorescent cell staining dye. As carboxyfluorescein diacetate succinimidyl ester (CFDA-SE), which is nonfluorescent, it enters the cytoplasm of cells, where intracellular esterases remove the acetate groups and convert the molecule to the fluorescent ester. CFSE is retained within cells and covalently couples, via its succinimidyl group, to intracellular molecules, for extremely long periods. Also, due to this stable linkage, once incorporated within cells, the dye is not transferred to adjacent cells. CFSE was originally developed as a fluorescent dye that could be used to stably label lymphocytes and track their migration within animals for many months. Subsequent studies revealed that the dye can be used to monitor lymphocyte proliferation, both *in vitro* and *in vivo*, due to the progressive halving of CFSE fluorescence within daughter cells following each cell division. The only limitation is that CFSE at high concentrations can be toxic for cells. However, when CFSE labeling is performed optimally, approximately 7–8 cell divisions can be identified before the CFSE fluorescence is too low to be distinguished above the autofluorescence background. Thus, CFSE represents an extremely valuable fluorescent dye for immunological studies, allowing lymphocyte proliferation, migration, and positioning to be simultaneously monitored. By the use of fluorescent antibodies against different lymphocyte cell surface markers, it is also possible to follow the proliferation behavior of different lymphocyte subsets. In addition, unlike other methods, CFSE-labeled viable cells can be recovered for further analysis.

Chemotaxis: A phenomenon whereby cells, bacteria, and other single-cell or multicellular organisms direct their movements according to the concentration gradients of selected chemicals in their environment. Positive chemotaxis occurs if the movement is toward a higher concentration of the chemical in question, which is therefore called a *chemoattractor*. Conversely, negative chemotaxis occurs if the movement is in the opposite direction, and the relative chemical substance is called *chemorepellent*. Such chemical-induced directional migration is important for bacteria to find food (for example, glucose), or to flee from poisons (for example, phenol). In multicellular organisms, chemotaxis is critical to early development (e.g., movement of sperm toward the egg during fertilization) and subsequent phases of development (e.g., migration of neurons or lymphocytes) as well as in normal function. In addition, it has been recognized that mechanisms that allow chemotaxis in animals can be subverted during cancer metastasis.

Compartmentalization: Representation of a simulating individual through a set of discrete objects, reproducing its specific subcompartments.

Contact inhibition: Inhibition of cell locomotion due to the overexpression of cell–cell adhesive contacts in regions with high cellular density.

Dulbecco's minimal essential medium (DMEM): A cell culture medium used to maintain cells in tissue culture. It contains vitamins (folic acid, nicotinamide, riboflavin), salts (calcium chloride, potassium chloride, magnesium sulfate, sodium chloride, and monosodium phosphate), amino acids, glucose, iron, and phenol red. DMEM is suitable for most types of cells, including human, monkey, hamster, rat, mouse, chicken, and fish cells.

Fetal calf serum (FCS, or fetal bovine serum): The portion of plasma remaining after coagulation of blood, as during this process the plasma protein fibrinogen is converted to fibrin and remains behind in the clot. FCS is the most widely used serum supplement for the *in vitro* cell culture of eukaryotic cells. This is because it has a very low level of antibodies and containing more growth factors, allowing for versatility in many different cell culture applications.

Fingers: Tongues of cells protruding from the main tumor mass, a consequence of the instability due to both cell proliferation and cell haptotactic movement.

Fisher equation: Also known as the Fisher–Kolmogorov equation and the Fisher–KPP equation, named after R. A. Fisher and A. N. Kolmogorov, is the partial differential equation

$$\frac{\partial u}{\partial t} = Au - Bu^2 + D\frac{\partial^2 u}{\partial x^2},$$

originally derived for the simulation of propagation of a gene in a population, where $u = u(x,t)$ is a real-valued function of the one-dimensional space-variable x and of the time t, and D, A, and B are positive parameters. The Fisher equation has two stationary states: $u \equiv 0$ is unstable, whereas $u \equiv 1$ is stable. It indeed admits traveling wave solutions of the form

$$u(x,t) = v(x \pm ct) \equiv v(z),$$

where c depends on the parameters and v is an increasing function characterized by $\lim_{z\to-\infty} v(z) = 0$ and $\lim_{z\to\infty} v(z) = 1$. That is, the solution switches from the equilibrium state $u = 0$ to the equilibrium state $u = 1$.

G1 phase (Gap 1 phase, or post-mitotic phase): A period in the cell cycle during interphase, before the S phase. For most cell lines, this phase is the major period of cell growth during its lifespan. During this stage, new organelles are being synthesized. Therefore, the cell requires both structural proteins and enzymes, resulting in a great amount of protein synthesis and a high rate of metabolism in the cell.

Glycosaminoglycan (GAGs, or mucopolysaccharides): Long unbranched polysaccharides formed by repeated disaccharide units, which consist of a hexose (six-carbon sugar) or a hexuronic acid, linked to a hexosamine (six-carbon sugar containing nitrogen). Protein cores are made in the rough endoplasmic reticulum and are posttranslationally modified by glycosyltransferases in the Golgi apparatus, where GAG disaccharides are added to them to yield proteoglycans; the exception is the GAG hyaluronan, which is uniquely synthesized without a protein core and is "spun out" by enzymes at the cell surface directly into the extracellular space. This family of carbohydrates is essential or important for life, as it forms an important component of connective tissues.

Gompertz function: Gompertz curve, named after Benjamin Gompertz, is a sigmoid function. It is a type of mathematical model for a time series, where growth is slowest at the start and end of a time period. The right-hand or future value asymptote of the function is approached much more gradually by the curve than the left-hand or lower-valued asymptote, in contrast to the logistic function in which both asymptotes are approached by the curve symmetrically. Its general formula is

$$y(t) = ae^{be^{ct}},$$

where a is the upper asymptote, since

$$ae^{be^{-\infty}} = ae^0 = a,$$

and b and c are negative numbers. In particular, b sets the x displacement and c sets the growth rate (x scaling). Examples of uses for Gompertz curves include modeling of growth of populations in a confined space, as birth rates first increase and then slow as resource limits are reached. In the 1960s, A.K. Laird for the first time successfully used the Gompertz curve to fit data of growth of tumors. In fact, tumors are cellular populations growing in a confined space where the availability of nutrients is limited.

Hamiltonian: Functional describing the effective energy of the system.

Haptotaxis: A directional motility or outgrowth of cells, e.g., in the case of axonal outgrowth, up a gradient of cellular adhesion sites or substrate-bound chemical sources. These gradients are naturally present in the extracellular matrix (ECM) of the body during processes such as angiogenesis or artificially present in biomaterials where gradients are established by altering the concentration of adhesion sites on a polymer substrate.

Hepatocyte growth factor (HGF)/ scatter factor (SF): Growth factor eliciting multiple processes, such as mitogenesis, motility, morphogenesis, cell dissociation, migration through the extracellular matrix, acquisition of polarity, and tubule formation.

Internal state vector: Vector characterizing an object whose components represent the internal factors (i.e., nutrients, proteins, growth factors, gene expressions, ...) influencing its properties and behavior.

Lattice: Regular repeated graph $\subseteq \mathbb{R}^d$ formed by identical d-dimensional closed grid sites, where $d = 1, 2, 3$.

Matrigel: Gelatinous protein mixture that resembles the complex extracellular environment found in many tissues and is used by cell biologists as a substrate for cell cultures. The ability of Matrigel to stimulate complex cell behavior is a consequence of its heterogeneous composition. The chief components of Matrigel are structural proteins such as laminin, entactin, and collagen, which present cultured cells with the adhesive peptide sequences that they would encounter in their natural environment. Also present are growth factors that promote differentiation and proliferation of many cell types. Matrigel contains numerous other proteins in small amounts, and its exact composition can vary from in each case.

Met: HGF/SF receptor.

MLP-29: Mouse liver progenitor cell.

Nested model: Mathematical framework where the information flows in a vertical fashion between submodels operating at different scales. Specifically, the parameters of the CPM are obtained as output of subcellular models related to protein and genetic networks.

Hybrid model: Mathematical framework characterized by the copresence of both continuous and discrete models usually operating in the same spatial domain. Specifically, CPM parameters may be linked to the evolution of concentration of chemicals diffusing in the extracellular environment.

Kolmogorov Smirnov test (K-S test): In statistics is a nonparametric test for the equality of continuous, one-dimensional probability distributions that can be used to compare either a sample with a reference probability distribution (one-sample KS test), or to two samples (two-sample KS test). In particular, in the first case, the Kolmogorov-Smirnov statistic quantifies the distance between the empirical distribution function of the sample and the cumulative distribution function of the reference distribution, whereas in the second case, the distance between the empirical distribution functions of two samples is quantified. The null distribution of this statistic is calculated under the null hypothesis that the sample is drawn from the reference distribution (in the one-sample case) or that the samples are drawn from the same distribution (in the two-sample case). In both cases, the distributions considered under the null hypothesis are continuous (i.e., otherwise they are unrestricted). In particular, the two-sample K-S test is one of the most useful and general nonparametric methods for comparing two samples, as it is sensitive to differences in both location and shape of the empirical cumulative distribution functions of the two samples. In the special case of testing for the normality of the distribution, samples are standardized and compared with a standard normal distribution. This is equivalent to setting the mean and variance of the reference distribution equal to the sample estimates, and it is known that using these to define the specific reference distribution changes the null distribution of the test statistic.

Monte Carlo Step: Basic unit of time of the CPM.

Nitric oxide (NO): Also known as nitrogen monoxide, it is an important cellular signaling molecule involved in many physiological and pathological processes. Nitric oxide, known also as the "endothelium-derived relaxing factor", or EDRF, is biosynthesized endogenously from L-arginine, oxygen, and NADPH by various nitric oxide synthase (NOS) enzymes. Reduction of inorganic nitrate may also serve to make nitric oxide. The endothelium (inner lining) of blood vessels uses nitric oxide to signal the surrounding smooth muscle to relax, thus resulting in vasodilation and increasing blood flow. Nitric oxide is highly reactive (having a lifetime of a few seconds), yet diffuses freely across membranes. These attributes make nitric oxide ideal for a transient paracrine (between adjacent cells) and autocrine (within a single cell) signaling molecule. It is a powerful vasodilator with a short half-life of a few seconds in the blood. Long-known pharmaceuticals like nitroglycerine and amyl nitrite were discovered, more than a century after their first use in medicine, to be active through the mechanism of being precursors to nitric oxide. Low levels of nitric oxide production are important in protecting organs such as the liver from ischemic damage. Chronic expression of NO is associated with various carcinomas and inflammatory conditions including Type-1 diabetes, multiple sclerosis, arthritis and ulcerative colitis.

Discrete object: Finite, spatially extended lattice subdomain of contiguous sites denoted with the same spin.

Continuous object: Field, i.e., spatiotemporal evolution of a microscopic entity.

Overdamped force-velocity response: Relation stating that the local velocity of an individual, and not its acceleration, is proportional to the local force.

Parallel computing: A form of computation in which many calculations are carried out simultaneously, operating on the principle that large problems can be divided into smaller ones, which are then solved concurrently ("in parallel"). There are several different forms of parallel computing: bit-level, instruction level, data, and task parallelism. In recent years, parallel computing has become the dominant paradigm in computer architecture, mainly in the form of multicore processors. Parallel computers can be roughly classified according to the level at which the hardware supports parallelism, with multi-core and multi-processor computers having multiple processing elements within a single machine, while clusters, MPPs, and grids use multiple computers to work on the same task. Specialized parallel computer architectures are sometimes used alongside traditional processors, for accelerating specific tasks. Communication and synchronization between the different subtasks are typically some of the greatest obstacles to getting good parallel program performance. Optimally, the speed-up from parallelization would be linear doubling the number of processing elements should halve the runtime, and doubling it a second time should again halve the runtime. However, very few parallel algorithms achieve optimal speed-up. Most of them have a near-linear speed-up for small numbers of processing elements, which flattens out into a constant value for large numbers of processing elements. The potential speed-up of an algorithm on a parallel computing platform is given by Amdahl's law, originally formulated by Gene Amdahl in the 1960s. It states that a small portion of the program that cannot be parallelized will limit the overall speed-up available from parallelization. A program solving a large mathematical or engineering problem will typically consist of several parallelizable parts and several non-parallelizable (sequential) parts. If α is the fraction of running time a sequential program spends on non-parallelizable parts, then $S = \frac{1}{\alpha}$ is the maximum speed-up with parallelization of the program. Gustafson's law is another law in computing, closely related to Amdahl's law. It states that the speed-up with P processors is $S(P) = P - \alpha(P-1)$. Amdahl's law assumes a fixed problem size and that the running time of the sequential section of the program is independent of the number of processors, whereas Gustafson's law does not make these assumptions.

Phosphate buffered saline (PBS): A water-based salt solution containing sodium chloride, sodium phosphate, and, in some formulations, potassium chloride and potassium phosphate. The buffer's phosphate groups help to maintain a constant pH. The osmolarity and ion concentrations of the solution usually match those of the human body

PKH26GL: Red fluorescent cell linker Kit for general cell membrane labeling.

Plug-ins: Energy functions, lattice monitors, or stored user-assigned data that CompuCell3D uses internally to initially configure the simulation. Examples are plug-ins defining objects' types, biophysical properties, or location and amplitude of external constant forces.

Rac: A subfamily of the Rho family of GTPases that regulates a wide range of cellular functions including migration and cytoskeletal reorganization.

RPMI: A series of media utilizing a bicarbonate buffering system and alterations in the amounts of amino acids and vitamins used for the culture of human normal and neoplastic leukocytes. RPMI, when properly supplemented, has demonstrated wide applicability for supporting the growth of many other types of cultured cells also.

Spin flip: Change in lattice configuration.

Steppables: CompuCell3D modules that are called every Monte Carlo Step (MCS). More precisely, they are called after all the spin copy attempts performed in a given MCS have been carried out. The steppables modules employ all functions that need to be done every MCS, such as updating the lattice configuration and recalculating the objects' attributes.

Student's t-test: Any statistical hypothesis test in which the test statistic follows a Student's t distribution if the null hypothesis is supported. It is most commonly applied when the test statistic would follow a normal distribution if the value of a scaling term in the test statistic were known.

When the scaling term is unknown and is replaced by an estimate based on the data, the test statistic (under certain conditions) follows a Student's t distribution. Among the most frequently used t-tests are

- A one-sample location test of whether the mean of a normally distributed population has a value specified in a null hypothesis.
- A two-sample location test of the null hypothesis that the means of two normally distributed populations are equal. All such tests are usually called Student's t-tests, though strictly speaking that name should only be used if the variances of the two populations are also assumed to be equal; the form of the test used when this assumption is dropped is sometimes called Welch's t-test. These tests are often referred to as "unpaired" or "independent samples" t-tests, as they are typically applied when the statistical units underlying the two samples being compared are nonoverlapping.
- A test of the null hypothesis that the difference between two responses measured on the same statistical unit has a mean value of zero. For example, suppose we measure the size of a cancer patient's tumor before and after a treatment. If the treatment is effective, we expect the tumor size for many of the patients to be smaller following the treatment. This is often referred to as the "paired" or "repeated measures" t-test.
- A test of whether the slope of a regression line differs significantly from 0.

Most t-test statistics have the form $T = \frac{Z}{s}$, where Z and s are functions of the data. Typically, Z is designed to be sensitive to the alternative hypothesis (i.e. its magnitude tends to be larger when the alternative hypothesis is true), whereas s is a scaling parameter that allows the distribution of T to be determined. As an example, in the one-sample t-test $Z = \frac{\bar{X}}{\frac{\sigma}{\sqrt{n}}}$, where $\bar{X}$ is the sample mean of the data, n is the sample size, and σ is the population standard deviation of the data. In particular, in the one-sample t-test, s is $\hat{\sigma}/\sigma$, where $\hat{\sigma}$ is the sample standard deviation. The assumptions underlying a t-test are that

- Z follows a standard normal distribution under the null hypothesis.
- ps^2 follows a χ^2 distribution with p degrees of freedom under the null hypothesis, where p is a positive constant.
- Z and s are independent.

Vascular endothelial growth factor (VEGF): A signal protein family produced by cells that stimulates vasculogenesis and angiogenesis. The most important member is VEGF-A. Other members are placenta growth factor (PGF), VEGF-B, VEGF-C and VEGF-D. The latter ones were discovered later than VEGF-A, and, before their discovery, VEGF-A was called just VEGF. A number of VEGF-related proteins have also been discovered encoded by viruses (VEGF-E) and in the venom of some snakes (VEGF-F). It is the part of the system that restores the oxygen supply to tissues when blood circulation is inadequate. Serum concentration of VEGF is high in bronchial asthma and low in diabetes mellitus. VEGF's normal function is to create new blood vessels during embryonic development, new blood vessels after injury, muscle following exercise, and new vessels (collateral circulation) to bypass blocked vessels. When VEGF is overexpressed, it can contribute to disease. Solid cancers cannot grow beyond a limited size without an adequate blood supply; cancers that can express VEGF are able to grow and metastasize. Overexpression of VEGF can cause vascular disease in the retina of the eye and other parts of the body. Drugs such as bevacizumab can inhibit VEGF and control or slow those diseases. VEGF is a subfamily of growth factors, to be specific, the platelet-derived growth factor family of cystine-knot growth factors. They are important signaling proteins involved in both vasculogenesis (the de novo formation of the embryonic circulatory system) and angiogenesis (the growth of blood vessels from preexisting vasculature). All members of the VEGF family stimulate cellular responses by binding to tyrosine kinase receptors (the VEGFRs) on the cell surface, causing them to dimerize and become activated through transphosphorylation, although to different sites, times, and extents. The VEGF receptors have an extracellular portion consisting of 7 immunoglobulin-like domains, a single transmembrane spanning region, and an intracellular portion containing a split tyrosine-kinase domain. VEGF-A binds to VEGFR-1 (Flt-1) and VEGFR-2 (KDR/Flk-1). VEGFR-2 appears to mediate almost all of the known cellular responses to VEGF. The function of VEGFR-1 is less well defined, although it is thought to modulate VEGFR-2 signaling. Another function of VEGFR-1 may be to act as a dummy/decoy receptor, sequestering VEGF from VEGFR-2 binding (this appears to be particularly important during vasculogenesis in the embryo). VEGF-C and VEGF-D, but not VEGF-A, are ligands for a third receptor (VEGFR-3), which mediates lymphangiogenesis.

Wound healing assay: Experimental model able to evaluate the migratory phenotype of a cell population. It consists in the incubation of the cell culture of interest toward confluence, and in the subsequent creation of an artificial scratch with a sharp object (for example, a pipette tip). The removal of cells from the wounded area acts as a stimulus for the remaining mass to invade and fill the open space, with a characteristic traveling-wave-like behavior. The rate of advance of the wound edge, i.e., the quantification of the area recolonized, gives a measure of the migratory capacity of the overall population. In particular, this technique is widely used to compare the motility properties of a cell line either in resting conditions (i.e., in the absence of any external stimuli) and in response to specific chemical stimulations or modifications of expressions of molecules putatively involved in the migratory processes. The wound healing experiment has been reproduced by several mathematical models, most of which are based on the Fisher equation. For instance, these methods have described the cell population as a density, while its motility has been prescribed by a diffusive flux, where the diffusivity has been considered either a constant or a function of an external chemical factor.

Appendix C

Parameter Values

TABLE C.1: Cell Scatter of ARO Aggregates. ($*$) indicates parameters that vary in the different sets of simulations.

Parameter	Description	Value	Reference
$A_C^{surface}$	Surface of ARO cells	200 [μm^2]	[101, 349]
$A_C^{perimeter}$	Perimeter of ARO cells	96 [μm]	[101, 349]
T	Motility of ARO cells	50*	[101, 349]
$\lambda_C^{surface}$	Compressibility of ARO cells	25 [μm^{-2}]	[349]
$\lambda_C^{perimeter}$	Stiffness of ARO cells	25 [μm^{-1}]	[349]
$J_{C,C}$	Cell–cell adhesion	60*	[101, 349]
$J_{C,M}$	Cell–matrix adhesion	20	[101, 349]

TABLE C.2: Cell Scatter of MLP-29 Aggregates. ($*$) indicates parameters that vary in the different sets of simulations.

Parameter	Description	Value	Reference
$A_C^{surface}$	Surface of MLP-29 cells	160 [μm^2]	[101, 349]
$A_C^{perimeter}$	Perimeter of MLP-29 cells	45 [μm]	[101, 349]
T	Motility of MLP-29 cells	50	[101, 349]
$\lambda_C^{surface}$	Compressibility of MLP-29 cells	25 [μm^{-2}]	[349]
$\lambda_C^{perimeter}$	Stiffness of MLP-29 cells	5 [μm^{-1}]	[349]
μ_C^{pers}	Persistence strength of MLP-29 cells	20*	[349]
μ_C^{chem}	Chemical strength of MLP-29 cells	20 [μM^{-1}]	[349]
$J_{C,C}$	Cell–cell adhesion	20	[101, 349]
$J_{C,M}$	Cell–matrix adhesion	20	[101, 349]
D_c	HGF diffusion constant	10 [μm^{2}s^{-1}]	[349]
ε_c	HGF uptake rate	$5{\cdot}10^{-4}$* [s^{-1}]	[349]
α_c	HGF addition rate	$5{\cdot}10^{-4}$* [s^{-1}]	[349]
λ_c	HGF decay rate	10^{-6} [s^{-1}]	[349]

TABLE C.3: OvCa Single Cell Transmigration. (*) indicates parameters that vary in the different sets of simulations.

Parameter	Description	Value	Reference
T_C	Motility of cancer cells	20	[162]
T_M	Motility of mesothelial cells	10	[162]
T_E	Motility of ECM fibers	0	[162]
$A_C^{surface}$	Target surface of cancer cells	175 [px]	[162]
$A_M^{surface}$	Target surface of mesothelial cells	243 [px]	[162]
$A_C^{perimeter}$	Target perimeter of cancer cells	70 [px]	[162]
$A_M^{perimeter}$	Target perimeter of mesothelial cells	72 [px]	[162]
$\lambda_C^{surface}$	Compressibility of cancer cells	3 [px^{-2}]	[162]
$\lambda_M^{surface}$	Compressibility of mesothelial cells	10 [px^{-2}]	[162]
$\lambda_C^{perimeter}$	Stiffness of cancer cells	5 [px^{-1}]	[162]
$\lambda_M^{perimeter}$	Stiffness of mesothelial cells	10 [px^{-1}]	[162]
$J_{C,F}$	Adhesion between cancer cells and extracellular fluid	15	[162]
$J_{M,F}$	Adhesion between mesothelial cells and extracellular fluid	5	[162]
$J_{E,F}$	Adhesion between matrix fibers and extracellular fluid	6	[162]
$J_{C,E}$	Adhesion between cancer cells and matrix fibers	1	[162]
$J_{C,M}$	Adhesion between cancer cells and mesothelial cells	1*	[162]
$J_{M,E}$	Adhesion between mesothelial cells and matrix fibers	1	[162]
$J_{M,M}$	Adhesion between mesothelial cells	0.3*	[162]
$J_{C,S}$	Adhesion between cancer cells and Petri dish	0.5	[162]
μ_C^{chem}	Chemotactic strength of cancer cells	30 [μM^{-1}]	[162]
D_m	MMPs diffusion rate	0.02 [px^2 MCS^{-1}]	[408]
λ_m	MMPs decay rate	0.01 [MCS^{-1}]	[408]
α_m	MMPs secretion rate	0.25* [MCS^{-1}]	[408]
D_c	Chemoattractant diffusion rate	0.1 [px^2 MCS^{-1}]	[162]
λ_c	Chemoattractant decay rate	0.001 [MCS^{-1}]	[162]
α_c	Chemoattractant secretion rate	0.2 [MCS^{-1}]	[162]

TABLE C.4: OvCa Multicellular Spheroid Invasion. (∗) indicates parameters that vary in the different sets of simulations.

Parameter	Description	Value	Reference
$A_C^{surface}$	Target surface of cancer cells	56 [px]	[234]
$\lambda_C^{surface}$	Compressibility of cancer cells	5 [px^{-2}]	[162]
$J_{C,C}$	Adhesion between cancer cells	0.1*	[162]
$J_{C,M}$	Adhesion between cancer cells and mesothelial cells	9	[234]
α_m	MMPs secretion rate	0.3* [MCS^{-1}]	[162]
g	Gravitational potential	9.8	

TABLE C.5: Wound Healing Assay of ARO Cells in Response to HGF.(∗) indicates parameters that vary in the different sets of simulations.

Parameter	Description	Value	Reference
$A_{\eta,N}^{surface}$	Surface of cell nucleus	50 [μm^2]	[101, 349]
$A_{\eta,N}^{perimeter}$	Perimeter of cell nucleus	43 [μm]	[101, 349]
$A_{\eta,C}^{surface}$	Surface of cell cytosol	150 [μm^2]	[101, 349]
$A_{\eta,C}^{perimeter}$	Perimeter of cell cytosol	90 [μm]	[101, 349]
$A_{\eta,M}^{surface}$	Surface of cell plasmamembrane	60 [μm^2]	[101, 349]
$A_{\eta,M}^{perimeter}$	Perimeter of cell plasmamembrane	60 [μm]	[101, 349]
$\lambda_{\eta,N,C,M}^{surface}$	Cell compressibility	20	[352]
$\lambda_{\eta,N,C,M}^{perimeter}$	Cell stiffness	20	[352]
$J_{C,N}^{int}$	Generalized adhesion between nucleus and cytoplasm	-20	[352]
$J_{C,M}^{int}$	Generalized adhesion between cytoplasm and plasmamembrane	-20	[352]
J_0	Basal adhesive strength	4.5	[101, 349]
k	Coefficient for $J_{E,E}^{ext}$	1/2 [μM^{-2}]	[352]
T_0	Basal cell motility	0.4	[101]

TABLE C.5: (Continued.)

Parameter	Description	Value	Reference
h	Michaelis–Menten constant for T	1/2	[352]
$T_{\eta,N}$	Generalized motility of the nucleus	0.25	[352]
D_h	HGF diffusion constant	10^{-13} [m^2s^{-1}]	[349]
λ_h	HGF on-rate degradation constant	10^{-4} [s^{-1}]	[349]
ϕ_h	HGF on-rate addition constant	10^{-4} [s^{-1}]	[349]
z	Coefficient of the amount of HGF internalized	1 [s^{-1}]	[352]
nr	Total number of Met receptors	311.200	[352]
D_p	PI3K diffusion constant	5 [$\mu\text{m}^2\text{s}^{-1}$]	[243]
λ_p	PI3K on-rate degradation constant	1 [s^{-1}]	[243]
μ_p	Maximal rate of Met-dependent PI3K activation	10 [μMs^{-1}]	[60]
$\mu_{p,0}$	Constant for Met-dependent PI3K activation	2 [μM]	[60]
D_k	MAPK diffusion constant	2 [$\mu\text{m}^2\text{s}^{-1}$]	[210, 255]
λ_k	MAPK on-rate degradation constant	1 [s^{-1}]	[210, 255]
μ_k	Maximal rate of Met-dependent MAPK activation	10 [μMs^{-1}]	[60]
$\mu_{k,0}$	Constant for Met-dependent MAPK activation	2 [μM]	[60]
D_r	Rac diffusion constant	10^5 [nm^2s^{-1}]	[254]
λ_r	Rac on-rate degradation constant	1 [s^{-1}]	[254]
μ_{p-r}	Rate of PI3K-dependent Rac activation	4.5 [s^{-1}]	[352]
μ_{k-r}	Rate of MAPK-dependent Rac activation	4.5 [s^{-1}]	[352]
μ_{c-r}	Rate of Cdc42-dependent Rac activation	4.5 [s^{-1}]	[254]
D_c	Cdc42 diffusion constant	10^5 [nm^2s^{-1}]	[254]
λ_c	Cdc42 on-rate degradation constant	1 [s^{-1}]	[254]
μ_{p-c}	Rate of PI3K-dependent Cdc42 activation	2.5 [s^{-1}]	[352]
μ_{k-c}	Rate of MAPK-dependent Cdc42 activation	3 [s^{-1}]	[352]

TABLE C.6: Effect of Calcium-Related Pathways on Single Cell Motility. (∗) indicates parameters that vary in the different sets of simulations.

Parameter	Description	Value	Reference
$A_{\eta,N}^{volume}$	Volume of cell nucleus	900 [μm^3]	[390]
$A_{\eta,N}^{surface}$	Surface of cell nucleus	314 [μm^2]	[390]
$A_{\eta,C}^{volume}$	Volume of cell cytosol	32000 [μm^2]	[390]
$A_{\eta,C}^{surface}$	Surface of cell cytosol	1110 [μm^2]	[390]
$\lambda_{\eta,N,C}^{volume}$	Cell compressibility	20	[351, 354]
$\lambda_{\eta,N}^{surface}$	Stiffness of cell nucleus	20	[351, 354]
λ_0^{surf}	Intrinsic stiffness of cell cytosol	12	[351, 354]
k	Michaelis–Menten constant for $\lambda_{\eta,C}^{surface}$	1	[351, 354]
T_0	Basal cell motility	3.5	[309]
$T_{\eta,N}$	Generalized motility of the nucleus	0.25	[351, 354]
h	Michaelis–Menten constant for T	1/2	[351, 354]
$J_{N,C}^{int}$	Generalized adhesion between sub-cellular units	-20	[351, 354]
μ_0^{ch}	basal chemotactic strength	0.2 [μM^{-1}]	[351, 354]
j	Michaelis–Menten constant for for $\mu_{\eta,C}^{ch}$	1/2	[351, 354]
μ_0^{pers}	Basal inertia strength	0.8	[309]
D_v	Diffusion constant of VEGF	10 [$\mu m^2 s^{-1}$]	[360]
λ_v	On-rate constant of VEGF degradation	$1.8 \cdot 10^{-4}$ [s^{-1}]	[360]
ϕ_v	On-rate constant of VEGF secretion	0.78 [h^{-1}]	[351, 360]
z	Coefficient of the amount of bound VEGF	1 [h^{-1}]	[351, 354]
D_a	Diffusion constant of AA	10 [$\mu m^2 s^{-1}$]	[283]
λ_a	On-rate constant of AA degradation	30 [s^{-1}]	[283]
k_a	Maximal rate of VEGF-dependent AA release	30* [$\mu M s^{-1}$]	[283]
K_a	Michaelis–Menten constant for VEGF-induced AA release	1 [μM]	[283]
v_c	Maximal rate of Ca-dependent AA release	0.75* [$\mu M s^{-1}$]	[351, 354]
c_a	Michaelis–Menten constant for for Ca-induced NO release	0.3 [μM]	[351, 354]
D_n	Diffusion constant of NO	3300 [$\mu m^2 s^{-1}$]	[226, 283]
λ_n	On-rate constant of NO degradation	0.1 [s^{-1}]	[283]
k_n	Maximal rate of VEGF-induced NO release	30* [$\mu M s^{-1}$]	[351, 354]

TABLE C.6: (Continued.)

Parameter	Description	Value	Reference
K_n	Michaelis–Menten constant for VEGF-induced NO release	1 [μM]	[351, 354]
v_{ca}	Maximal rate of NO release	1.5* [μMs^{-1}]	[37, 274]
c_n	Dissociation constant between Ca and eNOS	0.3 [μM]	[312]
a_a	Dissociation constant between AA and eNOS	0.2 [μM]	[283]
c_0	Basal intracellular calcium concentration	0.05 [μM]	[33, 283]
D_c	Diffusion constant of Ca	220* [μm^{2}s^{-1}]	[132, 212, 283]
k_{Ca}	Maximal rate of calcium efflux	24.7 [μMs^{-1}]	[283, 399]
Ca_{out}	Limit for calcium extrusion	0.32 [μM]	[283, 399]
$F_{AA,max}$	Maximal rate of AA-induced calcium influx	6 [s^{-1}]	[274, 283]
$F_{NO,max}$	Maximal rate of NO-induced calcium influx	4.5 [s^{-1}]	[274, 283]
q_a	Michaelis–Menten constant for AA-activated channels	2 [μM]	[274, 283]
q_n	Michaelis–Menten constant for NO-activated channels	5 [μM]	[274, 283]
K_{off}	Dissociation constant between buffers and Ca$_c$	300 [s^{-1}]	[24]
K_{on}	On-rate constant of Ca$_c$ buffering	100 [μM^{-1}s^{-1}]	[24]
b_T	Total buffer concentration	450 [μM]	[132, 283]
$c_{ext,0}$	Extracellular calcium level	2000 [μM]	[33, 283]

TABLE C.7: Tumor-Derived Tubulogenesis. (∗) indicates parameters that vary in the different sets of simulations.

Parameter	Description	Value	Reference
$A_{\eta,N}^{surface}$	Surface of cell nucleus	130 [μm^2]	[133]
$A_{\eta,N}^{perimeter}$	Perimeter of cell nucleus	35 [μm]	[133]
$A_{\eta,C}^{surface}$	Surface of cell cytosol	1150 [μm^2]	[133]
$A_{\eta,C}^{perimeter}$	Perimeter of cell cytosol	150 [μm]	[133]
J_0	Basal adhesion strength	5 [μm^{-1}]	[135]
p	Coefficient for $J_{E,E}^{ext}$	1/2	

TABLE C.8: Different Morphologies of Tumor Invasion Fronts. (∗) indicates parameters that vary in the different sets of simulations.

Parameter	Description	Value	Reference
$A^{surface}_{\eta,N}$	Surface of cell nucleus	50 [μm^2]	[49]
$A^{perimeter}_{\eta,N}$	Perimeter of cell nucleus	43 [μm]	[49]
$A^{surface}_{\eta,C}$	Surface of cell cytosol	150 [μm^2]	[49]
$A^{perimeter}_{\eta,C}$	Perimeter of cell cytosol	90 [μm]	[49]
$T_{\eta,N}$	Generalized motility of the nucleus	1	[355]
T_0	Basal cell motility	4.5	[355]
h	Michaelis–Menten constant for T	1/3 [355]	
$\lambda^{surface}_{\eta,N,C}$	Cell compressibility	10	[355]
$\lambda^{perimeter}_{\eta,C}$	Stiffness of cell cytosol	5*	[355]
$\lambda^{perimeter}_{\eta,N}$	Stiffness of cell nucleus	10	[355]
$J^{int}_{N,C}$	Generalized adhesion between nucleus and cytosol	-20	[355]
$J^{ext}_{T,T}$	Cell–cell adhesive strength	5*	[355]
$J^{ext}_{T,M}$	Cell–matrix adhesive strength	2*	[355]
μ^{hapt}	Haptotaxis strength	5* [μM^{-1}]	[355]
D_n	Diffusion constant of nutrients	10^3 [μm^{2}s^{-1}]	[139]
λ_n	On-rate constant of nutrient decay	$2 \cdot 10^{-4}$ [s^{-1}]	[138, 139]
χ_n	On-rate constant of nutrient uptake	1 [h^{-1}]	[138, 139]
ϕ_n	On-rate constant of growth factor production	0.78 [h^{-1}]	[138, 139]
n_{max}	Maximal growth factor consumption	0.08 [pg/cell/h]	[140]
$n_{ext,0}$	Initial extracellular level of growth factors	10 [μM]	[355]
n_0	Basal intracellular level of growth factors	0.2 [μM]	[67, 138]
λ_p	On-rate constant of ECM protein decay	$1.3 \cdot 10^{-4}$ [s^{-1}]	[286]
χ_p	On-rate constant of ECM protein degradation	3* [h^{-1}]	[355]
$p_{ext,0}$	Initial extracellular level of ECM protein	4 [μM]	[355]
D_m	Diffusion constant of MMPs	5^{-4} [μm^{2}s^{-1}]	[286]
λ_m	On-rate constant of MMP decay	$2 \cdot 10^{-3}$ [s^{-1}]	[286]
π_m	On-rate constant of MMP production	$5 \cdot 10^{-3}$ [s^{-1}]	[286]

TABLE C.9: Early Stages of Tumor Spheroid Growth. (*) indicates parameters that vary in the different sets of simulations.

Parameter	Description	Value	Reference
A_C^{volume}	Cell volume	3380 [μm^3]	[50]
$A_C^{surface}$	Cell surface	1130 [μm^2]	[50]
T_0	Basal cell motility	4.5 [356]	
h	Michaelis–Menten constant for T	1/3	[356]
λ_C^{volume}	Cell compressibility	10	[356]
k	Rate of cell loose volume	1/5	[356]
$\lambda_C^{surface}$	Cell stiffness	5	[356]
$J_{C,C}$	Cell–cell adhesive strength	2.5*	[356]
$J_{C,M}$	Cell–matrix adhesive strength	5	[356]
μ	Haptotaxis strength	5 [μM^{-1}]	[356]
D_n	Diffusion constant of nutrients	10^3 [$\mu m^2 s^{-1}$]	[139]
λ_n	On-rate constant of nutrient decay	$2 \cdot 10^{-4}$ [s^{-1}]	[138, 139]
χ_n	On-rate constant of nutrient uptake	1 [h^{-1}]	[138, 139]
ϕ_n	On-rate constant of growth factor production	0.78 [h^{-1}]	[138, 139]
n_{max}	Maximal growth factor consumption	0.08 [pg/cell/h]	[140]
$n_{ext,0}$	Initial extracellular level of growth factors	0.4 [μM]	[356]
n_0	Basal intracellular level of growth factors	0.5 [μM]	[29, 67, 138]
n_l	Threshold level for necrotic transition	0.2 [μM]	[29, 67, 138]
χ_p	On-rate constant of ECM protein degradation	3 [h^{-1}]	[286]
$p_{ext,0}$	Initial extracellular level of ECM protein	4 [μM]	[356]
D_m	Diffusion constant of MMPs	5^{-4} [$\mu m^2 s^{-1}$]	[286]
λ_m	On-rate constant of MMP decay	$2 \cdot 10^{-3}$ [s^{-1}]	[286]
π_m	On-rate constant of MMP production	$5 \cdot 10^{-3}$ [s^{-1}]	[286]

TABLE C.10: Cell Migration in Matrix Environments. (∗) indicates parameters that vary in the different sets of simulations.

Parameter	Description	Value	Reference
$A_{\eta,N}^{surface}$	Surface of cell nucleus in 2D	34.7 [μm^2]	[7]
$A_{\eta,N}^{perimeter}$	Perimeter of cell nucleus in 2D	20.8 [μm]	[7]
$A_{\eta,C}^{surface}$	Surface of cell cytosol in 2D	104.3 [μm^2]	[7]
$A_{\eta,C}^{perimeter}$	Perimeter of cell cytosol in 2D	62.6 [μm]	[7]
$A_{\eta,N}^{volume}$	Volume of cell nucleus in 3D	155.2 [μm^3]	[7]
$A_{\eta,N}^{surface}$	Surface of cell nucleus in 3D	138.8 [μm^2]	[7]
$A_{\eta,C}^{volume}$	Volume of cell cytosol in 3D	1076.3 [μm^3]	[7]
$A_{\eta,C}^{surface}$	Surface of cell cytosol in 3D	694.6 [μm^2]	[7]
$A_{F}^{surface}$	Surface of matrix fibers	20 [μm^2]	[7]
$A_{F}^{perimeter}$	Length of matrix fibers	20 [μm]	[162]
$T_{\eta,N}$	Motility of cell nucleus	0.5	[353]
$T_{\eta,C}$	Motility of cell cytosol	9	[353]
T_F	Motility of matrix fibers	0	[353]
$\lambda_{\eta,N,C}^{volume}$	Cell compressibility in 3D	10	[353]
$\lambda_{\eta,N,C}^{surface}$	Cell compressibility in 2D	10	[353]
$\lambda_{\eta,N}^{surface}$	Stiffness of cell nucleus in 3D	8.5*	[353]
$\lambda_{\eta,N}^{perimeter}$	Stiffness of cell nucleus in 2D	8.5*	[353]
$\lambda_{\eta,C}^{surface}$	Stiffness of cell cytosol in 3D	0.8	[353]
$\lambda_{\eta,C}^{perimeter}$	Stiffness of cell cytosol in 2D	0.8	[353]
$\lambda_{F}^{surface}$	Compressibility of matrix fibers	11*	[353]
$\lambda_{F}^{perimeter}$	Stiffness of matrix fibers	11	[353]
$J_{N,C}^{int}$	Generalized adhesion between nucleus and cytosol	-20	[353]
$J_{E,E}^{ext}$	Cell–cell adhesive strength	12	[353]
$J_{E,M}^{ext}$	Cell–medium adhesive strength	6.5	[353]
$J_{E,F}^{ext}$	Cell–fiber adhesive strength	4.5*	[353]
D_m	Diffusion constant of MMPs	5^{-4} [μm^{2}s^{-1}]	[286]
λ_m	On-rate constant of MMP decay	$2 \cdot 10^{-3}$ [s^{-1}]	[286]
π_m	On-rate constant of MMP production	$5 \cdot 10^{-3}$ [s^{-1}]	[286]

TABLE C.11: Cancer Cell Migration in Matrix Microchannels. (∗) indicates parameters that vary in the different sets of simulations.

Parameter	Description	Value	Reference
$A^{volume}_{\eta,N}$	Volume of cell nucleus	523 [μm^3]	[28, 332]
$A^{surface}_{\eta,N}$	Surface of cell nucleus	314 [μm^2]	[28, 332]
$A^{volume}_{\eta,C}$	Volume of cell cytosol	6542 [μm^3]	[28, 332]
$A^{surface}_{\eta,C}$	Surface of cell cytosol	3140 [μm^2]	[28, 332]
$T_{\eta,N}$	Motility of cell nucleus	0.25	[357]
$T_{\eta,C}$	Motility of cell cytosol	9	[357]
$\lambda^{volume}_{\eta,N,C}$	Cell compressibility	15	[357]
$\lambda^{surface}_{\eta,N}$	Stiffness of cell nucleus	15 ; 0.5*	[357]
$\lambda^{surface}_{\eta,C}$	Stiffness of cell cytosol	15 ; 0.9*	[357]
$J^{int}_{N,C}$	Generalized adhesion between nucleus and cytosol	-20	[357]
$J^{ext}_{E,E}$	Cell–cell adhesive strength	12	[357]
$J^{ext}_{E,M}$	Cell–medium adhesive strength	0	[350]
$J^{ext}_{E,P}$	Cell–fibronectin gel adhesive strength	0	[350]
$\mu_{pers,0}$	Basal persistence strength	0.8	[357]

Appendix D

Color Insert

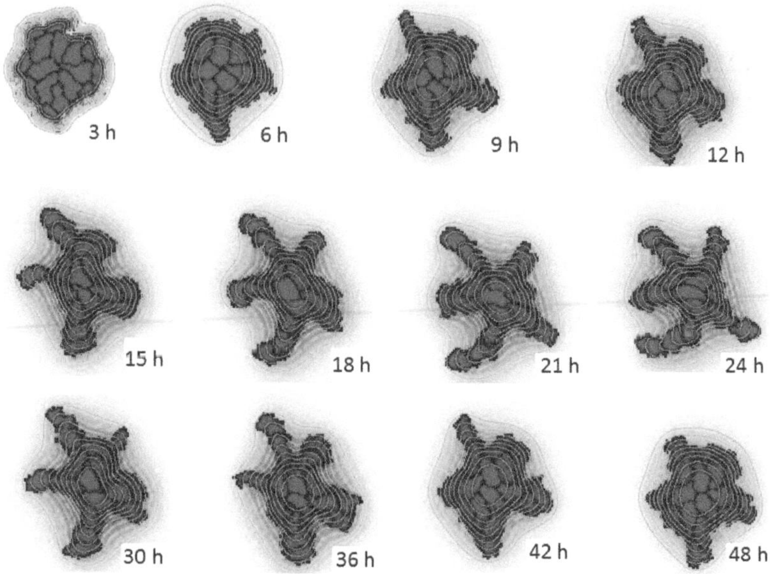

COLOR FIGURE 2.9

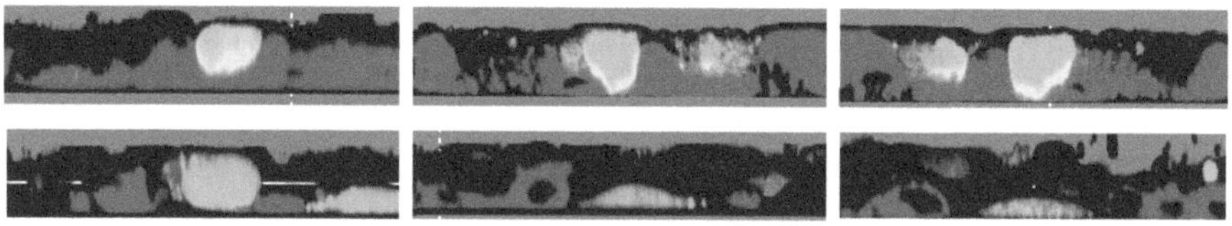

COLOR FIGURE 3.3

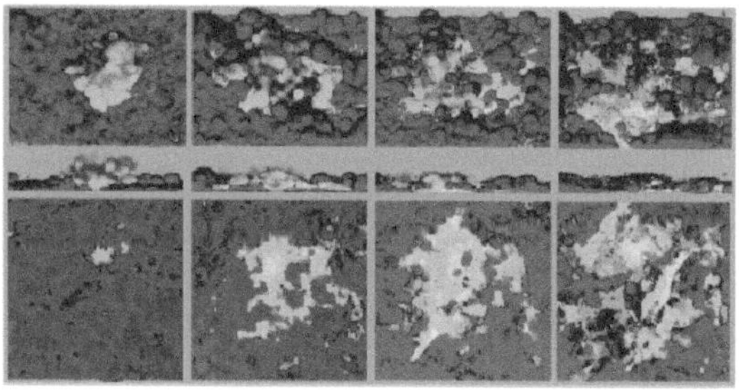

COLOR FIGURE 3.4

FIGURE D.1: Color Figure 1

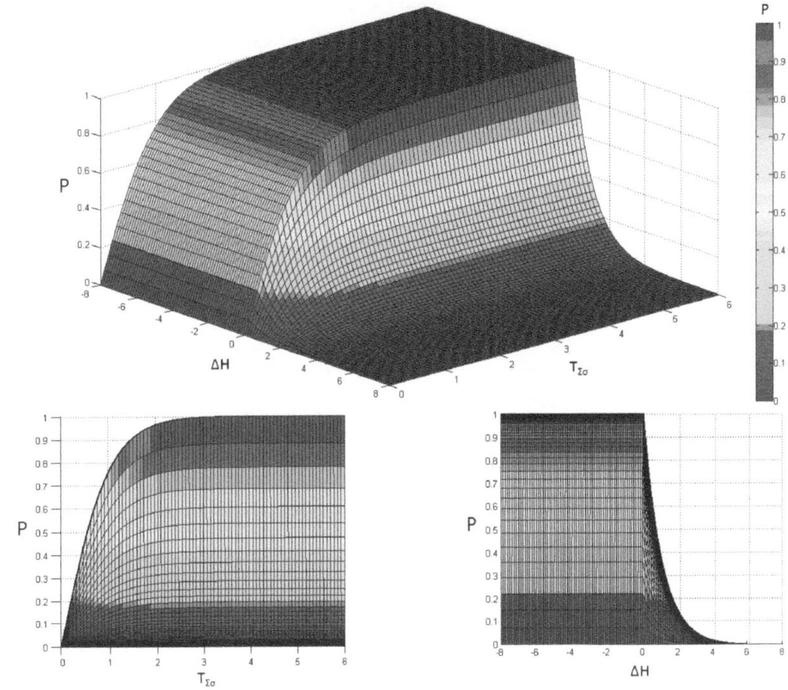

COLOR FIGURE 4.5

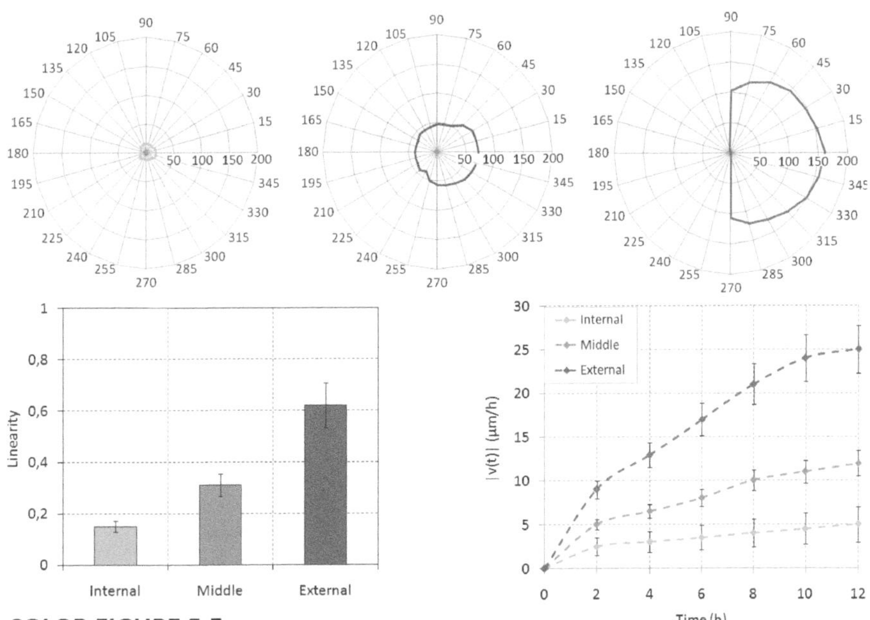

COLOR FIGURE 5.5

FIGURE D.2: Color Figure 2

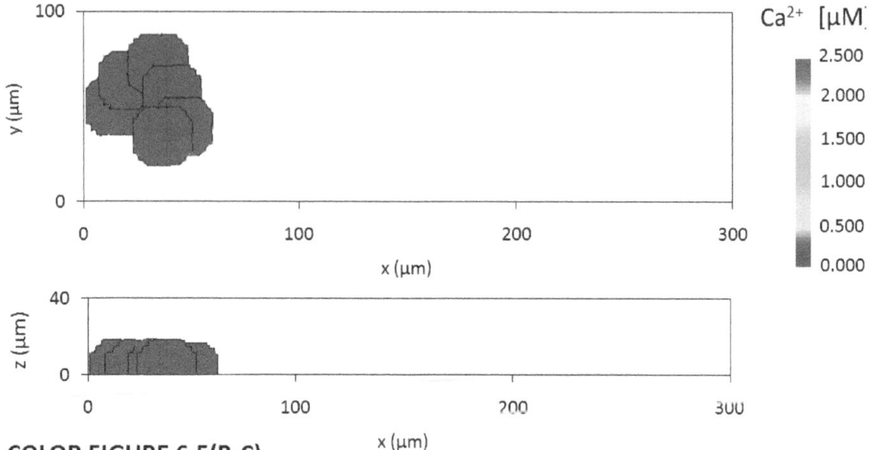

COLOR FIGURE 6.5(B-C)

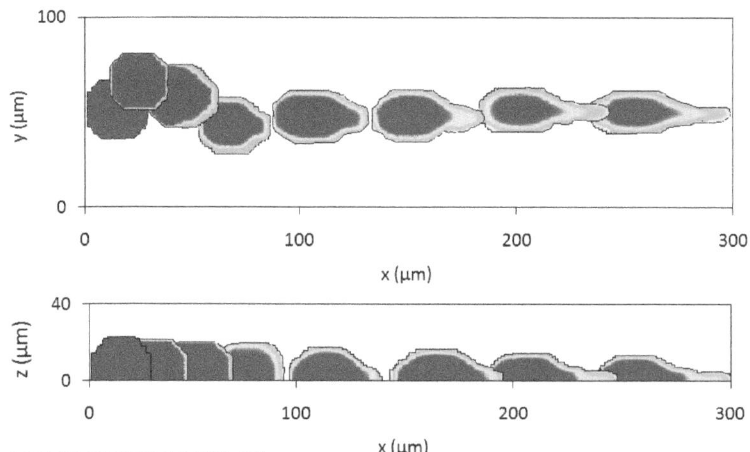

COLOR FIGURE 6.7(A-B)

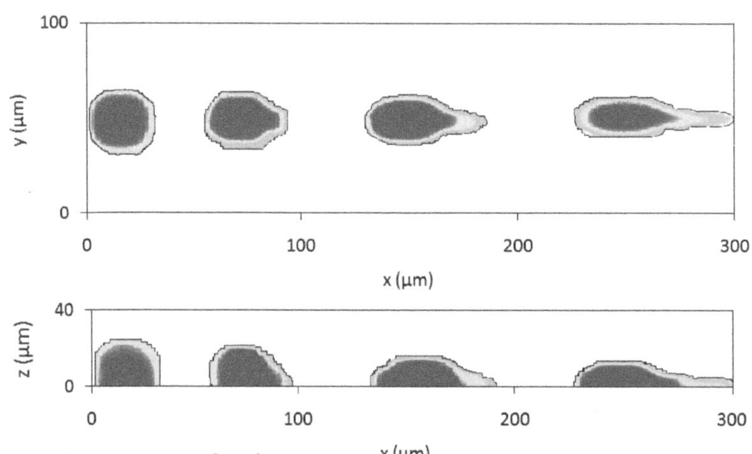

COLOR FIGURE 6.8(B-C)

FIGURE D.3: Color Figure 3

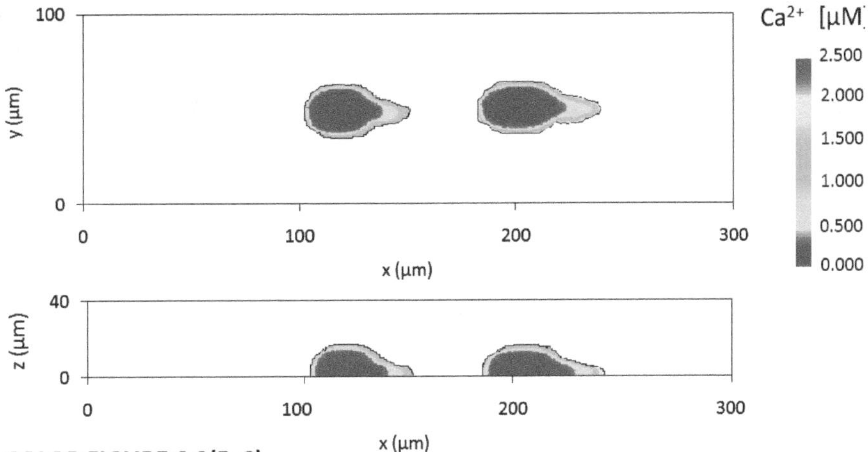

COLOR FIGURE 6.9(B-C)

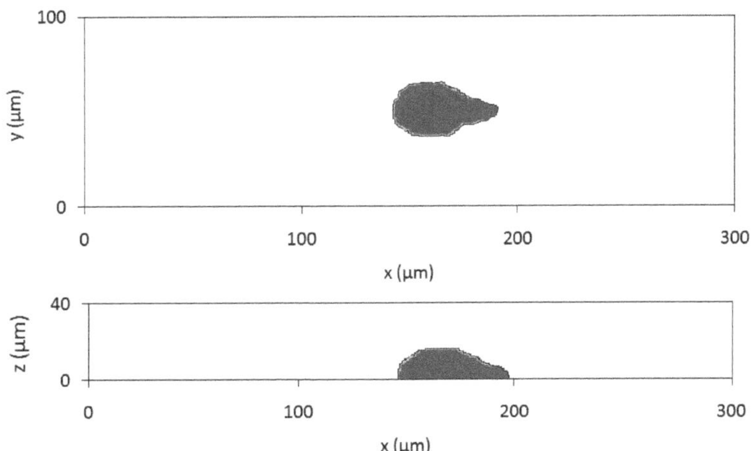

COLOR FIGURE 6.10(B-C)

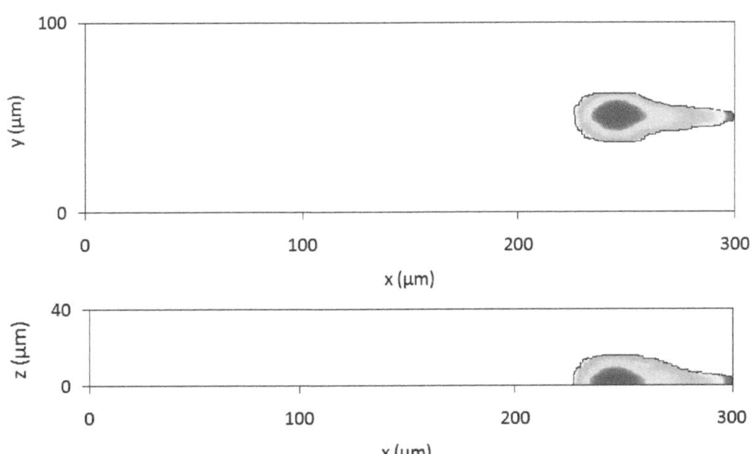

COLOR FIGURE 6.11(B-C)

FIGURE D.4: Color Figure 4

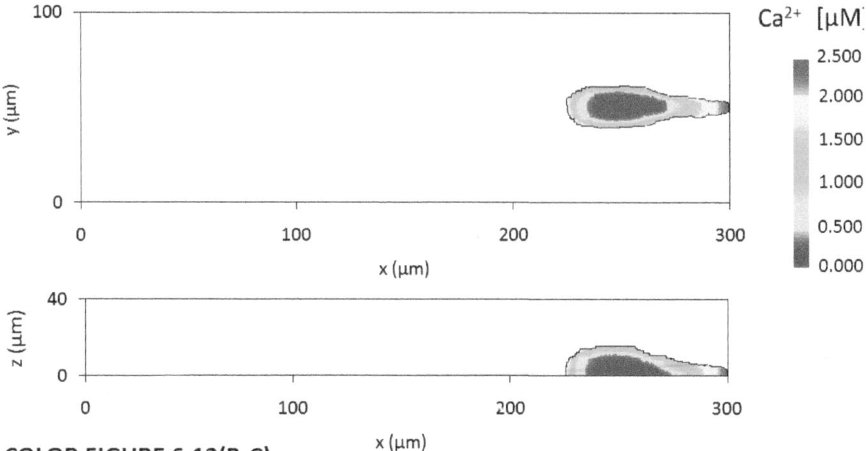

COLOR FIGURE 6.12(B-C)

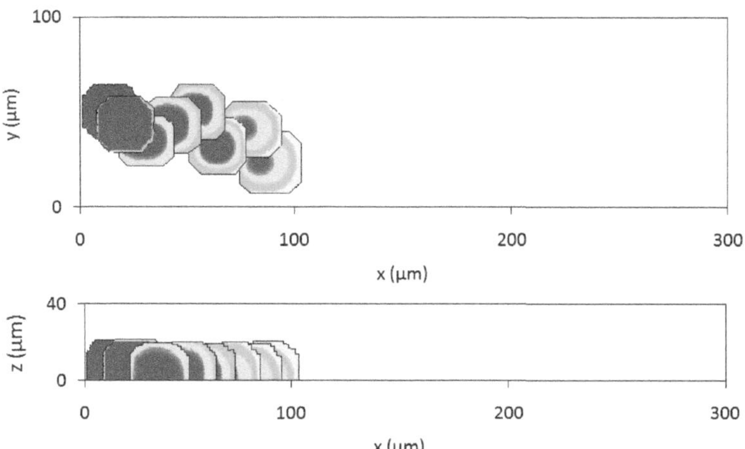

COLOR FIGURE 6.13(B-C)

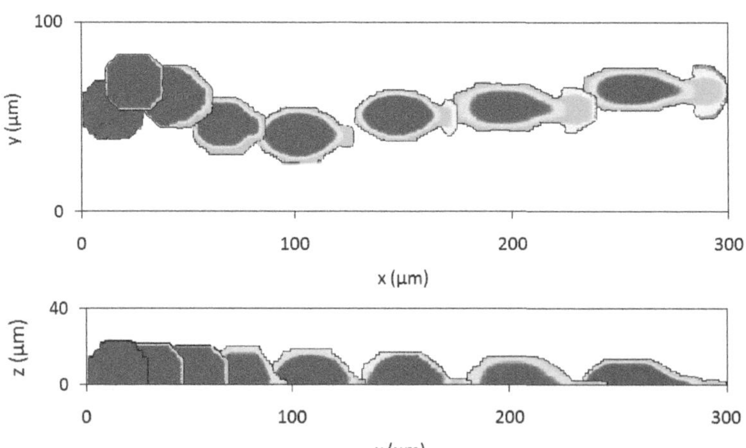

COLOR FIGURE 6.14(B-C)

FIGURE D.5: Color Figure 5

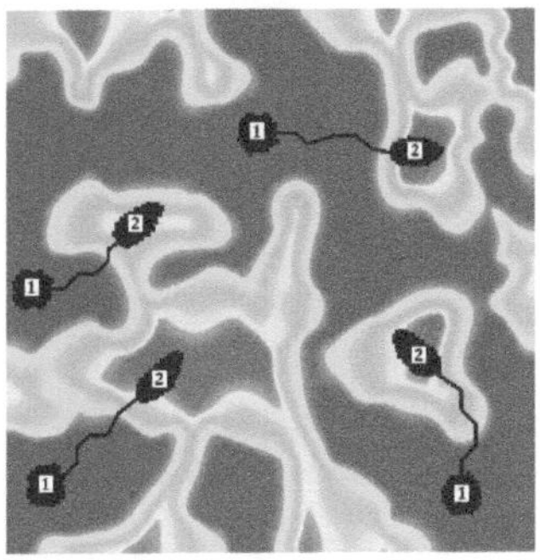

COLOR FIGURE 7.5

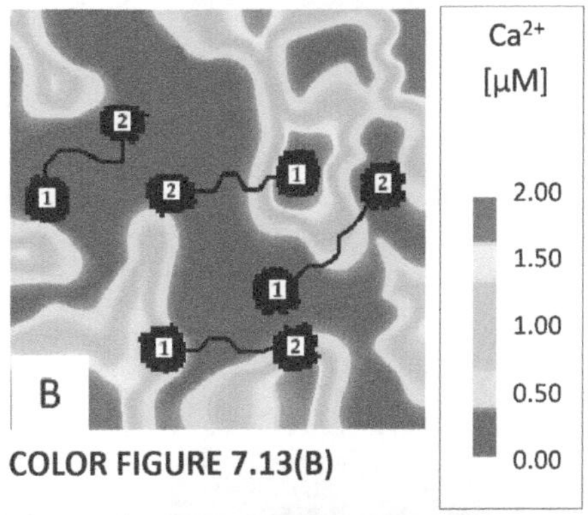

COLOR FIGURE 7.13(B)

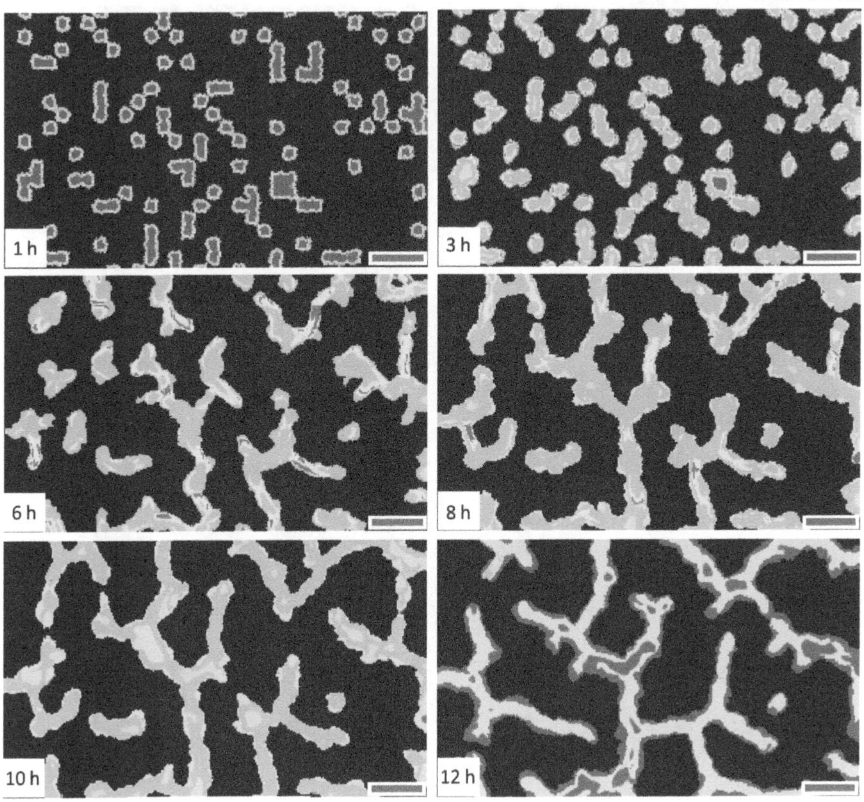

COLOR FIGURE 7.6

FIGURE D.6: Color Figure 6

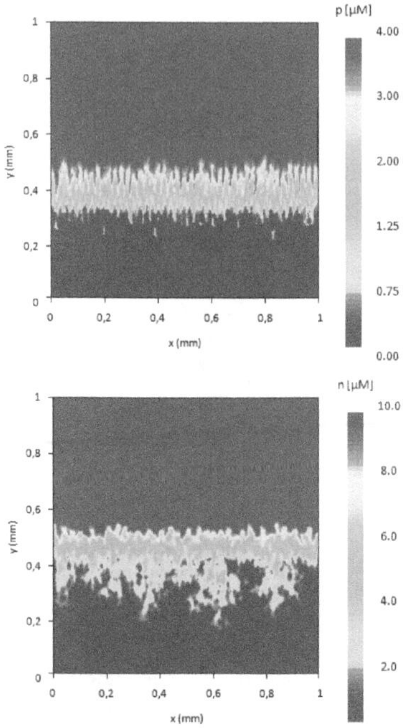

COLOR FIGURE 8.4(C-D)

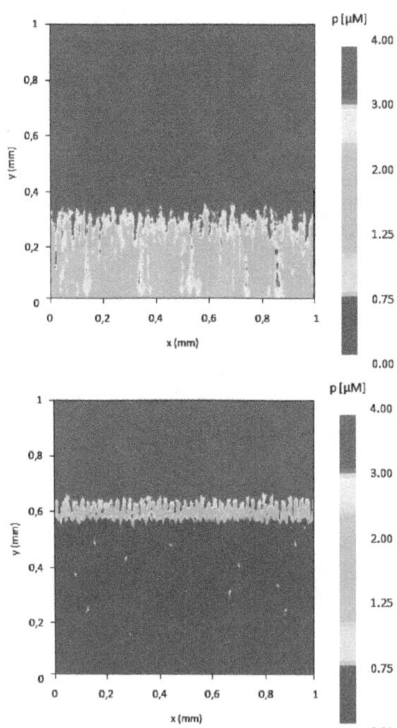

COLOR FIGURE 8.8(C-D)

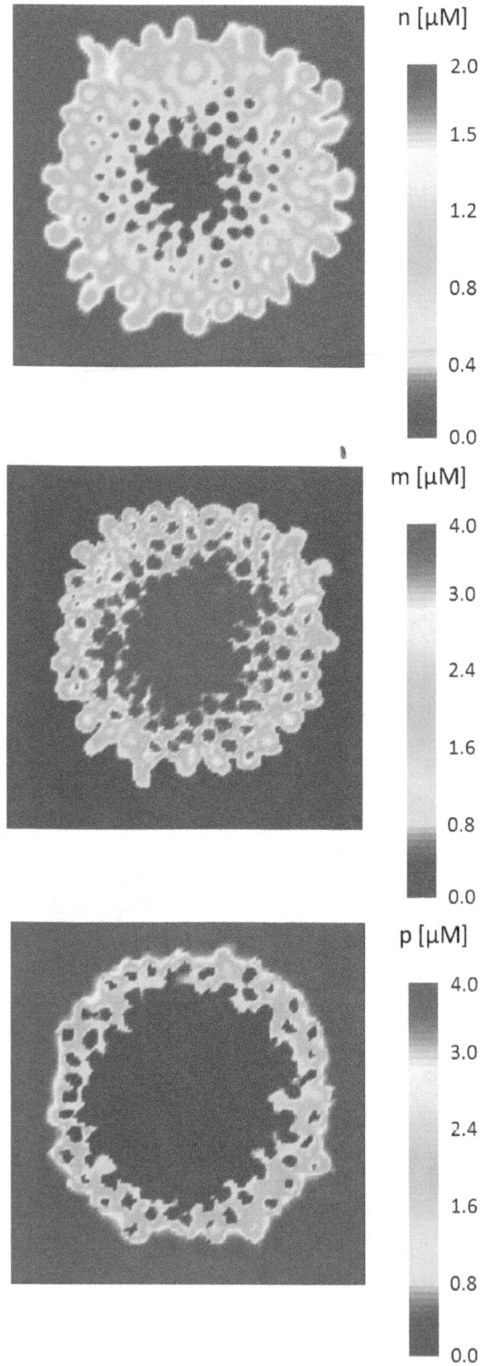

COLOR FIGURE 9.4

FIGURE D.7: Color Figure 7

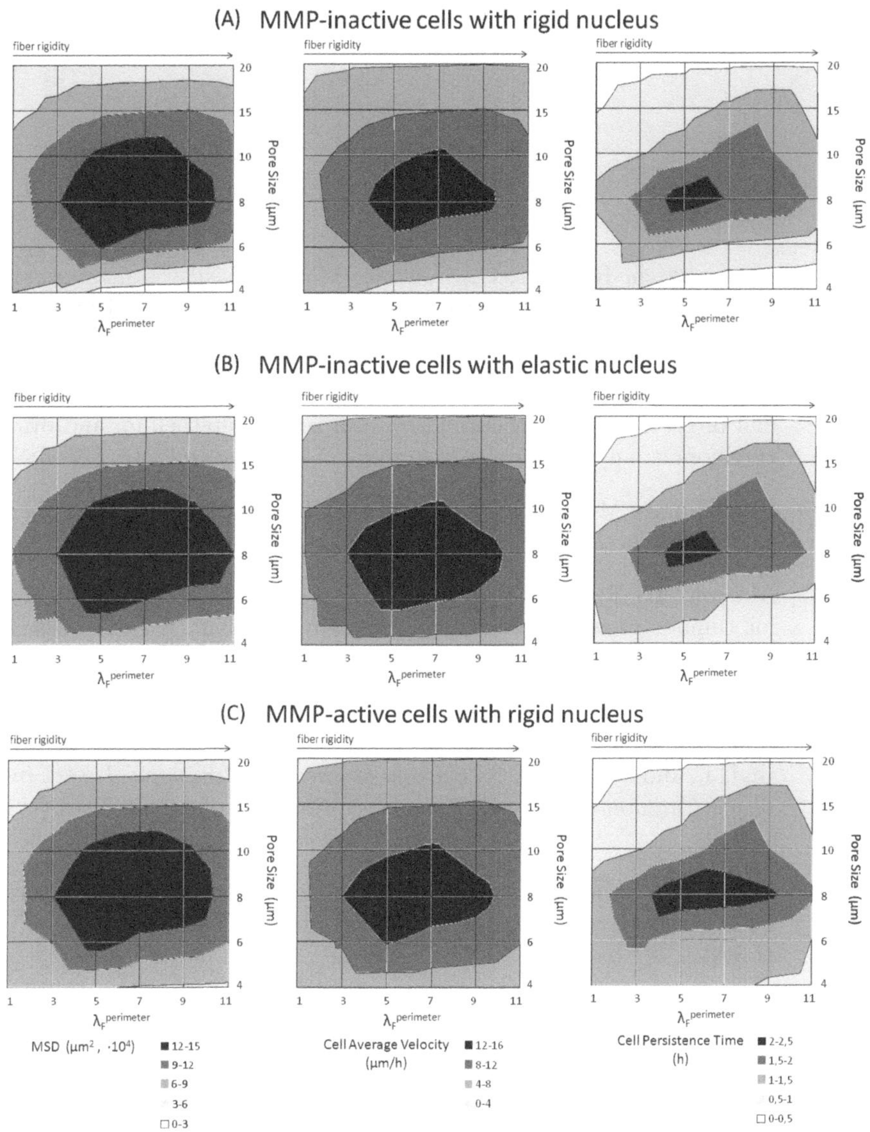

COLOR FIGURE 10.10

FIGURE D.8: Color Figure 8

Bibliography

[1] Abercrombie, M., 1980. The crawling movement of cells. *Proc R Soc London B*, 207, 129–147.

[2] Abramovitch, R., Meir, G., Neeman, M., 1995. Neovascularization induced growth of implanted C6 glioma multicellular spheroids: magnetic-resonance microimaging. *Cancer Res*, 55, 1956–1962.

[3] Ahmed, N., Thomson, E. W., Quinn, M. A., 2007. Epithelial - mesenchymal interconversions in normal ovarian surface epithelium and ovarian carcinomas: an exception to the norm. *J Cell Physiol*, 213, 581–588.

[4] Ahmedin, J., Murray, T., Samuels, A., Ghafoor, A., War, E., Thun, M. J., 2003. Cancer statistics. *Cancer J Clin*, 53, 5–26.

[5] Alarcon, T., Byrne, H., Maini, P., 2003. A cellular automaton model for tumour growth in inhomogeneous environment. *J Theor Biol*, 225, 257–274.

[6] Alber, M. S., Kiskowski, M. A., Glazier, J. A., and Jiang, Y., 2004. On cellular automation approaches to modeling biological cells. In Rosenthal, J., and Gilliam, D. S., editors, *Mathematical Systems Theory in Biology, Communication and Finance.* Springer-Verlag, New York (USA), 1–40.

[7] Alberts, B., Johnson, A., Lewis, J., Raff, M., Roberts, K., Walter, P., 2002. *Molecular Biology of the Cell*, 4th edn. Garland Science, New York (USA).

[8] Albrecht, M. A., Colegrove, S., L., Friel, D. D., 2002. Differential regu lation of ER Ca2+ uptake and release rates accounts for multiple modes of Ca2+-induced Ca2+ release. *J Gen Physiol*, 119 (3), 211–233.

[9] Ambrosi, D., Gamba, A., Serini, G., 2004. Cell directional persistence and chemotaxis in vascular morphogenesis. *Bull Math Biol*, 66 (6), 1851–1873.

[10] Ambrosi, D., Bussolino F., Preziosi, L., 2005. A review of vasculogenesis models. *J Theor Med*, 6, 1–19.

[11] Anderson, A. R. A., Chaplain, M. A. J., Rejniak, K. A., editors, 2007. *Single-Cell-Based Models in Biology and Medicine*, Mathematics and Biosciences in Interactions. Birkaüser, Basel (Switzerland).

[12] Anderson, A. R. A., Weaver, A., Commmings, P., Quaranta, V., 2006. Tumor morphology and phenotypic evolution driven by selective pressure from the microenvironment. *Cell*, 127, 905–915.

[13] Anderson, A. R. A., 2005. A hybrid mathematical model of solid tumour invasion: the importance of cell adhesion. *Math Med Biol*, 22, 163–186.

[14] Anderson, A. R. A., Chaplain, M. A. J., 1998. Continuous and discrete mathematical models of tumor-induced angiogenesis. *Bull Math Biol*, 60, 857–899.

[15] Araujo, R., McElwain, D., 2004. A history of the study of solid tumour growth: The contribution of mathematical modelling. *Bull Math Biol*, 66, 1039–1091.

[16] Armstrong, P. B., Parenti, D., 1972. Cell sorting in the presence of cytochalasin b. *J Cell Biol*, 55 (3), 542–553.

[17] Arnold, M., Hirschfeld-Warneken, V. C., Lohmüller, T., Heil, P., Blümmel, J., Cavalcanti-Adam, E. A., López-García, M., Walther, P., Kessler, H., Geiger, B., Spatz, J. P., 2008. Induction of cell polarization and migration by a gradient of nanoscale variations in adhesive ligand spacing. *Nano Lett*, 8 (7), 2063–2069.

[18] Baird, D., Feng, Q., Cerione, R. A., 2005. The Cool-2/alpha-Pix protein mediates a Cdc42-Rac signaling cascade. *Curr Biol*, 15 (1), 1–10.

[19] Bajénoff, M., Egen, J. G., Koo, L. Y., Laugier, J. P., Brau, F., Glaichenhaus, N., Germain, R. N., 2006. Stromal cell networks regulate lymphocyte entry, migration, and territoriality in lymph nodes. *Immunity*, 25 (6), 989–1001.

[20] Balter, A., Merks, R., M., H., Poplawski, N., J., Swat, M., Glazier, J., A., 2007. The Glazier-Graner-Hogeweg model: Extensions, future directions, and opportunities for further study. In Anderson, A. R. A., Chaplain, M. A. J., and K. A. Rejniak, editors, *Single-Cell-Based Models in Biology and Medicine*, Mathematics and Biosciences in Interactions. Birkaüser, Basel (Switzerland), 151–167.

[21] Baluk, P., Morikawa, S., Haskell, A., Mancuso, M., McDonald, D. M., 2003. Abnormalities of basement membrane on blood vessels and endothelial sprouts in tumors. *Am J Pathol*, 163, 1801–1815.

[22] Barbiero, G., Munaron, L., Antoniotti, S., Baccino, F. M., Bonelli, G., Lovisolo, D., 1995. Role of mitogen-induced calcium influx in the control of the cell cycle in Balb-c 3T3 fibroblasts. *Cell Calcium*, 18 (6), 542–556.

[23] Batschelet, E., 1981. *Circular Statistics in Biology.* Academic Press, London (United Kingdom).

[24] Bayley, P., Ahlstrom, P., Martin, S. R., Forsen S., 1984. The kinetics of calcium binding to calmodulin: Quin 2 and ANS stopped-flow fluorescence studies. *Biochem Biophys Res Commun*, 120 (1), 185–191.

[25] Bauer, A. L., Jackson, T. L., Jiang, Y., 2007. A cell-based model exhibiting branching and anastomosis during tumor-induced angiogenesis. *Biophys J*, 92 (9), 3105–3121.

[26] Bauer, K. S., Cude, K. J., Dixon, S. C., Kruger, E. A., Figg, W. D., 2000. Carboxyamido-triazole inhibits angiogenesis by blocking the calcium-mediated nitric-oxide synthase-vascular endothelial growth factor pathway. *J Pharmacol Exp Ther*, 292 (1), 31–37.

[27] Beadle, C., Assanah, M. C., Monzo, P., Vallee, R., Rosenfeld, S. S., Canoll, P., 2008. The role of myosin ii in glioma invasion of the brain. *Mol Biol Cell*, 19, 3357–3368.

[28] Beil, M., Micoulet, A., Wichert, G., Paschke, S., Walther, P. et al., 2003. Sphingosylphosphorylcholine regulates keratin network architecture and viscoelastic properties of human cancer cells. *Nat Cell Biol*, 5, 803–811.

[29] Bell, H. S., Whittle, I. R., Walker, M., Leaver, H. A., Wharton, S. B., 2001. The development of necrosis and apoptosis in glioma: experimental findings using spheroid culture systems. *Neuropathol Appl Neurobiol*, 27, 291–304.

[30] Bellomo, N., Li, N. K., Maini, P. K., 2008. On the foundations of cancer modelling: Selected topics, speculations, and perspectives. *Math Models Methods Appl Sci*, 18, 593–646.

[31] Bellomo, N., Bellouquid, A., Nieto, J., Soler, J., Maini, P. K., 2007. Multicellular biological growing systems: Hyperbolic limits towards macroscopic description. *Math Models Methods Appl Sci*, 17, 1675–1692.

[32] Bennett, J., Weeds, A., 1986. Calcium and the cytoskeleton. *Br Med Bull*, 42, 385–390.

[33] Berridge, M. J., Bootman, M. D., Roderick, H. L., 2003. Calcium signalling: Dynamics, homeostasis and remodelling. *Nat Rev Mol Cell Biol*, 4 (7), 517–529.

[34] Berridge, M. J., 1995. Calcium signalling and cell proliferation. *Bioessays*, 17 (6), 491–500.

[35] Berridge, M. J., Lipp, P., Bootman, M. D., 2000. The versatility and universality of calcium signalling. *Nat Rev Mol Cell Biol*, 1 (1), 11–21.

[36] Berridge, M. J., Bootman, M. D., Lipp, P., 1998. Calcium–a life and death signal. *Nature*, 395 (6703), 645–648.

[37] Blatter, L. A., Taha, Z., Mesaros, S., Shacklock, P. S., Wier, W. G., Malinski, T., 1995. Simultaneous measurements of Ca2+ and nitric oxide in bradykinin-stimulated vascular endothelial cells. *Circ Res*, 76 (5), 922–924.

[38] Bock, J. M., Sinclair, L. L., Bedford, N. S., Jackson, R. E., Lee, J. H., Trask, D. K., 2008. Modulation of cellular invasion by VEGF-C expression in squamous cell carcinoma of the head and neck. *Arch Otolaryngol Head Neck Surg*, 134 (4), 355–362.

[39] Boonstra, J., Van Rossum, G. S., 2003. The role of cytosolic phospholipase A2 in cell cycle progression. *Prog Cell Cycle Res*, 5, 181–190.

[40] Bootman, M. D., Lipp, P., Berridge, M. J., 2001. The organisation and functions of local Ca2+ signals. *J Cell Sci*, 114 (12), 2213–2222.

[41] Bortz, A. B., Kalos, M. H., Lebowitz, J. L., 1975. A new algorithm for Monte Carlo simulation of Ising spin systems. *J Comp Phys*, 17 (1), 10–18.

[42] Bosse, T., Ehinger, J., Czuchra, A., Benesch, S., Steffen, A., Wu, X., Schloen, K., Niemann, H. H., Scita, G., Stradal, T. E. B., Brakebusch, C., Rottner, K., 2007. Cdc42 and PI3K drive Rac-mediated actin polymerization downstream of c-Met in distinct and common pathways. *Mol Cell Biol*, 27 (19), 6615–6628.

[43] Brembeck, F. H., Schwartz-Romond, T., Bakkers, J., Wilhelm, S., Hammerschmidt, M., Birchmeier, W., 2004. Essential role of BCL9-2 in the switch between beta-catenin's adhesive and transcriptional functions. *Genes Dev*, 18, 2225–2230.

[44] Brightman, A. O., Rajwa, B. P., Sturgis, J. E., McCallister, M. E., Robinson, J. P., Voytik-Harbin, S. L., 2000. Time-lapse confocal reflection microscopy of collagen fibrillogenesis and extracellular matrix assembly in vitro. *Biopolymers*, 54, 222–234.

[45] Brinkmann, V., Foroutan, H., Sachs, H., Weidner, K. M., Birchmeier, W., 1995. Hepatocyte growth factor/scatter factor induces a variety of tissue specific morphogenic programs in epithelial cells. *J Cell Biol*, 131 (6 Pt 1), 1573–1586.

[46] Brock, A., Chang, E., Ho, C. C., LeDuc, P., Jiang, X., Whitesides, G. M., Ingber, D. E., 2003. Geometric determinants of directional cell motility revealed using microcontact printing. *Langmuir*, 19 (5), 1611–1617.

[47] Brown, J. M., 2002. Tumor microenvironment and the response to anticancer therapy. *Cancer Biol Ther*, 1, 453–458.

[48] Bru, A., Albertos, S., Subiza, J. L., Garca-Asenjo, J. L., Bru, I., 2003. The universal dynamics of tumor growth. *Bioph J*, 85 (5), 2948–2961.

[49] Burgess, P. K., Kulesa, P. M., Murray, J. D., Alvord, E. C., 1997. The interaction of growth rates and 544 diffusion coefficients in a three-dimensional mathematical model of gliomas. *J Neuropathol Exp Neurol*, 56, 704–713.

[50] Burgess, B. T., Myles, J. L., Dickinson, R. B., 2000. Quantitative analysis of adhesion-mediated cell migration in three-dimensional gels of RGD-grafted collagen. *Ann Biomed Eng*, 28 (1), 110–118.

[51] Burleson, K. M., Casey, R. C., Skubitz, K. M., Parmabuccian, S. E., Oegema Jr, T. R., Skubitz, A. P., 2004. Ovarian carcinoma ascites spheroids adhere to extracellular matrix components and mesothelial cell monolayers. *Gynec Oncol*, 93, 170–181.

[52] Burleson, K. M., Boente, M. P., Parmabuccian, S. E., and A. Skubitz, A. P., 2004. Ovarian carcinoma spheroids disaggregate on type I collagen and invade live mesothelial cell monolayers. *Clin Exp Metastasis*, 21, 685–697.

[53] Burleson, K. M., Boente, M. P., Parmabuccian, S. E., and A. Skubitz, A. P., 2004. Disaggregation and invasion of ovarian carcinoma ascites spheroids. *J Transl Med*, 4, 1–16.

[54] Bussolati, B., Deregibus, M. C., Camussi, G., 2010. Characterization of molecular and functional alterations of tumor endothelial cells to design anti-angiogenic strategies. *Curr Vasc Pharmacology*, 8 (2), 220–232.

[55] Bussolati, B., Deambrosis, I., Russo, S., Deregibus, M. C., Camussi, G., 2003. Altered angiogenesis and survival in human tumor-derived endothelial cells. *FASEB J*, 17 (9), 1159–1161.

[56] Bussolati, B., Grange, C., Camussi, G., 2011. Tumor exploits alternative strategies to achieve vascularization. *FASEB J*, 25 (9), 2874–2882.

[57] Bussolino, F., Arese, M., Audero, E., Giraudo, E., Marchio, S., Mitola, S., Primo, L., Serino, G., 2003. Biological aspects in tumor angiogenesis. In Preziosi, L., editor, *Cancer Modeling and Simulation*, Mathematical Biology and Medicine Sciences. Chapman and Hall/CRC Press, London (United Kingdom), 1–16.

[58] Byrne, H. M., Owen, M. R., Alarcon, T., Murphy, J., Maini, P. K., 2006. Modelling the response of vascular tumours to chemotherapy: A multiscale approach. *Math Mod Meth in Appl Sci*, 16 (7), 1219–1241.

[59] Byrne, H., Alarcon, T., Owen, M., Webb, S., Maini, P., 2006. Modeling aspects of cancer dynamics: A review. *Philos Trans R Soc A*, 364, 1563–1578.

[60] Cai, W., Rook, S. L., Jiang, Z. Y., Takahara, N., Aiello, L. P., 2000. Mechanisms of hepatocyte growth factorinduced retinal endothelial cell migration and growth. *Invest Ophthalmol Vis Sci*, 41 (7), 1885–1893.

[61] Cai, A. Q., Landman, K. A., Hughes, B. D., 2007. Multi-scale modeling of a wound-healing cell migration assay. *J Theor Biol*, 245, 576–594.

[62] Cannistra, S. A., 1993. Cancer of the ovary. *N Engl J Med*, 329, 1550–1559.

[63] Capito, R. M., Spector, M., 2003. Scaffold-based articular cartilage repair. *IEEE Eng Med Biol Mag*, 22, 42–50.

[64] Carmeliet, P., Jain, R. K., 2000. Angiogenesis in cancer and other diseases. *Nature*, 407, 249–257.

[65] Carmeliet, P., 2005. Angiogenesis in life, disease and medicine. *Nature*, 438, 932–936.

[66] Carmeliet, P., 2005. VEGF as a key mediator of angiogenesis in cancer. *Oncology*, 69, 4–10.

[67] Casciari, J. J., Sotirchos, S. V., Sutherland, R. M., 1992. Variations in tumor-cell growth-rates and metabolism with oxygen concentration, glucose-concentration, and extracellular pH. *J Cell Physiol*, 151, 386–394.

[68] Casey, R. C., Burleson, K. M., Skubitz, K. M., Parmabuccian, S. E., Oegema, T. J., Ruff, L. E., Skubitz, A. P., 2001. $\beta 1$–integrins regulate the formation and adhesion of ovarian carcinoma multicellular spheroids. *Am J Pathol*, 159, 2071–2080.

[69] Castro, M. A. A., Klamt, F., Grieneisen, V. A., Grivicich, I., Moreira, J. C. F., 2003. Gompertzian growth pattern correlated with phenotypic organization of colon carcinoma, malignant glioma and non-small cell lung carcinoma cell lines. *Cell Prolif*, 36, 65–73.

[70] Cavalcanti-Adam, E. A., Volberg, T., Micoulet, A., Kessler, H., Geiger, B., Spatz, J. P., 2007. Cell spreading and focal adhesion dynamics are regulated by spacing of integrin ligands. *Biophys J*, 92 (8), 2964–2974.

[71] Cavallaro, U., Christofori, G., 2001. Cell adhesion in tumor invasion and metastasis: Loss of the glue is not enough. *Biochim Biophys Acta*, 1552, 39–45.

[72] Cavallaro, U., Christofori, G., 2004. Cell adhesion and signalling by cadherins and Ig-CAMs in cancer. *Nat Rev Cancer*, 4, 118–132.

[73] Cercato, F. P., Mombach, J. C. M., Cavalheiro, G. G. H., 2006. High performance simulations of the cellular Potts model. *20th International Symposium on High-Performance Computing in an Advanced Collaborative Environment*, 28.

[74] Chaplain, M. A. J., Anderson, A. R. A., 2003. Mathematical modelling of tissue invasion. In Preziosi, L., editor, *Cancer Modelling and Simulation*. Chapman Hall/CRC Press, 269–297.

[75] Chaturvedi, R., Huang, C., Izaguirre, J. A., Newman, S. A., Glazier, J. A., Alber, M. S., 2004. A hybrid discrete-continuum model for 3-D skeletogenesis of the vertebrate limb. *Lect Notes Comput Sci*, 3305, 543–552.

[76] Chaturvedi, R., Huang, C., Kazmierczak, B., Schneider, T., Izaguirre, J. A., Glimm, T., Hentschel, H. G., Glazier, J. A., Newman, S. A., Alber, M. S., 2005. On multiscale approaches to three-dimensional modelling of morphogenesis. *J R Soc Interface*, 2 (3), 237–253.

[77] Chauviere, A., Preziosi, L., and Verdier, C., editors, 2010. *Cell Mechanics: From Single Scaled-Based Models to Multscale Modeling*, Mathematical and Computational Biology Series. Chapman Hall/CRC Press, London (United Kingdom).

[78] Chauviere, A., Hillen, T., Preziosi, L., 2007. A continuum model for mesenchymal motion in a fibrous network. *Netw Heterog Media*, 2, 333–357.

[79] Chauviere, A., Hillen, T., Preziosi, L., 2008. Modelling the motion of a cell population in the extracellular matrix. *Discr Cont Dyn Sys B*, Supplement Volume, 250–259.

[80] Chen, N., Glazier, J. A., Izaguirre, J. A., Alber, M. S., 2007. A parallel implementation of the cellular Potts model for simulation of cell-based morphogenesis. *Comput Phys Commun*, 176 (11-12), 670–681.

[81] Chen, R. R., Silva, E. A., Yuen, W. W., Brock, A. A., Fischbach, C., Lin, A. S., Guldberg, R. E., Mooney, D. J., 2007. Integrated approach to designing growth factor delivery system. *FASEB J*, 21 (14), 3896–3903

[82] Chilibeck, P. D., Paterson, D. H., Cunningham, D. A., Taylor, A. W., Noble, E. G., 1997. Muscle capillarization O_2 diffusion distance, and VO_2 kinetics in old and young individuals. *J Appl Physiol*, 82 (1), 63–69.

[83] Christofori, G., 2003. Changing neighbours, changing behaviour: cell adhesion molecule-mediated signalling during tumour progression. *EMBO J*, 22, 2318–2323.

[84] Cickovski, T., Huang, C., Chaturvedi, R., Glimm, T., Hentschel, H. G. E., Alber, M. S., Glazier, J. A., Newman, S. A., Izaguirre, J. A., 2005. A framework for three-dimensional simulation of morphogenesis. *IEEE/ACM Trans Computational Biology and Bioinformatics*, 2 (4), 273–288.

[85] Cipra, B. A., 1987. An introduction to the Ising model. *Am Math Monthly*, 94, 937–959.

[86] Clapham, D. E., 2007. Calcium signalling. *Cell*, 131 (6), 1047–1058

[87] Clairambault, J., 2009. Modelling physiological and pharmacological control on cell proliferation to optimise cancer treatments. *Math Model Nat Phenom*, 4, 12–67.

[88] Clauss, M., and Breier, G., editors, 2005. *Mechanisms of Angiogenesis*. Birkhaüser-Verlag, Basel (Switzerland).

[89] Coatesworth, W., Bolsover, S., 2008. Calcium signal transmission in chick sensory neurones is diffusion based. *Cell Calcium*, 43 (3), 236–249.

[90] Cohen, J. E., 2004. Mathematics is biology's next microscope, only better; biology is mathematics' next physics, only better. *PLoS Biol*, 2 (12), e439.

[91] Comoglio, P. M., Boccaccio, C., 2000. Scatter factors and invasive growth. *Semin Cancer Biol*, 11 (2), 153–165.

[92] Comoglio, P. M., Stella, M. C., 1999. HGF: A multifunctional growth factor controlling cell scattering. *Int J Biochem Cell Biol*, 31 (12), 1357–1362.

[93] Condeelis, J., Segall, J. E., 2003. Intravital imaging of cell movement in tumours. *Nat Rev Cancer*, 3, 921–930.

[94] Cristini, V., Frieboes, H., Gatenby, R., Caserta, S., Ferrari, M., Sinek, J., 2005. Morphologic instability and cancer invasion. *Clin Cancer Res*, 11, 6772–6779.

[95] Cristini, V., Lowengrub, J., Nie, Q., 2003. Nonlinear simulation of tumor growth. *J Math Biol*, 46, 191–224.

[96] Cross, S. S., 1997. Fractals in pathology. *J Pathol*, 182, 1–8.

[97] D' Alessandro, L., 2006. *High-throughput approaches to explore the transcriptional network leading to invasive growth*. PhD Thesis in Complex systems applied to post-genomic biology, Torino (Italy).

[98] Dale, P. D., Maini, P. K., Sherratt, J. A., 1994. Mathematical modeling of corneal epithelial wound healing. *Math Biosci*, 124, 127–147.

[99] Dan, D., Mueller, C., Chen, K., Glazier, J. A., 2005. Solving the advection-diffusion equations in biological contexts using the cellular Potts model. *Phys Rev E Stat Nonlin Soft Matter Phys*, 72 (4 Pt 1), 041909.

[100] De Bock, K., Cauwenberghs, S., Carmeliet, P., 2011. Vessel abnormalization: Another hallmark of cancer? Molecular mechanisms and therapeutic implications. *Curr Opin Genet Dev*, 21 (1), 73–79.

[101] De Luca, A., Arena, N., Sena, L. M., Medico, E., 1999. Met overexpression confers HGF-dependent invasive phenotype to human thyroid carcinoma cells in vitro. *J Cell Physiol*, 180 (3), 365–371.

[102] Dembo, M., Bell, G. I., 1987. The thermodynamics of cell adhesion. In Bronner F., and Kleinzeller, A., editors, *Current Topics in Membranes and Transport*, Vol. 29. Academic Press (USA), 71–89.

[103] Deutsch, A., Dormann, S., 2005. *Cellular automaton modeling of biological pattern formation: Characterization, applications, and analysis*, Modeling and Simulation in Science, Engeneering and Technology, Birkäuser, Basel (Switzerland).

[104] Di Renzo, M. F., Oliviero, M., Narsimhan, R. P., Bretti, S., Giordano, S., Medico, E., Gaglia, P., Zara, P., Comoglio, P. M., 1991. Expression of the Met/HGF receptor in normal and neoplastic human tissues. *Oncogene*, 6 (11), 1997–2003.

[105] Dickinson, R. B., Guido, S., Tranquillo, R. T., 1994. Biased cellmigration of fibroblasts exhibiting contact guidance in oriented collagen gels. *Ann Biomed Eng*, 22, 342–356.

[106] DiMilla, P. A., Stone, J. A., Quinn, J. A., Albelda, S. M., Lauffenburger, D. A., 1993. Maximal migration of human smooth-muscle cells on fibronectin and type-IV collagen occurs at an intermediate attachment strength. *J Cell Biol*, 122, 729–737.

[107] DiMilla, P. A., Barbee, K., Lauffenburger, D. A., 1991. Mathematical model for the effects of adhesion and mechanics on cell migration speed. *Biophys J*, 60 (1), 15–37.

[108] Dolmetsch, R. E., Xu, K., Lewis, R. S., 1998. Calcium oscillations increase the efficiency and specificity of gene expression. *Nature*, 392 (6679), 933–936.

[109] Doumic, M., 2007. Analysis of a population model structured by the cells molecular content. *Math Model Nat Phenom*, 2, 121–152.

[110] Doyle, A. D., Wang, F. W., Matsumoto, K., Yamada, K. M., 2009. One-dimensional topography underlies three-dimensional fibrillar cell migration. *J Cell Biol*, 184, 481–490.

[111] Drasdo, D., Hohme, S., 2003. Individual-based approaches to birth and death in avascular tumors. *Math Comp Model*, 37 (11), 1163–1175.

[112] Drasdo, D., Hohme, S., 2005. A single-cell-based model of tumor growth in vitro: Monolayers and spheroids. *Phys Biol*, 2, 133–147.

[113] Drasdo, D., 2005. On selected individual-based approaches to the dynamics of multicellular systems. In Alt, W., and Griebel, M., editors, *Multiscale Modeling.* Birkhaüser, Basel (Switzerland), 169–203.

[114] Drasdo, D. and Forgacs, G., 2000. Modeling the interplay of generic and genetic mechanisms in cleavage, blastulation, and gastrulation. *Dev Dyn*, 219 (2), 182–191.

[115] Drasdo, D., 2005. Coarse graining in simulated cell populations. *Adv Complex Syst*, 8, 319–363.

[116] Drake, C. J., LaRue, A., Ferrara, N., Little, C. D., 2000. VEGF regulates cell behavior during vasculogenesis. *Dev Biol*, 224 (2), 178–188.

[117] Duan, X., Sheardown, H., 2005. Crosslinking of collagen with dendrimers. *J Biomed Mater Res A*, 75 (3), 510–518.

[118] Duband, J.-L., Dufour, S., Thiery, J. P., 1989. Migration-promoting effect of fibronectin is mimicked by antibodies to integrins with moderate affinity. *J Cell Biol*, 109, 76a.

[119] Dubey, N., Letourneau, P. C., Tranquillo, R. T., 2001. Neuronal contact guidance in magnetically aligned fibrin gels: Effect of variation in gel mechano-structural properties. *Biomaterials*, 22, 1065–1075.

[120] Dunn, G. A., Ebendal, T., 1978. Contact guidance on oriented collagen gels. *Exp Cell Res*, 111, 475–479.

[121] Dunn, G. A. 1983. Characterizing a kinesis response: Time-averaged measures of cell speed and directional persistence. *Agents Actions Suppl*, 12, 14–33.

[122] Egeblad, M., Werb, Z., 2002. New functions for the matrix metalloproteinases in cancer progression. *Nature*, 2, 2071–2080.

[123] Engler, A., Bacakova, L., Newman, C., Hategan, A., Griffin, M., Discher, D., 2004. Substrate compliance versus ligand density in cell on gel responses. *Biophys J*, 86, 617–628.

[124] Erler, J. T., Weaver, V. M., 2009. Three-dimensional context regulation of metastasis. *Clin Exp Metastasis*, 26, 35–49.

[125] Etienne-Manneville, S., 2004. Cdc42the centre of polarity. *J Cell Sci*, 117 (Pt 8), 1291–1300.

[126] Feeley, K. M., Wells, M., 2001. Precursor lesions of ovarian epithelial malignancy. *Histopathology*, 38, 87–95.

[127] Feki, A., Berardi, P., Bellingan, G., Major, A., Krause, K. H., Petignat, P., et al., 2009. Dissemination of intraperitoneal ovarian cancer: Discussion of mechanisms and demonstration of lymphatic spreading in ovarian cancer model. *Crit Rev Oncol Hematol*, 72, 1–9.

[128] Ferrara, N., 2002. VEGF and the quest for tumour angiogenesis factors. *Nat Rev Cancer*, 2 (10), 795–803.

[129] Ferrara, N., Kerbel, R. S., 2005. Angiogenesis as a therapeutic target. *Nature*, 438 (7070), 967–974.

[130] Feske, S., Giltnane, J., Dolmetsch, R., Staudt, L. M., Rao, A., 2001. Gene regulation mediated by calcium signals in T lymphocytes. *Nat Immunol*, 2 (4), 316–324.

[131] Fichthorn, K. A., Weinberg, W. H., 1991. Theoretical foundations of dynamical Monte Carlo simulations. *J Chem Phys*, 95 (2), 1090–1096.

[132] Fink, C. C., Slepchenko, B., Moraru, I. I., Watras, J., Schaff, J. C., Loew, L. M., 2000. An image-based model of calcium waves in differentiated neuroblastoma cells. *Biophys J*, 79 (1), 163–183.

[133] Fiorio Pla, A., Grange, C., Antoniotti, S., Tomatis, C., Merlino, A., Bussolati, B., Munaron, L., 2008. Arachidonic acid-induced Ca2+ entry is involved in early steps of tumor angiogenesis. *Mol Cancer Res*, 6 (4), 535–545.

[134] Fiorio Pla, A., Munaron, L., 2001. Calcium influx, arachidonic acid, and control of endothelial cell proliferation. *Cell Calcium*, 30 (4), 235–244.

[135] Fiorio Pla, A., Genova, T., Pupo, E., Tomatis, C., Genazzani, A., Zaninetti, R., Munaron, L., 2010. Multiple roles of protein kinase a on arachidonic acidmediated Ca2+ entry and tumor-derived human endothelial cells migration. *Mol Cancer Res*, doi:10.1158/1541-7786.MCR-10-0002, in press.

[136] Fishman, D. A., Liu, Y., Ellerbroek, S. M., Stack, S. M., 2001. Lysophosphatidic acid promotes matrix metalloproteinase (MMP) activation and MMP-dependent invasion in ovarian cancer cells. *Cancer Res*, 61, 3194–3199.

[137] Fleming, I., and Busse, R., 1999. Signal transduction of eNOS activation. *Cardiovasc Res*, 43, 532–541.

[138] Freyer, J. P., Sutherland, R. M., 1986. Regulation of growth saturation and development of necrosis in EMT6-Ro multicellular spheroids by the glucose and oxygen-supply. *Cancer Res*, 46, 3504-3512.

[139] Freyer, J. P., Sutherland, R. M., 1983. Determination of diffusion constants for metabolites in multicell tumor spheroids. *Adv Exp Med Biol*, 159, 463–475.

[140] Freyer, J. P., 1994. Rates of oxygen consumption for proliferating and quiescent cells isolated from multicellular tumor spheroids. *Adv Exp Med Biol*, 345, 335–342.

[141] Friedl, P., Brocker, E. B., 2000. The biology of cell locomotion within three-dimensional extracellular matrix. *Cell Mol Life Sci*, 57 (1), 41–64.

[142] Friedl, P., Wolf, K. 2003. Tumour-cell invasion and migration: Diversity and escape mechanisms. *Nat Rev Cancer*, 3 (5), 362–374.

[143] Friedl, P., Maaser, K., Klein, C. E., Niggemann, B., Krohne, G., Zänker, K.S.,1997. Migration of highly aggressive MV3 melanoma cells in 3-dimensional collagen lattices results in local matrix reorganization and shedding of alpha2 and beta1 integrins and CD44. *Cancer Res*, 57 (10), 2061–2070.

[144] Friedl, P., Wolf, K., Lammerding, J., 2011. Nuclear mechanics during cell migration. *Curr Opin Cell Biol*, 23 (2), 253.

[145] Friedl, P., Weigelin, B., 2008. Interstitial leukocyte migration and immune function. *Nat Immunol*, 9 (9), 960–969.

[146] Friedl, P., Wolf, K., 2009. Plasticity of cell migration: A multiscale tuning model. *J Cell Biol*, 188 (1), 11–19.

[147] Friedman, A., 2009. Free boundary problems associated with multiscale tumor models. *Math Model Nat Phenom*, 4, 134–155.

[148] Fukumura, D., Duda, D. G., Munn, L. L., Jain, R. K., 2010. Tumor microvasculature and microenvironment: Novel insights through intravital imaging in pre-clinical models. *Microcirculation*, 17 (3), 206–225.

[149] Funamoto, S., Meili, R., Lee, S., Parry, L., Firtel, R. A., 2002. Spatial and temporal regulation of 3-phosphoinositides by PI 3-kinase and PTEN mediates chemotaxis. *Cell*, 109 (5), 611–623.

[150] Funaro, A., Ortolan, E., Bovino, P., Lo Buono, N., Nacci, G., Parrotta, E., Ferrero, E., Malavasi, F., 2009. Ectoenzymes and innate immunity: The role of human CD157 in leukocyte trafficking. *Front Biosci*, 14, 929–943.

[151] Fusco, A., Grieco, M., Santoro, M., Berlingieri, M. T., Pilotti, S., Pierotti, M. A., Dalla Porta, G., Vecchio, G., 1987. A new oncogene in human thyroid papillary carcinomas and their lymph-nodal metastases. *Nature*, 328, 170–172.

[152] Gabhann, F. M., Popel, A. S., 2004. Model of competitive binding of vascular endothelial growth factor and placental growth factor to VEGF receptors on endothelial cells. *Am J Physiol Heart Circ Physiol*, 286 (1), H153–H164.

[153] Gallagher, H. C., Bacon, C. L., Odumeru, O. A., Gallagher, K. F., Fitzpatrick, T., Regan, C. M., 2004. Valproate activates phosphodiesterase-mediated cAMP degradation: Relevance to C6 glioma G1 phase progression. *Neurotoxicol Teratol*, 26, 73–81.

[154] Gamba, A., Ambrosi, D., Coniglio, A., de Candia, A., di Talia, S., Giraudo, E., Serini, G., Preziosi, L., Bussolino, F., 2003. Percolation, morphogenesis, and Burgers dynamics in blood vessel formation. *Phys Rev Letters*, 90 (11), 101–118.

[155] Gatenby, R., Smallbone, K., Maini, P., Rose, F., Averill, J., Nagle, R., Worrall, L., Gillies, R., 2007. Cellular adaptations to hypoxia and acidosis during somatic evolution of breast cancer. *Br J Cancer*, 97, 646–653.

[156] Gaudet, C., Marganski, W., Kim, S., Brown, C. T., Gunderia, V., Dembo, M., Wong, J., 2003. Influence of type I collagen surface density on fibroblast spreading, motility, and contractility. *Biophys J*, 85, 3329–3335.

[157] Gerhardt, H., Golding, M., Fruttiger, M., Ruhrberg, C., Lundkvist, A., Abramsson, A., Jeltsch, M., Mitchell, C., Alitalo, K., Shima, D., Betsholtz, C., 2003. VEGF guides angiogenic sprouting utilizing endothelial tip cell filopodia. *J Cell Biol*, 161 (6), 1163–1177.

[158] Gerlee, P., Anderson, A., 2007. An evolutionary hybrid cellular automaton model of solid tumor growth. *J Theor Biol*, 246, 583–603.

[159] Gerlee, P., Anderson, A., 2007. Stability analysis of a hybrid cellular automaton model of cell colony growth. *Phys Rev E*, 75, 051911.

[160] Gerlitz, G., Bustin, M., 2011. The role of chromatin structure in cell migration. *Trends Cell Biol*, 21 (1), 6–11.

[161] Girton, T. S., Oegema, T. R., Tranquillo, R. T., 1999. Exploiting glycation to stiffen and strengthen tissue equivalents for tissue engineering. *J Biomed Mater Res*, 46 (1), 87–92.

[162] Giverso, C., Scianna, M., Preziosi, L., Lo Buono, N., Funaro, A., 2010. Individual cell-based model for in-vitro mesothelial invasion of ovarian cancer. *Math Model Nat Phenom*, 5 (1), 203–223.

[163] Glazier, J. A., Balter, A., Merks, R. M. H., Poplawski, N. J., Swat, M., 2007. The Glazier-Graner-Hogeweg model: Extension, future direction, and opportunitites for further study. *Single-Cell-Based Model Biol Med*, 4, 157–167.

[164] Glazier, J. A., Balter, A., Poplawski, N. J., 2007. Magnetization to morphogenesis: A brief history of the Glazier-graner-Hogeweg model. In Anderson, A. R. A., Chaplain, M. A. J., and Rejniak, K. A., editors, *Single-Cell-Based Models in Biology and Medicine*, Mathematics and Biosciences in Interactions. Birkaüser, Basel (Switzerland), 79–106.

[165] Glazier, J. A., Graner, F., 1993. Simulation of the differential adhesion driven rearrangement of biological cells. *Phys Rev E Stat Phys Plasmas Fluids Relat Interdiscip Topics*, 47 (3), 2128–2154.

[166] Goto, Y., Miura, M., Iijima, T., 1996. Extrusion mechanisms of intracellular Ca2+ in human aortic endothelial cells. *Eur J Pharmacol*, 314 (1-2), 185–192.

[167] Goodman, S. L., Risse, G., Vondermark, K., 1989. The E8 subfragment of laminin promotes locomotion of myoblasts over extracellular matrix. *J Cell Biol*, 109 (2), 799–809.

[168] Grakoui, A., Bromley, S. K., Sumen, C., Davis, M. M., Shaw, A. S., Allen, P. M., Dustin, M. L., 1999. The immunological synapse: A molecular machine controlling T cell activation. *Science*, 285 (5425), 221–227.

[169] Graner, F., Glazier, J. A., 1992. Simulation of biological cell sorting using a two dimensional extended Potts model. *Phys Rev Lett*, 69 (13), 2013–2017.

[170] Grange, C., Bussolati, B., Bruno, S., Fonsato, V., Sapino, A., Camussi, G., 2006. Isolation and characterization of human breast tumor-derived endothelial cells. *Oncol Rep*, 15 (2), 381–386.

[171] Grinnell, F., Geiger, B., 1986. Interaction of fibronectin-coated beads with attached and spread fibroblasts: Binding, phagocytosis, and cytoskeletal reorganization. *Exp Cell Res*, 162 (2), 449–446.

[172] Guck, J., Lautenschläger, F., Paschke, S., Beil, M., 2010. Critical review: Cellular mechanobiology and amoeboid migration. *Integr Biol*, 2 (11), 575–583.

[173] Guo, C., Kaufman, L. J., 2007. Flow and magnetic field induced collagen alignment. *Biomaterials*, 28, 1105–1123.

[174] Gusatto, E., Mombach, J. C. M., Cercato, F. P., Cavalheiro, G. H., 2005. An efficient parallel algorithm to evolve simulations of the cellular Potts model. *Parallel Process Lett*, 15, 199–208.

[175] Guyton, A. C., Hall, J. E., 2000. *Textbook of medical physiology, 10th edition*. W. B. Sauders, Philadelphia (USA).

[176] Hamed, G., Rodriguez, F., 1975. Gelation of dilute collagen solutions by ultraviolet light. *J Appl Polym Sci*, 19 (12), 3299–3313.

[177] Harley, B. A. C., Kim, H., Zaman, M. H., Yannas, I. V., Lauffenburger, D. A., Gibson, L. J., 2008. Microarchitecture of three-dimensional scaffolds influences cell migration behavior via junction interactions. *Biophys J*, 95 (8), 4013–4024.

[178] Harley, B. A., Spilker, M. H., Wu, J. W., Asano, K., Hsu, H. P., Spector, M., Yannas, I. V., 2004. Optimal degradation rate for collagen chambers used for regeneration of peripheral nerves over long gaps. *Cells Tissues Organs*, 176, 153–165.

[179] Hatzikirou, H., Deutsch, A., Schaller, C., Simon, M., Swanson, K., 2005. Mathematical modeling of glioblastoma tumour development: A review. *Math Models Methods Appl Sci*, 15, 1779–1794.

[180] Haugh, J. M., Codazzi, F., Teruel, M., Meyer, T., 2000. Spatial sensing in fibroblasts mediated by 3' phosphoinositides. *J Cell Biol*, 151 (6), 1269–1280.

[181] Hawkins, R., Piel, M., Faure-Andre, G., Lennon-Dumenil, A., Joanny, J. et al., 2009. Pushing off the walls: A mechanism of cell motility in confinement. *Phys Rev Lett*, 102, 4.

[182] Hegedus, B., Marga, F., Jakab, K., Sharpe-Timms, K. L., Forgacs, G., 2006. The interplay of cell–cell and cell–matrix interactions in the invasive properties of brain tumors. *Biophys J*, 91, 2708–2716.

[183] Hentschel, H. G. E., Glimm, T., Glazier J. A., Newman, S. A., 2004. Dynamical mechanisms for skeletal pattern formation in the vertebrate limb. *Proc Biol Sci*, 271 (1549), 1713–1722.

[184] Higgs, H. N., Pollard, T. D., 2000. Activation by Cdc42 and PIP2 of Wiskott-Aldrich syndrome protein (WASp) stimulates actin nucleation by Arp2/3 complex. J Cell Biol, 150 (6), 1311–1320.

[185] Hogea, C., Murray, B., Sethian, J., 2006. Simulating complex tumor dynamics from avascular to vascular growth using a general level-set method. *J Math Biol*, 53, 86–134.

[186] Hogeweg, P., 2000. Evolving mechanisms of morphogenesis: On the interplay between differential adhesion and cell differentiation. *J Theor Biol*, 203 (4), 317–333.

[187] Hong, D., Jaron, D., Buerk, D. G., Barbee, K. A., 2008. Transport-dependent calcium signaling in spatially segregated cellular caveolar domains. *Am J Physiol Cell Physiol*, 294 (3), C856–C866.

[188] Horwitz, A. R., Parsons, J. T., 1999. Cell biology: Cell migration-movin'on. *Science*, 286 (5442), 1102–1103.

[189] Hotary, K. B., Allen, E. D., Brooks, P. C., Datta, N. S., Long, M. W., Weiss, S. J., 2003. Membrane type I matrix metalloproteinase usurps tumor growth control imposed by the three-dimensional extracellular matrix. *Cell*, 114, 33–45.

[190] Howard, J., 2001. *Mechanics of motor proteins and the cytoskeleton.* Sinauer Associates Inc., Sunderland (USA).

[191] Hryshko, L., V., Philipson, K., D., 1997. Sodium-calcium exchange: Recent advances. *Basic Res Cardiol*, 92 (1), 45–51.

[192] Huang, S., Brangwynne, C. P., Parker, K. K., Ingber, D. E., 2005. Symmetry-breaking in mammalian cell cohort migration during tissue pattern formation: Role of random-walk persistence. *Cell Motil Cytoskeleton*, 61, 201–213.

[193] Huang, S. Ingber, D. E., 1999. The structural and mechanical complexity of cell-growth control. *Nat Cell Biol*, 1, 131–138.

[194] Hunding, A., Ipsen, M., 2003. Simulation of waves in calcium models with 3D spherical geometry. *Math Biosci*, 182 (1), 45–66.

[195] Inoue, N., Bessho, M., Furuta, M., Kojima, T., Okuda, S., Hara, M., 2006. A novel collagen hydrogel cross-linked by gamma-ray irradiation in acidic pH conditions. *J Biomater Sci Polym Ed*, 17 (8), 837–858.

[196] Ilina, O., Bakker, G. J., Vasaturo, A., Hofmann, R. M., Friedl, P., 2011. Two-photon laser-generated microtracks in 3D collagen lattices: Principles of MMP-dependent and -independent collective cancer cell invasion. *Phys Biol*, 8 (1), 015010.

[197] Irimia, D., Charras, G., Agrawal, N., Mitchison, T., Toner, M., 2007. Polar stimulation and constrained cell migration in microfluidic channels. *Lab on a Chip*, 7, 1783.

[198] Irimia, D., Toner, M., 2009. Spontaneous migration of cancer cells under conditions of mechanical confinement. *Integr Biol*, 1, 506–512.

[199] Ising, E., 1925. Beitrag zur theorie des ferromagnetismus. *Z Physik*, 31, 253.

[200] Jafri, M. S., Keizer, J., 1995. On the roles of Ca2+ diffusion, Ca2+ buffers, and the endoplasmic reticulum in IP3-induced Ca2+ waves. *Biophys J*, 69 (5), 2139–2153.

[201] Jain, R. K., 2003. Molecular regulation of vessel maturation. *Nat Med*, 9, 685–693.

[202] Jiang, Y., Pjesivac-Grbovic, J., Cantrell, C., Freyer, J., 2005. A multiscale model for avascular tumor growth. *Biophys J*, 89, 3884–3894.

[203] Jiang, Y., 1998. *Cellular pattern formation.* PhD Thesis, University of Notre Dame (USA).

[204] Kahl, C. R., Means, A. R., 2003. Regulation of cell cycle progression by calcium/calmodulin-dependent pathways. *Endocr Rev*, 24 (6), 719–736.

[205] Kaufman, L. J., Brangwynne, C. P., Kasza, K. E., Filippidi, E., Gordon, V. D., Deisboeck, T. S., Weitz, D. A., 2005. Glioma expansion in collagen I matrices: Analyzing collagen concentration-dependent growth and motility patterns. *Biophys J*, 89, 635–650.

[206] Kelm, J. M., Timmins, N. E., Brown, C. J., Fussenegger, M., Nielsen, L. K., 2003. Method for generation of homogeneous ulticellular tumor spheroids applicable to a wide variety of cell types. *Biotechnol Bioeng*, 83, 173–180.

[207] Kenny, H. A., Kaur, S., Coussens, L. M., Lengyel, E., 2008. The initial steps of ovarian cancer cell metastasis are mediated by MMP-2 cleavage of vitronectin and fibronectin. *J Clin Invest*, 118, 1367–1379.

[208] Khaitan, D., Chandna, S., Arya, M. B., Dwarakanath, B. S., 2006. Establishment and characterization of multicellular spheroids from a human glioma cell line: Implications for tumor therapy. *J Transl Med*, 4, 12–25.

[209] Khanapure, S. P., Garvey, D. S., Janero, D. R., Letts, L., G., 2007. Eicosanoids in inflammation: Biosynthesis, pharmacology, and therapeutic frontiers. *Curr Top Med Chem*, 7 (3), 311–340.

[210] Kholodenko, B. N., 2003. Four-dimensional organization of protein kinase signaling cascades: The roles of diffusion, endocytosis and molecular motors. *J Exp Biol*, 206 (Pt 12), 2073–2082.

[211] Kimura, H., Esumi, H., 2003. Reciprocal regulation between nitric oxide and vascular endothelial growth factor in angiogenesis. *Acta Biochim Pol*, 50 (1), 49–59.

[212] Klingauf, J., Neher, E., 1997. Modeling buffered Ca^{2+} diffusion near the membrane: Implications for secretion in neuroendocrine cells. *Biophys J*, 72, 674–690.

[213] Kohn, E. C., Sandeen, M. A., Liotta, L. A., 1992. In vivo efficacy of a novel inhibitor of selected signal transduction pathways including calcium, arachidonate, and inositol phosphates. *Cancer Res*, 52 (11), 3208–3212.

[214] Kohn, E. C., Liotta, L. A., 1990. L651582: A novel antiproliferative and antimetastasis agent. *J Natl Cancer Inst*, 82 (1), 54–60.

[215] Kowalczyk, R., 2005. Preventing blow-up in a chemotaxis model. *J Math Anal App*, 305 (2), 566–588.

[216] Kowanetz, M., Ferrara, N.,2006. Vascular endothelial growth factor signaling pathways: Therapeutic perspective. *Clin Cancer Res*, 12 (17), 5018–5022.

[217] Kraus, M., Wolf, B., Wolf, B., 1996. Crosstalk between cellular morphology and calcium oscillation patterns. Insights from a stochastic computer model. *Cell Calcium*, 19 (6), 461–472.

[218] Kumar. S., Weaver, V. M., 2009. Mechanics, malignancy, and metastasis: The force journey of a tumor cell. *Cancer Metastasis Rev*, 28, 113–127.

[219] Kuntz, R. M., Saltzman, W. M., 1997. Neutrophil motility in extracellular matrix gels: Mesh size and adhesion affect speed of migration. *Biophys J*, 72 (3), 1472–1480.

[220] Kurosaka, S., Kashina, A., 2008. Cell biology of embryonic migration. *Birth Defects Res C Embryo Today*, 84 (2), 102–122.

[221] Kwan, H. Y., Huang, Y., Yao, X., 2007. TRP channels in endothelial function and dysfunction. *Biochim Biophys Acta*, 1772, 907–914.

[222] Lachowicz, M., 2002. From microscopic to macroscopic description for generalized kinetic models. *Math Models Methods Appl Sci*, 12, 985–1005.

[223] Lahat, N., Sheinfeld, M., Sobel, E., Kinarty, A., Kraiem, Z., 1992. Divergent effects of cytokines on human leukocyte antigen-DR antigen expression of neoplastic and non-neoplastic human thyroid cells. *Cancer*, 69 (7), 1799–1807.

[224] Lamers, E., van Horssen, R., Te Riet, J., Van Delft, F. C., Luttge, R., Walboomers, X. F., Jansen, J. A., 2010. The influence of nanoscale topographical cues on initial osteoblast morphology and migration. *Eur Cell Mater*, 9 (20), 329–343.

[225] Lämmermann, T., Bader, B. L., Monkley, S. J., Worbs, T., Wedlich-Soldner, R., et al., 2008. Rapid leukocyte migration by integrin-independent flowing and squeezing. *Nature*, 453, 51–55.

[226] Lancaster, J. R. Jr, 1997. A tutorial on the diffusibility and reactivity of free nitric oxide. *Nitric Oxide*, 1 (1), 18–30.

[227] Landini, G., Hirayama, Y., Li, T. J., Kitano, M., 2000. Increased fractal complexity of the epithelial connective tissue interface in the tongue of 4NQ0-treated rats. *Pathol Res Pract*, 196, 251–258.

[228] Lauffenburger, D. A., Lindermann, J. J., 1996. *Receptors: Models for binding, trafficking, and signaling.* Oxford University Press, London (United Kingdom).

[229] Lauffenburger, D. A., Horwitz, A. F., 1996. Cell migration: A physically integrated molecular process. *Cell*, 84 (3), 359–369.

[230] Lautenschläger, F., Paschke, S., Schinkinger, S., Bruel, A., Beil, M., Guck J., 2009. The regulatory role of cell mechanics for migration of differentiating myeloid cells *Proc Natl Acad Sci USA*, 106 (37), 15696–15701.

[231] Lee, K.-C., 1995. Rejection-free Monte Carlo technique. *J Phys A*, 28, 4835.

[232] Lehnert, D., Wehrle-Haller, B., David, C., Weiland, U., Ballestrem, C., Imhof, B. A., Bastmeyer, M., 2004. Cell behaviour on micropatterned substrata: Limits of extracellular matrix geometry for spreading and adhesion. *J Cell Sci*, 117 (1), 41–52.

[233] Lesko, E., Majka, M., 2008. The biological role of HGF- MET axis in tumor growth and development of metastasis. *Front Biosci*, 13, 1271–1280.

[234] Lessan, K., Aguiar, D. J., Oegema, T. J., Siebenson, L., Skubitz, A. P., 1999. CD44 and β1–integrin mediate ovarian carcinoma cell adhesion to peritoneal mesothelial cells. *Am J Pathol*, 154, 1525–1537.

[235] Li, X., Cristini, V., Nie, Q., Lowengrub, J., 2007. Nonlinear three-dimensional simulation of solid tumor growth. *Discrete Dyn Continuous Dyn Syst B*, 7, 581–604.

[236] Lin, S., Fagan, K. A., Li, K.-X., Shaul, P. W., Cooper, D. M. F., Rodman, D. M., 2000. Sustained endothelial nitric oxide synthase activation requires capacitative Ca2+ entry. *J Biol Chem*, 275 (24), 17979–17985.

[237] Lo, C. M., Wang, H. B., Dembo, M., Wang, Y. L., 2000. Cell movement is guided by the rigidity of the substrate. *Biophys J*, 79, 144–152.

[238] Lo Buono, N., Scianna, M., Bovino, P., Morone, S., Panotta, R., Ferrero, E., Giverso, C., Preziosi, L., Malavasi, F., Funaro, A., Ortolan, E., 2011. CD157 expression enhances invasiveness of ovarian cancer cells and induces a mesenchymal phenotype. *FEBS J*, 278, 218–219.

[239] Lowengrub, J., Cristini, V., Frieboes, H. B., Li, X., Macklin, P., Sanga, S., Wise, S. M., Zheng, X., 2011. Nonlinear modeling and simulation of tumor growth. In Bellomo, N., Chaplain, M. A. J., and De Angelis, E., editors, *Modeling and Simulation in Science.* Birkaüser, Basel (Switzerland).

[240] Lowengrub, J., Cristini, V., 2010. *Multiscale modeling of cancer: An integrated experimental and mathematical modeling approach.* Cambridge University Press, Cambridge (United Kingdom).

[241] Lowengrub, J., Frieboes, H. B., Jin, F., Chuang, Y.-L., Li, X., Macklin, P., Wise, S. M., Cristini, V., 2010. Nonlinear modeling of cancer: Bridging the gap between cells and tumors. *Nonlinearity*, 23 (1), R1–R91.

[242] Lutolf, M. P., Lauer-Fields, J. L., Schmoekel, H. G., Metters, A. T., Weber, F. E., Fields, G. B., Hubbell, J. A., 2003. Synthetic matrix metalloproteinase-sensitive hydrogels for the conduction of tissue regeneration: Engineering cell-invasion characteristics. *Proc Natl Acad Sci USA*, 100 (9), 5413–5418.

[243] Ma, L., Janetopoulos, C., Yang, L., Devreotes, P. N., Iglesias, P. A., 2004. Two complementary, local excitation, global inhibition mechanisms acting in parallel can explain the chemoattractant-induced regulation of PI(3,4,5)P3 response in Dictyostelium cells. *Biophys J*, 87 (6), 3764–3774.

[244] Macklin, P., Lowengrub, J., 2006. An improved geometry-aware curvature discretization for level set methods: Application to tumor growth. *J Comput Phys*, 215, 392–401.

[245] Macklin, P., Lowengrub, J., 2007. Nonlinear simulation of the effect of microenvironment on tumor growth. *J Theor Biol*, 245 (4), 677–704.

[246] Mahoney, A. W., Smith, B. G., Flann, N. S., Podgorski, G. J., 2008. Discovering novel cancer therapies: A computational modeling and search approach. *IEEE Conference on Computational Intelligence in Bioinformatics and Bioengineering*, 233–240.

[247] Maini, P. K., Olsen, L., Sherratt, J. A., 2002. Mathematical models for cell-matrix interactions during dermal wound healing. *Int J Bifurcations Chaos*, 12 (9), 2021–2029.

[248] Maini, P.K., McElwain, D.L.S., Leavesley, D., 2004. Travelling waves in a wound healing assay. *Appl Math Lett*, 17, 575–580.

[249] Maini, P.K., McElwain, D.L.S., Leavesley, D., 2004. Travelling wave model to interpret a wound-healing cell migration assay for human peritoneal mesothelial cells. *Tissue Eng*, 10, 475–482.

[250] Manoussaki, D., Lubkin, S. R., Vernon, R. B., Murray, J. D., 1996. A mechanical model for the formation of vascular networks in vitro. *Acta Biotheor*, 44 (3-4), 271–282.

[251] Mardia, K. V., Jupp, P. E., 2000. *Directional Statistics.* John Wiley and Sons, Chichester (United Kingdom).

[252] Marée, A. F. M., Grieneisen, V. A., Hogeweg, P., 2007. The cellular Potts model and biophysical properties of cells, tissues and morphogenesis. In Anderson, A. R. A., Chaplain, M. A. J., and Rejniak, K. A., editors, *Single-Cell-Based Models in Biology and Medicine*, Mathematics and Biosciences in Interactions. Birkaüser, Basel (Switzerland), 107–136.

[253] Marée, A. F. M., Hogeweg, P., 2001. How amoeboids self-organize into a fruiting body: Multicellular coordination in Dictyostelium discoideum. *Proc Natl Acad Sci USA*, 98, 3879–3883.

[254] Marée, A. F. M., Jilkine, A., Dawes, A., Grieneisen, V. A., Edelstein-Keshet, L., 2006. Polarization and movement of keratocytes: A multiscale modelling approach. *Bull Math Biol*, 68 (5), 1169–1211.

[255] Markevich, N. I., Tsyganov, M. A., Hoek, J. B., Kholodenko, B. N., 2006. Long-range signaling by posphoprotein waves arising from bistability in protein kinase cascades. *Mol Sys Biol*, 2 (61), doi:10.1038/msb4100108.

[256] Martin, M. L., Regan, C. M., 1991. The anticonvulsant valproate teratogen restricts the glial cell cycle at a defined point in the mid-G1 phase. *Brain Res*, 554, 223–228.

[257] Matsumoto, K., Nakamura, T., 2008. NK4 gene therapy targeting HGF-Met and angiogenesis. *Front Biosci*, 13, 1943–1951.

[258] Meili, R., Firtel, R. A., 2003. Two poles and a compass. *Cell*, 114 (2), 153–156.

[259] Merks, R. M. H., Glazier, J. A., 2006. Dynamic mechanisms of blood vessel growth. *Nonlinearity*, 19 (1), C1–C10.

[260] Merks, R. M. H., Brodsky, S. V., Goligorksy, M. S., Newman, S. A., Glazier, J. A., 2006. Cell elongation is key to in silico replication of in vitro vasculogenesis and subsequent remodeling. *Dev Biol*, 289 (1), 44–54.

[261] Merks, R. M. H., Perryn, E. D., Shirinifard, A., Glazier, J. A., 2008. Contact-inhibited chemotactic motility: Role in de novo and sprouting blood vessel growth. *PLoS Comput Biol*, 4 (9), e1000163.

[262] Merks, R. M. H., Koolwijk, P., 2009. Modeling morphogenesis in silico and in vitro: Towards quantitative, predictive, cell-based modeling. *Math Model Nat Phenom*, 4 (4), 149–171.

[263] Metropolis, N., Rosenbluth, A., Rosenbluth, M. N., Teller, A. H., Teller, E., 1953. Equation of state calculations by fast computing machines. *J Chem Phys*, 21 (6), 1087–1092.

[264] Mittara, S., Ulyatt, C., Howell, G. J., Bruns, A. F., Zachary, I., Walker, J. H., Ponnambalam, S., 2009. VEGFR1 receptor tyrosine kinase localization to the Golgi apparatus is calcium-dependent. *Exp Cell Res*, 315 (5), 877–889.

[265] Moccia, F., Berra-Romani, R., Baruffi, S., Spaggiari, S., Signorelli, S., Castelli, L., Magistretti, J., Taglietti, V., Tanzi, F., 2002. Ca2+ uptake by the endoplasmic reticulum Ca2+-ATPase in rat microvascular endothelial cells. *Biochem J*, 364 (1), 235–244.

[266] Mochizuki, A., 2002. Pattern formation of the cone mosaic in the zebrafish retina: A cell rearrangement model. *J Theor Biol*, 215 (3), 345–361

[267] Mogilner, A., Oster, G., 2003. Polymer motors: Pushing out the front and pulling up the back. *Curr Biol*, 13 (18), R721–R733.

[268] Mombach, J. C. M., Glazier, J. A., 1996. Single cell motion in aggregates of embryonic cells. *Phys Rev Lett*, 76 (16), 3032–3035.

[269] Mombach, J. C. M, de Almeida, R. M. C., Thomas, G. L., Upadhyaya, A., Glazier, J. A., 2001. Bursts and cavity formation in Hydra cells aggregates: Experiments and simulations. *Physica A*, 297 (3-4), 495–508.

[270] Mombach, J. C. M, de Almeida, R. M. C., Iglesias, G. L., 1993. Mitosis and growth in biological tissues. *Phys Rev E Stat Phys Plasmas Fluids Relat Interdiscip Topics*, 48 (1), 598–608.

[271] Mongiovi, A. M., Huff, J., Jelinek, M. A., Follenzi, A., Gaudino, G., Parsons, J. T., Comoglio, P. M., Medico, E., 1996. The tyrosine kinase receptors Ron and Sea control "scattering" and morphogenesis of liver progenitor cells in vitro. *Mol Biol Cell*, 7 (4), 495–504.

[272] Montero, E., Abreu, C., Tonino, P., 2007. Relationship between VEGF and p53 expression and tumor cell proliferation in human gastrointestinal carcinomas. *J Cancer Res Clin Oncol*, 134 (2), 193–201.

[273] Moreira, J., Deutsch, A., 2005. Pigment pattern formation in zebrafish during late larval stages: A model based on local interactions. *Develop Dynam*, 232 (1), 33–42.

[274] Mottola, A., Antoniotti, S., Lovisolo, D., Munaron, L., 2005. Regulation of noncapacitative calcium entry by arachidonic acid and nitric oxide in endothelial cells. *FASEB J*, 19 (14), 2075–2077.

[275] Mueller-Klieser, W., 2000. Tumor biology and experimental therapeutics. *Crit Rev Oncol Hematol*, 36, 123–139.

[276] Munaron, L., 2002. Calcium signalling and control of cell proliferation by tyrosine kinase receptors (review). *Int J Mol Med*, 10 (6), 671–676.

[277] Munaron, L., 2006. Intracellular calcium, endothelial cells and angiogenesis. *Recent Patents Anticancer Drug Discov*, 1 (1), 105–119.

[278] Munaron, L., Antoniotti, S., Distasi, C., Lovisolo, D., 1997. Arachidonic acid mediates calcium influx induced by basic fibroblast growth factor in Balb-c 3T3 fibroblasts. *Cell Calcium*, 22 (3), 179–188.

[279] Munaron, L., Antoniotti, S., Fiorio Pla, A., Lovisolo, D., 2004a. Blocking Ca2+ entry: A way to control cell proliferation. *Curr Med Chem*, 11 (12), 1533–1543.

[280] Munaron, L., Antoniotti, S., Lovisolo, D., 2004b. Intracellular calcium signals and control of cell proliferation: How many mechanisms? *J Cell Mol Med*, 8 (2), 161–168.

[281] Munaron, L., Fiorio Pla, A., 2000. Calcium influx induced by activation of tyrosine kinase receptors in cultured bovine aortic endothelial cells. *J Cell Physiol*, 185 (3), 454–463.

[282] Munaron, L., Tomatis, C., Fiorio Pla, A., 2008. The secret marriage between calcium and tumor angiogenesis. *Technol Cancer Res Treat*, 7 (4), 335–339.

[283] Munaron, L., 2009. A tridimensional model of proangiogenic calcium signals in endothelial cells. *The Open Biology J*, 2, 114-129.

[284] Munaron, L., Fiorio Pla, A., 2000. Endothelial calcium machinery and angiogenesis: Understanding physiology to interfere with pathology. *Current Medicinal Chemistry*, 16, 4691–4703.

[285] Munaron, L., Scianna, M., 2012. Multiscale complexity of calcium signaling: Modeling angiogenesis (review). *World J Biol Chem*, 3(6), 121–126.

[286] Murphy, G., Gavrilovic, J., 1999. Proteolysis and cell migration: Creating a path? *Curr Opin Cell Biol*, 11, 614–621.

[287] Mutsaers, S. E., 2002. Mesothelial cells: Their structure, function and role in serosal repair. *Respirology*, 7, 171–191.

[288] Nakamura, T., Teramoto, H., Ichihara, A., 1986. Purification and characterization of a growth factor from rat platelets for mature parenchymal hepatocytes in primary cultures. *Proc Natl Acad Sci USA*, 83, 6489–6493.

[289] Namy, P., Ohayon, J., Traqui, P., 2004. Critical conditions for pattern formation and in vitro tubulogenesis driven by cellular traction fields. *J Theor Biol*, 227 (1), 103–120.

[290] Naora, H., Montell, D. J., 2005. Ovarian cancer metastasis: Integrating insights from disparate model organisms. *Nat Rev Cancer*, 5, 355–366.

[291] Niedbala, M. J., Crickard, K., Bernacki, R. J., 1985. Interactions of human ovarian tumor cells with human mesothelial cells grown on extracellular matrix. An in vitro model system for studying tumor cell adhesion and invasion. *Exp Cell Res*, 160, 499–513.

[292] Nilius, B., Droogmans, G., 2001. Ion channels and their functional role in vascular endothelium. *Physiol Rev*, 81, 1415–1459.

[293] Nobes, C. D., Hall, A., 1999. Rho GTPases control polarity, protrusion, and adhesion during cell movement. *J Cell Biol*, 144 (6), 1235–1244.

[294] O'Brien, F. J., Harley, B. A., Yannas, I. V., Gibson, L., 2004. Influence of freezing rate on pore structure in freeze-dried collagen-GAG scaffolds. *Biomaterials*, 25 (6), 1077–1086.

[295] Olde Damink, L. H. H., Dijkstra, P. J., Luyn, M. J. A., Wachem, P. B., Nieuwenhuis, P., Feijen, J., 1995. Glutaraldehyde as a crosslinking agent for collagen-based biomaterials. *J Mater Sci Mater Med*, 6 (8), 460–472.

[296] Orban, J. M., Wilson, L. B., Kofroth, J. A., El-Kurdi, M. S., Maul, T. M., Vorp, D. A., 2004. Crosslinking of collagen gels by transglutaminase. *J Biomed Mater Res A*, 68 (4), 756–762.

[297] Osada, H., Takahashi, T., 2002. Genetic alterations of multiple tumor suppressors and oncogenes in the carcinogenesis and progression of lung cancer. *Oncogene*, 21, 7421–7434.

[298] Ouchi, N. B., Glazier, J. A., Rieu, J. P., Upadhyaya, A., Sawada, J., 2003. Improving the realism of the cellular Potts model in simulations of biological cells. *Physica A*, 329 (3-4), 451–458.

[299] Owen, M. R., Alarcón, T., Maini, P. K., Byrne, H. M., 2009. Angiogenesis and vascular remodelling in normal and cancerous tissues. *J Math Biol*, 58, 689–721.

[300] Pacher, P., Beckman, J. S., Liaudet, L., 2007. Nitric oxide and peroxynitrite in health and disease. *Physiol Rev*, 87 (1), 315–424.

[301] Paik, D. C., Dillon, J., Galicia, E., Tilson, M. D., 2001. The nitrite/collagen reaction: Nonenzymatic nitration as a model system for age-related damage. *Connect Tissue Res*, 42 (2), 111–122.

[302] Palecek, S. P., Loftus, J. C., Ginsberg, M. H., Lauffenburger, D. A., Horwitz, A. F., 1997. Integrin-ligand binding properties govern cell migration speed through cell-substratum adhesiveness. *Nature*, 385 (6616), 537–540.

[303] Parker, B. S., Argani, P., Cook, B. P., Liangfeng, H., Chartrand, S. D., Zhang, M., Saha, S., Bardelli, A., Jiang, Y., St Martin, T. B., Nacht, M., Teicher, B. A., Klinger, K. W., Sukumar, S., Madden, S. L., 2004. Alterations in vascular gene expression in invasive breast carcinoma. *Cancer Res*, 64 (21), 7857–7866.

[304] Paszek, M. J., Zahir, N., Johnson, K. R., et al., 2005. Tensional homeostasis and the malignant phenotype. *Cancer Cell*, 8, 241–254.

[305] Patel, S., Madan, P., Getsios, S., Bertrand, M. A., Maccalman, C. D., 2003. Cadherin switching in ovarian cancer progression. *Int J Cancer*, 106, 172–177.

[306] Patton, A. M., Kassis, J., Doong, H., Kohn, E. C., 2003. Calcium as a molecular target in angiogenesis. *Curr Pharm Des*, 9, 543–551.

[307] Pedersen, J. A., Swartz, M. A., 2005. Mechanobiology in the third dimension. *Ann Biomed Eng*, 33 (11), 1469–1490.

[308] Pedeux, R., Ythier, D., Duperray, A., 2010. Cell motility and tumor suppressor genes. In Chauvier, A., Preziosi, L., and Verdier, C., editors, *Cell Mechanics: From Single Scale-Based Models to Multiscale Modeling*, Mathematical and Computational Biology Series. Chapman and Hall/CRC Press, London (United Kingdom), 67–88.

[309] Perryn, E. D., Czirok, A., Little, C. D., 2008. Vascular sprout formation entails tissue deformations and VE-cadherin dependent cell-autonomous motility. *Dev Biol*, 313, 545–555.

[310] Peyton, S. R., Putnam, A. J., 2005. Extracellular matrix rigidity governs smooth muscle cell motility in a biphasic fashion. *J Cell Physiol*, 204, 198–209.

[311] Pollard, T. D., Borisy, G. G., 2003. Cellular motility driven by assembly and disassembly of actin filaments. *Cell*, 112 (4), 453–465.

[312] Pollock, J. S., Förstermann, U., Mitchell, J. A., Warner, T. D., Schmidt, H. H. H. W., Nakane, M., Murad, F., 1991. Purification and characterization of particulate endothelium-derived relaxing factor synthase from cultured and native bovine aortic endothelial cells. *Proc Natl Acad Sci USA*, 88 (23), 10480–10484.

[313] Poplawski, N. J., Shirinifard, A., Swat, M., Glazier, J. A., 2008. Simulation od single-species bacterical-biofilm growth using the Glazier–Graner–Hogeweg model and the CompuCell3D modeling environment. *Math Biosci Eng*, 5 (2), 355–388.

[314] Poplawski, N. J., Swat, M., Gens, J. S., Glazier, J. A., 2007. Adhesion between cells, diffusion of growth factors, and elasticity of the AER produce the paddle shape of the chick limb. *Physica A*, 373, 521–532.

[315] Potts, R. B., 1952. Some generalized order-disorder transformations. *Proc Camb Phil Soc*, 48, 106–109.

[316] Preziosi, L., editor, 2003. *Cancer Modeling and Simulation*, Mathematical Biology and Medicine Sciences. Chapman and Hall/CRC Press, London (United Kingdom).

[317] Preziosi, L., Tosin, A., 2009. Muntiphase and multiscale trends in cancer modelling. *Math Model Nat Phenom*, 4, 1–11.

[318] Preziosi, L., Tosin, A., 2007. Multiphase modeling of tumor growth and extracellular matrix interaction: Mathematical tools and applications. *J Math Biol*, 58 (4-5), 625–656.

[319] Provenzano, P. P., Inman, D. R., Eliceiri, K. W., Knittel, J. G., Yan, L., Rueden, C. T., White, J. G., Keely, P. J., 2008. Collagen density promotes mammary tumor initiation and progression. *BMC Med*, 6–11.

[320] Puiffe, M. L., Le Page, C., Filali-Mouhim, A., Zietarska, M., Ouellet, V., Tonin, P. N., Chevrette, M., Provencher, D. M., Mes-Masson, A. M., 2007 Characterization of ovarian cancer ascites on cell invasion, proliferation, spheroid formation, and gene expression in an in vitro model of epithelial ovarian cancer. *Neoplasia*, 9, 820–829.

[321] Quaranta, V., Weaver, A., Cummings, P., Anderson, A. R. A., 2005. Mathematical modeling of cancer: The future of prognosis and treatment. *Clin Chim Acta*, 357, 173–179.

[322] Raeber, G. P., Lutolf, M. P., Hubbell, J. A., 2005. Molecularly engineered PEG hydrogels: A novel model system for proteolytically mediated cell migration. *Biophys J*, 89 (2), 1374–1388.

[323] Rafelski, S. M., Theriot, J. A., 2004. Crawling toward a unified model of cell motility: Spatial and temporal regulation of actin dynamics. *Annu Rev Biochem*, 73, 209–239.

[324] Ramis-Conde, I., Drasdo, D., Anderson, A. R. A., Chaplain, M. A. J., 2008. Modeling the influence of E-cadherin-beta-catenin pathway in cancer cell invasion: A multiscale approach. *Biophys J*, 95, 155–165.

[325] Raub, C. B., Suresh, V., Krasieva, T., Lyubovitsky, J., Mih, J. D., Putnam, A. J., Tromberg, B. J., George, S. C., 2007. Noninvasive assessment of collagen gel microstructure and mechanics using multiphoton microscopy. *Biophys J*, 92, 2212–2222.

[326] Rejniak, K. A., Dillon, R. H., 2007. A single cell-based model of the ductal tumor microarchitecture. *Comp Math Meth Med*, 8 (1), 51–69.

[327] Ribba, B., Sautb, O., Colinb, T., Breschc, D., Grenierd, E., Boissel, J.P., 2006. A multiscale mathematical model of avascular tumor growth to investigate the therapeutic benefit of anti-invasive agents. *J Theor Biol*, 243, 532–541.

[328] Ridley, A. J., Paterson, H. F., Johnston, C. L., Diekmann, D., Hall, A., 1992. The small GTPbinding protein Rac regulates growth factor-induced membrane ruffling. *Cell*, 70 (3), 401–410.

[329] Ridley, A. J., 2001. Rho GTPases and cell migration. *J Cell Sci*, 114, 2713–2722.

[330] Rizzuto, R., Duchen, M. R., Pozzan, T., 2004. Flirting in little space: The ER/mitochondria Ca2+ liaison. *Sci STKE*, 2004 (215), re1.

[331] Roeder, B. A., Kokini, K., Sturgis, J. E., Robinson, J. P., Voytik-Harbin, S. L., 2002. Tensile mechanical properties of three-dimensional type I collagen extracellular matrices with varied microstructure. *J Biomech Eng*, 124, 214–222.

[332] Rolli C. G., Seufferlein, T., Kemkemer, R., Spatz, J. P., 2010. Impact of tumor cell cytoskeleton organization on invasiveness and migration: A microchannel-based approach. *PLoS ONE*, 5, e8726.

[333] Ross S.M., 2003. *Probabilitá e statistica per l'ingegneria e le scienze.* Apogeo, Milano (Italy).

[334] Rubenstein, B. M., Kaufman, L. J., 2008. The role of extracellular matrix in glioma invasion: A cellular Potts model approach. *Biophys J*, 95 (12), 5661–5680.

[335] Rupp, P. A., Czirk, A., Little, C. D., 2004. $\alpha v\beta 3$ integrin-dependent endothelial cell dynamics in vivo. *Development*, 131, 2887–2897.

[336] Sabeh, F., Shimizu-Hirota, R., Weiss, S. J., 2009. Protease-dependent versus -independent cancer cell invasion programs: Three-dimensional amoeboid movement revisited. *J Cell Biol*, 185 (1), 11–19.

[337] Sahai, E., 2007. Illuminating the metastatic process. *Nat Rev Cancer*, 7 (10), 737–749.

[338] Sakata, H., Stahl, S. J., Taylor, W. G., Rosenberg, J. M., Sakaguchi, K., Wingfield, P. T., Rubin, J. S. , 1997. Heparin binding and oligomerization of hepatocyte growth factor/scatter factor isoforms. *J Biol Chem*, 272, 9457–9463.

[339] Sanga, S., Sinek, J., Frieboes, H., Ferrari, M., Fruehauf, J., Cristini, V., 2006. Mathematical modeling of cancer progression and response to chemotherapy. *Expert Rev Anticancer Ther*, 6, 1361–1376.

[340] Santella, L., Ercolano, E., Nusco, G. A., 2005. The cell cycle: A new entry in the field of Ca2+ signaling. *Cell Mol Life Sci*, 62 (21), 2405–2413

[341] Saucerman, J. J., Zhang, J., Martin, J. C., Peng, L. X., Stenbit, A. E., Tsien, R. Y., McCulloch, A. D., 2006. Systems analysis of PKA-mediated phosphorylation gradients in live cardiac myocytes. *Proc Natl Acad Sci USA*, 103 (34), 12923–12928.

[342] Savill, N. J., Hogeweg, P., 1997. Modelling morphogenesis: From single cells to crawling slugs. *J Theor Biol*, 184, 118–124.

[343] Savill, N. J., Sherratt, J. A., 2003. Control of epidermal stem cell clusters by Notch-mediated lateral induction. *Dev Biol*, 258 (1), 141–153.

[344] Savla, U., Olson, L.E., Waters, C.M., 2004. Mathematical modeling of airway epithelial wound closure. *J Appl Physiol*, 96, 566–574.

[345] Sawada, K., Mitra, A. K., Radjabi, A. R., Bhaskar, V., Kistner, E. O., Tretiakova, M., Jagadeeswaran, S., Montag, A., Becker, A., Kenny, H. A., Peter, M. E., Ramakrishnan, V., Yamada, S. D., Lengyel, E., 2008. Loss of E-cadherin promotes ovarian cancer metastasis via α5-integrin, which is a therapeutic target. *Cancer Res*, 68, 2329–2339.

[346] Sawada, M., Shii, J., Akedo, H., Tanizawa, O., 1994. An experimental model for ovarian tumor invasion of cultured mesothelial cell monolayer. *Lab Invest*, 70, 333–338.

[347] Schinkinger, S., Lincoln, B., Wottawah, F., Ebert, S., Romeyke, M., Lenz, D., Erickson, H. M., Ananthakrishnan, R., Mitchell, D., Ks, J., Ulvick, S., Bilby, C., 2005. Optical deformability as an inherent cell marker for testing malignant transformation and metastatic competence. *Biophys J*, 88, 3689–3698.

[348] Schmidt, S., Friedl, P., 2010. Interstitial cell migration: Integrin-dependent and alternative adhesion mechanisms. *Cell Tissue Res*, 339 (1), 83–92.

[349] Scianna, M., Merks, R. M. H., Preziosi, L., Medico, E., 2009. Individual cell-based models of cell scatter of ARO and MLP-29 cells in response to hepatocyte growth factor. *J Theor Biol*, 260 (1), 151–160.

[350] Scianna, M., 2011. A multiscale hybrid model for pro-angiogenic calcium signals in a vascular endothelial cell. *Bull Math Biol*, 74 (6), 1253–1291.

[351] Scianna, M., Munaron, L., Preziosi, L., 2011. A multiscale hybrid approach for vasculogenesis and related potential blocking therapies. *Prog Biophys Mol Biol*, 106 (2), 450–462.

[352] Scianna, M., Preziosi, L., 2012. Multiscale developments of the cellular Potts model. *Multiscale Model Simul*, 10 (2), 342–382.

[353] Scianna, M., Preziosi, L., Wolf, K., 2013. A cellular Potts model simulating cell migration on and in matrix environments. *Math Biosci Eng*, 10 (1), 235–261.

[354] Scianna, M., Munaron, L., 2011. Multiscale model of tumor-derived capillary-like network formation. *Netw Heterog Media*, 6 (4), 597–624.

[355] Scianna, M., Preziosi, L., 2012. A hybrid cellular Potts model describing different morphologies of tumor invasion fronts. Math Model Nat Phenom, 7 (1), 78–104.

[356] Scianna, M., Preziosi, L., 2012. Hybrid cellular Potts model for solid tumor growth. In D'Onofrio, A., Cerrai, P., and Gandolfi, A., editors, *New Challenges for Cancer Systems Biomedicine*, SIMAI Springer Series. Springer-Verlag Italia, Milano (Italy), 209–228.

[357] Scianna, M., Preziosi, L., 2012. Modelling the influence of nucleus elasticity on cell invasion in fiber networks and microchannels. *J Theor Biol*, in press. doi: 10.1016/j.jtbi.2012.11.003.

[358] Scianna, M., Bell, C., Preziosi, L., 2012. A review of mathematical models for the formation of vascular networks. Submitted for publication.

[359] Seaman, S., Stevens, J., Yang, M. Y., Logsdon, D., Graff-Cherry, C., St Croix, B., 2007. Genes that distinguish physiological and pathological angiogenesis. *Cancer Cell*, 11 (6), 539–554.

[360] Serini, G., Ambrosi, D., Giraudo, E., Gamba, A., Preziosi, L., Bussolino, F., 2003. Modeling the early stages of vascular network assembly. *EMBO J*, 22 (8), 1771–1779.

[361] Seufferlein, T., Rozengurt, E., 1995. Sphingosylphosphorylcholine rapidly induces tyrosine phosphorylation of p125FAK and paxillin, rearrangement of the actin cytoskeleton and focal contact assembly. Requirement of p21rho in the signaling pathway. *J Biol Chem*, 270, 24343–24351.

[362] Sheu, M. T., Huang, J. C., Yeh, G. C., Ho, H. O., 2001. Characterization of collagen gel solutions and collagen matrices for cell culture. *Biomaterials*, 22 (13), 1713–1719.

[363] Shield, K., Ackland, M. L., Ahnmed, N., Rice, G. E., 2008. Multicellular spheroids in ovarian cancer metastases: Biology and pathology. *Gynec Oncol*, 113, 143–148.

[364] Shield, K., Riley, C., Quinn, M. A., Rice, G. E., Ackland, M. L., Ahnmed, N., 2007. $\alpha 2\beta 1$–integrin affects metastatic potential of ovarian carcinoma spheroids by supporting disaggregation and proteolysis. *J Carcinog*, 6, 6–11.

[365] Shirinifard, A., Gens, J. S., Zaitlen, B., L., Poplawski, N. J., Swat, M., Glazier, J. A., 2009. 3D multi-cell simulation of tumor growth and angiogenesis. *PLoS ONE*, 4 (10), e7190.

[366] Shojaei, F., Ferrara, N., 2007. Antiangiogenic therapy for cancer: An update. *The Cancer J*, 13 (6), 345–348.

[367] Siegel, R. C., 1976. Collagen cross-linking. Synthesis of collagen cross-links in vitro with highly purified lysyl oxidase. *J Biol Chem*, 251 (18), 5786–5792.

[368] Skubitz, P. N., Bast Jr, R. C., Wayner, E. A., Letourneau, P. C., Wilke, M. S., 1996. Expression of $\alpha 6$ and $\beta 4$ integrins in serous ovarian carcinoma correlates with expression of the basement membrane protein laminin. *Am J Pathol*, 148, 1445–1461.

[369] Smith, J. A., Martin, L., 1973. Do cells cycle? *Proc Natl Acad Sci USA*, 70, 1263–1267.

[370] Smolle, J., 1998. Fractal tumor stromal border in a nonequilibrium growth model. *Anal Quant Cytol Histol*, 20, 7–13.

[371] Sneyd, J., Keizer, J., Sanderson, M. J., 1995. Mechanisms of calcium oscillations and waves: A quantitative analysis. *FASEB J*, 9, 1463–1472.

[372] Sneyd, J., Keizer, J., 2009. *Mathematical Physiology.* In Interdisciplinary Applied Mathematics, Springer, USA.

[373] Spadaccio, C., Rainer, A., De Porcellinis, S., Centola, M., De Marco, F., Chello, M., Trombetta, M., Genovese, J. A., 2010. A G-CSF functionalized PLLA scaffold for wound repair: An in vitro preliminary study. *Conf Proc IEEE Eng Med Biol Soc*, 2010, 843–846.

[374] Starruss, J., Bley, T., Sgaard-Andersen, L., Deutsch, A., 2007. A new mechanism for collective migration in Myxococcus Xanthus, *J Stat Phys*, 128 (1-2), 269–286.

[375] Steele, I. A., Edmondson, R. J., Leung, H. Y., Davies, B. R., 2006. Ligands to FGF receptor 2-IIIb induce proliferation, motility, protection from cell death and cytoskeletal rearrangements in epithelial ovarian cancer cell lines. *Growth Factors*, 24 (1), 45–53.

[376] Stein, A. M., Demuth, T., Mobley, D., Berens, M., Sander, L. M., 2007. A mathematical model of glioblastoma tumor spheroid invasion in a three-dimensional in vitro experiment. *Biophys J*, 92, 356–365.

[377] Steinberg, M. S., 1963. Reconstruction of tissues by dissociated cells. Some morphogenetic tissue movements and the sorting out of embryonic cells may have a common explanation. *Science*, 141, 401–408.

[378] Steinberg, M. S., 1970. Does differential adhesion govern self-assembly processes in histogenesis? Equilibrium configurations and the emergence of a hierarchy among populations of embryonic cells. *J Exp Zool*, 173 (4), 395–433.

[379] Stetler-Stevenson, W. G., Aznavoorian, S., Liotta, L. A., 1993. Tumor cell interactions with the extracellular matrix during invasion and metastasis. *Ann Rev Cell Biol*, 9, 541–573.

[380] Stoker, M., Gherardi, E., Perryman, M., Gray, J., 1987. Scatter factor is a fibroblast-derived modulator of epithelial cell mobility. *Nature*, 21 (6119), 239–242.

[381] Straume, O., Salvesen, H. B., Akslen, L. A., 1999. Angiogenesis is prognostically important in vertical growth phase melanomas. *Int J Oncol*, 15, 595–599.

[382] Stupack, D. G., 2007. The biology of integrins. *Oncology*, 21 (9), 6–12.

[383] Su, J. L., Yang, P. C., Shih, J. Y., Yang, C. Y., Wei, L. H., Kuo, M. L., et al., 2006. The VEGF-C/Flt-4 axis promotes invasion and metastasis of cancer cells. *Cancer Cell*, 9, 209–223.

[384] Sundararaghavan, H. G., Monteiro, G. A., Lapin, N. A., Chabal, Y. J., Miksan, J. R., Shreiber, D. I., 2008. Genipin-induced changes in collagen gels: Correlation of mechanical properties to fluorescence. *J Biomed Mater Res A*, 87 (2), 308–320.

[385] Sundfeldt, K., 2003. Cell–cell adhesion in the normal ovary and ovarian tumors of epithelial origin; an exception to the rule. *Molecular and Cellular Endocrinology*, 202, 89–96.

[386] Szczur, K., Xu, H., Atkinson, S., Zheng, Y., Filippi, M.-D., 2006. Rho GTPase CDC42 regulates directionality and random movement via distinct MAPK pathways in neutrophils. *Blood*, 108 (13), 4205–4213.

[387] Takamizawa, K., Niu, S., Matsuda, T., 1997. Mathematical simulation of unidirectional tissue formation: In vitro transanastomotic endothelialization model. *J Biomater Sci Polym Ed*, 8, 323–334.

[388] Takai, Y., Sasaki, T., Matozaki, T., 2001. Small GTP-binding proteins. *Physiol Rev*, 81 (1), 153–208.

[389] Tamagnone, L., Comoglio, P. M., 1997. Control of invasive growth by hepatocyte growth factor (HGF) and related scatter factors. *Cytokine Growth Factor Rev*, 8 (2), 129–142.

[390] Tomatis, C., Fiorio Pla, A., Munaron, L., 2007. Cytosolic calcium microdomains by arachidonic acid and nitric oxide in endothelial cells. *Cell Calcium*, 41 (3), 261–269.

[391] Tosin, A., Ambrosi, D., Preziosi, L., 2006. Mechanics and chemotaxis in the morphogenesis of vascular networks. *Bull Math Biol*, 68, 1819–1836.

[392] Tracqui, P., 2009. Biophysical models of tumour growth. *Rep Prog Phys*, 72 (5), 056701.

[393] Tremel, A., 2006. *Cell migration and proliferation during monolayer formation.* Experimental Semester Thesis, Department of Chemical and Biomolecular Engineering, The University of Melbourne (Australia).

[394] Trusolino, L., Comoglio, P. M., 2002. Scatter-factor and semaphorin receptors: Cell signalling for invasive growth. *Nat Rev Cancer*, 2 (4), 289–300.

[395] Turner, S., Sherratt, J. A., 2002. Intercellular adhesion and cancer invasion: A discrete simulation using the extended Potts model. *J Theor Biol*, 216 (1), 85–100.

[396] Ulrich, T. A., Jain, A., Tanner, K., MacKay, J. L., Kumar, S., 2010. Probing cellular mechanobiology in three-dimensional culture with collagen-agarose matrices. *Biomaterials*, 31 (7), 1875–1884.

[397] Ulrich, T. A., De Juan Pardo, E. M., Kumar, S., 2009. The mechanical rigidity of the extracellular matrix regulates the structure, motility, and proliferation of glioma cells. *Cancer Res*, 69 (10), 4167–4174.

[398] Ura, H., Takeda, F., Okochi, H., 2004. An in vitro outgrowth culture system for normal human keratinocytes. *J Dermatol Sci*, 35, 19–28.

[399] Valant, P. A., Adjei, P. N., Haynes, D. H., 1992. Rapid Ca2+ extrusion via the Na+/Ca2+ exchanger of the human platelet. *J Membr Biol*, 130 (1), 63–82.

[400] Van Nieuw Amerongen, G. P., Koolwijk, P., Versteilen, A., Van Hinsbergh, V. W. M., 2003. Involvement of RhoA/Rho kinase signaling in VEGF-induced endothelial cell migration and angiogenesis in vitro. *Arteriosclerosis, Thrombosis, and Vascular Biology*, 23, 211–217.

[401] Vaupel, P., Hockel, M., 2000. Blood supply, oxygenation status and metabolic micromilieu of breast cancers: Characterization and therapeutic relevance (Review). *Int J Oncol*, 17, 869–879.

[402] Watson, E. L., Jacobson, K. L., Singh, J. C., Di Julio, D. H., 2004. Arachidonic acid regulates two Ca2+ entry pathways via nitric oxide. *Cell Signal*, 16, 157–165.

[403] Walther, T., Reinsch, H., Grosse, A., Ostermann, K., Deutsch, A., Bley, T., 2004. Mathematical modeling of regulatory mechanisms in yeast colony development. *J Theor Biol*, 229, 327–338.

[404] Wang, N., Tytell, J. D., Ingber, D. E., 2009. Mechanotransduction at a distance: Mechanically coupling the extracellular matrix with the nucleus. *Nat Rev Mol Cell Biol*, 10 (1), 75–82.

[405] Wang, F., So, J., Reierstad, S., Fishman, D. A., 2006. Vascular endothelial growth factorregulated ovarian cancer invasion and migration involves expression and activation of matrix metalloproteinases. *Int J Cancer*, 118, 879–888.

[406] Wang, H. S., Hung, Y., Su, C. H., Peng, S. T., Guo, Y. J., Lai, M. C., et al., 2005. CD44 cross-linking induces integrin-mediated adhesion and transendothelial migration in breast cancer cell line by up-regulation of LFA-1 ($\alpha L\beta 2$) and VLA-4 ($\alpha 4\beta 1$). *Exp Cell Res*, 304, 116–126.

[407] Weidner, K. M., Behrens, J., Vandekerckhove, J., Birchmeier, W., 1990. Scatter factor: molecular characteristics and effect on the invasiveness of epithelial cells. *J Cell Biol*, 111 (5 Pt 1), 2097–2108.

[408] Wolf, K., Wu, Y. I., Liu, Y., Geiger, J., Tam, E., et al., 2007. Multistep pericellular proteolysis controls the transition from individual to collective cancer cell invasion. *Nat Cell Biol*, 9, 893–904.

[409] Wolf, K., Alexander, S., Schacht, V., Coussens, L. M., von Andrian, U. H., van Rheenen, J., Deryugina, E., Friedl, P., 2009. Collagen-based cell migration models in vitro and in vivo. *Semin Cell Dev Biol*, 20 (8), 931–941.

[410] Wolf, K., Friedl, P., 2011. Extracellular matrix determinants of proteolytic and non-proteolytic cell migration. *Trends Cell Biol*, 21 (12), 736–744.

[411] Wolf, K., Mazo, I., Leung, H., Engelke, K., von Andrian, U. H., Deryugina, E. I., Strongin, A. Y., Bröcker, E. B., Friedl, P., 2003. Compensation mechanism in tumor cell migration mesenchymalamoeboid transition after blocking of pericellular proteolysis. *J Cell Biol*, 160 (2), 267–277.

[412] Wright, S. A., Plimpton, S. J., Swiler, T. P., Fye, R. M., Young, M. F., Holm, E. A., 1997. Potts-model grain growth simulations: Parallel algorithms and applications, *SAND Report*, 971925.

[413] Wu, P., Hoying, J. B.,Williams, S. K., Kozikowski, B. A., Lauffenburger, D. A., 1994. Integrin-binding peptide in solution inhibits or enhances endothelial cell migration, predictably from cell adhesion. *Ann Biomed Eng*, 22 (2), 144-152.

[414] Xu, Z., Lioi, J., Mu, J., Kamocka, M. M., Liu, X., Chen, D. Z., Rosen, E. D., Alber, M. S., 2010. A multiscale model of venous thrombus formation with surface-mediated control of blood coagulation cascade. *Biophys J*, 98 (9), 1723–1732.

[415] Xu, Z., Chen, N., Shadden, S., Marsden, J. E., Kamocka, M. M., Rosen, E. D., Alber, M. S., 2009. Study of blood flow impact on growth of thrombi using a multiscale model. *Soft Matter*, 5, 769–779.

[416] Xu, Z., Chen, N., Kamocka, M. M., Rosen, E. D., Alber, M. S., 2008. A multiscale model of thrombus development. *J R Soc Interface*, 5 (24), 705–722.

[417] Yang, Y. L., Kaufman, L. J., 2009. Rheology and confocal reflectance microscopy as probes of mechanical properties and structure during collagen and collagen/hyaluronan self-assembly. *Biophys J*, 96 (4), 1566–1585.

[418] Yannas, I. V., Lee, E., Orgill, D. P., Skrabut, E. M., Murphy, G. F., 1989. Synthesis and characterization of a model extracellular matrix that induces partial regeneration of adult mammalian skin. *Proc Natl Acad Sci USA*, 86, 933–937.

[419] Yung, S., Li, F. K., Chan, T. M., 2006. Peritoneal mesothelial cell culture and biology. *Perit Dial Int*, 26, 162–173.

[420] Zahm, J. M., Kaplan, H., Herard, A., Doriot, F., Pierrot, D., Somelette, P., Puchelle, E., 1997. Cell migration and proliferation during the in vitro repair of the respiratory epithelium. *Cell Motil Cytoskel*, 37, 33–43.

[421] Zaman, M. H., Matsudaira, P., Lauffenburger, D. A., 2007. Understanding effects of matrix protease and matrix organization on directional persistence and translational speed in three-dimensional cell migration. *Ann Biomed Eng*, 35 (1), 91–100.

[422] Zaman, M. H., Trapani, L. M., Sieminski, A. L., Mackellar, D., Gong, H., Kamm, R. D., Wells, A., Lauffenburger, D. A., Matsudaira, P., 2006. Migration of tumor cells in 3D matrices is governed by matrix stiffness along with cell-matrix adhesion and proteolysis. *Proc Natl Acad Sci USA*, 103 (29), 10889–10894.

[423] Zhang, Y., Vande Woude, G. F., 2003. HGF/SF-Met signaling in the control of branching morphogenesis and invasion. *J Cell Biochem*, 88 (2), 408–417.

[424] Zhdanov, V. P., Kasemo, B., 2004. Simulation of the growth and differentiation of stem cells on a heterogeneous scaffold. *Phys Chem Chem Phys*, 6, 4347–4350.

[425] Zhu, Y., Sunfeldt, K., 2007. Tight junction formation in epithelial ovarian adenocarcinoma. *Acta Obstetricia et Gynecologica*, 86, 1011–1019.

Index

activators, 58
algorithm
- checkerboard, 208
- kinetic Monte Carlo, 207
- Metropolis, 6, 205
- N-fold way, 207
- non-Metropolis Monte Carlo, 207
- Random-Walker (RW), 207, 208

amoeboid movement, 189, 191
angiogenesis, 115
anti-angiogenic therapies, 82, 119, 133
anti-tumoral therapies, 80, 137, 149, 161
arachidonic acid (AA), 82
average velocity, 11

BioSpice, 206
Boltzmann
- probability, 6
- temperature, 6

Boyden chamber, 195

cadherins, 8, 18, 34, 36, 42, 44, 71, 119, 132, 141, 146, 148, 158, 163, 168
calcium, 81
- signals, 126
- buffers, 91
- channels, 81, 91, 92
- efflux, 91
- influx, 92
- pathways, 82, 88, 118
- pumps, 81

cancer
- breast, 162
- glioblastoma multiforme (GBM), 159
- glioma, 162, 172, 184, 185
- invasion, 36, 37, 189
- melanoma, 172, 182
- of the cervix, 137
- ovarian, 33, 162
- prostate, 182, 184
- thyroid, 17
- urothelial, 137

carboxyamidotriazole (CAI), 101, 130
CD157, 42, 210
CD44, 35, 209
cell
- acute promyelocytic leukemia (APL), 198
- ARO, 17, 19, 68, 69
- blood, 170
- endothelial (EC), 81, 115
- epithelial, 169
- fibroblast, 169, 174, 178, 187, 200
- glioma, 148, 158, 164, 198
- human umbilical vein endothelial (HUVEC), 117, 124
- invasive, 193
- leukocyte, 200
- mesenchymal, 169
- mesothelial, 35
- MLP-29, 25, 27
- mouse fibroblast, 172, 184
- neuronal, 174
- ovarian cancer, 35
- pancreatic epithelial cancer (Panc-1), 192, 195, 200

penetrating, 193
permeative, 193
smooth muscle, 184
T-, 175, 179
tumor-derived endothelial (TEC), 82, 117
cell cytoskeleton, 84, 96, 131, 166, 180
cell cytosol, 50, 55, 56, 60, 70, 72, 82, 84, 85, 87, 95, 119, 139, 153, 167, 190
cell lysing, 157
cell migration, 67, 81, 165, 189
cell nucleus, 50, 55, 56, 70, 72, 82, 84, 85, 87, 94–96, 98, 119, 126, 139–141, 153, 167, 190
cell plasmamembrane (PM), 49, 54, 60, 70, 94
cell polarization, 18, 94
cell proliferation, 152
cell sorting, 14
Cell Type Map (CTM), 4
center of mass, 11
CFSE, 210
chemoattractant, 10
chemorepellent, 10
chemotaxis, 10, 15, 39, 85, 121, 132, 175, 211
compartmentalization approach, 51, 54, 60
CompuCell3D (CC3D), 205
contact guidance, 175, 195
contact inhibition, 17, 26, 79, 145, 161, 211

Differential Adhesion Hypothesis (DAH), 7
extension, 52, 85
directional guidance, 174
distributive computing, 207

ECM fibers, 38, 167, 189
ECM protein, 142, 158
collagen, 35, 36, 165, 168, 172, 174, 178, 190
fibronectin, 35, 36, 57, 141, 175, 181, 182, 184, 191
laminin, 35, 36, 141, 181
engulfment, 15
extracellular matrix (ECM), 139, 165, 189

focal adhesions (FAs), 96

Hamiltonian, xv, 6, 7, 49, 56, 206
haptotaxis, 10, 137, 141, 149
hepatocyte growth factor (HGF), 17, 68, 142

inhibitors, 58
instantaneous velocity, 11
integrin receptor saturation, 179
integrins, 8, 34–36, 42, 45, 141, 146, 158, 166, 168, 179, 180, 200
invasive growth, 17, 68

lamellipodium, 111
lattice site, 3
source, 6, 206
target, 6, 206

Matrigel, 70, 117, 184
matrix metalloproteinases (MMPs), 35, 39, 45, 137, 143, 152, 158, 166, 186
matrix microchannels, 189, 190
maximum transition probability, 64
mean square displacement (MSD), 11, 170
mean velocity, 11
mesenchymal movement, 17, 145, 195
mesenchymal transition, 17, 145, 175
Met receptor, 17, 68, 71
microlaser procedures, 190
model
cellular automata, xiii
continuous, xiii, 138
discrete, xiii, 138
hybrid, xiv, 3, 4, 58, 214
IBM, xiii
multiphase, 98, 138

nested, xiv, 56, 214
Monte Carlo Step (MCS), 7, 206
Monte Carlo-Boltzmann thermodynamics, 6
motility, 6, 60, 61

nitric oxide (NO), 82
non-capacitative calcium entry (NCCE), 82
non-store-operated calcium entry (NSOCE), 82
nonmuscle myosin II (NMMII), 185, 198
nuclear chromatin structure, 184
nuclear membrane, 184

object
  compartmentalized, 51
  continuous, 3, 57
  discrete, 3
overdamped force-velocity response, 62

parallel computing, 207, 208
percolative transition, 128
persistence, 25, 26, 30, 86, 132, 170, 191
persistence-random-walk (PRW) law, 170
phalloidin, 109, 131, 149
Potts parameter, 8, 10, 53, 54
pseudopodium, 29, 94
pug-ins, 205
push-and-pull movement, 202

rejection free dynamics, 207

scale
  macroscopic, xi, xiii
  mesoscopic, xi, xiv, xvi, 3, 4, 50, 58, 60
  microscopic, xi, xiv, xv, 4, 8, 50, 54, 57, 58, 60
scatter factor (SF), 17
scatter process, 17, 68
sliding m

sphingosylphosphorylcholine (SPC), 197, 198
spin, 3
  flip, 6, 61, 206
steppables, 205
Swiss cheese transition, 128
system biology approach, xi
Systems Biology Workbench, 206

tubulogenesis, 116
tumor metastasis, 14, 46, 152, 165, 189
tumor spheroid, 36, 156

vascular endothelial growth factor (VEGF), 73, 89, 118, 121, 142
vascularization, 81, 157
vasculogenesis, 115

wound healing assay, 23, 67, 218